Deontic Logic and Normative Systems

13th International Conference, DEON 2016, Bayreuth, Germany, July 18-21, 2016

Deontic Logic and Normative Systems

13th International Conference, DEON 2016, Bayreuth, Germany, July 18-21, 2016

Edited by

Olivier Roy,
Allard Tamminga
and
Malte Willer

ISBN 978-1-84890-215-2

College Publications
Scientific Director: Dov Gabbay
Managing Director: Jane Spurr

http://www.collegepublications.co.uk

Original cover design by Laraine Welch
Printed by Lightning Source, Milton Keynes, UK

Preface

This volume contains the proceedings of DEON 2016, the 13th International Conference on Deontic Logic and Normative Systems that was held at the University of Bayreuth (Germany) on 18–21 July 2016. The biennial DEON conferences are designed to promote interdisciplinary cooperation amongst scholars interested in linking the formal-logical study of normative concepts and normative systems with computer science, artificial intelligence, linguistics, philosophy, organization theory and law.

There have been twelve DEON conferences: Amsterdam 1991; Oslo 1994; Sesimbra 1996; Bologna 1998; Toulouse 2000; London 2002; Madeira 2004; Utrecht 2006; Luxembourg 2008; Fiesole 2010; Bergen 2012; Ghent 2014.

General Topics and Special Focus

DEON conferences focus on the following general topics:

- the logical study of normative reasoning, including formal systems of deontic logic, defeasible normative reasoning, logics of action, logics of time, and other related areas of logic
- the formal analysis of normative concepts and normative systems
- the formal specification of aspects of norm-governed multi-agent systems and autonomous agents, including (but not limited to) the representation of rights, authorization, delegation, power, responsibility and liability
- the normative aspects of protocols for communication, negotiation and multi-agent decision making
- the formal analysis of the semantics and pragmatics of deontic and normative expressions in natural language
- the formal representation of legal knowledge
- the formal specification of normative systems for the management of bureaucratic processes in public or private administration applications of normative logic to the specification of database integrity constraints
- game-theoretic aspects of deontic reasoning
- emergence of norms
- deontic paradoxes

In addition to these general topics, DEON 2016's special focus was "Reasons, Argumentation and Justification." Reasons play a prominent role in the normative study of action, belief, intention, and the emotions, as well as in everyday justification and argumentation. Recent years have seen numerous fruitful exchanges between deontic logicians, computer scientists, and philosophers on the nature of reasons and their role in practical and theoretical deliberation. There have also been multiple applications of formal frameworks for the study of reasons in areas of interest to linguists and philosophers of language. The goal of DEON 2016's special focus was to continue this positive trend by encouraging submissions that explore the significance of deontic logic for the study of reasons and their connection with justification and argumentation (and vice versa).

Topics of interest in this special theme included:

- the role of general (though perhaps defeasible) principles in justification and argumentation
- the relation between practical and epistemic reasons
- the formal analysis of reasons and of reasoning about reasons
- the role of justification in multi-agent systems
- the connection between justification and argumentation

Our call for papers attracted 44 submissions from a variety of research communities. All submitted papers were reviewed by three members of the Program Committee. In total, 19 papers were accepted for presentation at the conference and 18 are published in this volume.

Keynote Speakers

Our four keynote speakers were chosen with an eye to the conference's special focus. They were:

- John Broome (University of Oxford)
- Janice Dowell (Syracuse University)
- Xavier Parent (University of Luxembourg)
- Gabriella Pigozzi (Université Paris-Dauphine)

Titles and abstracts of the invited talks were the following:

John Broome, "A Linking Belief is Not Essential for Reasoning":

> Reasoning is a mental process through which some attitudes of yours – premise attitudes – give rise to a new attitude of yours – a conclusion attitude. Not all processes of this sort are reasoning, so what further conditions are essential for a process of this sort to be reasoning? A common view is that you must believe that the content of the conclusion attitude is implied by the contents of the premise attitudes. Call this a 'first-order linking belief'. A first-order linking belief is plausibly a necessary condition for some sorts of reasoning – specifically for theoretical reasoning that concludes in a belief. But it is not a necessary condition for other sorts of reasoning, such as practical reasoning that concludes in an intention. And it is not essential even for reasoning that concludes in a belief: it is not part of what makes a process reasoning.

Janice Dowell, “Methodology for Semantic Theorizing: The Case of Deontic Modals”:

Recent challenges to Kratzer-style contextualism about modals accord speakers' truth- and warrant-assessments a special evidential role: They presuppose that any adequate semantic hypothesis must vindicate those assessments. Here I challenge this presupposition, focusing on John MacFarlane's central challenge to contextualism about deontic modals. In order for our judgments about his challenge case to be reasonably accorded that evidential role, its characterization (of its discourse context and circumstances of evaluation) must be non-defective. However, his case does not meet this minimal constraint on characterizations. That characterization may be repaired to reveal data that properly plays the presumed evidential role. However, none of that data is data the contextualist cannot easily explain.

Xavier Parent, “Preference-based semantics for dyadic deontic logics in Hansson's tradition: a survey of axiomatisation results”:

I present and discuss a number of axiomatization results about so-called dyadic deontic logics in the preference-based semantics tradition. These rely on ranking possible worlds in terms of a Hanssonian binary preference relation of comparative goodness or betterness. In that framework the conditional obligation operator is defined in terms of best antecedent-worlds. The goal is to identify the different systems that can be obtained, depending on the special properties envisaged for the betterness relation, and depending on how the notion of “best” is understood (optimality vs. maximality, stringent vs. liberal maximization). If time allows, decidability issues will also be discussed.

Gabriella Pigozzi, “Changing norms: a framework for revising and contracting rules”:

In human societies as well as in artificial ones, norms change over time: new norms can be created to face changes in the society, and old norms can be retracted. Multiagent systems need mechanisms to model and reason about norm change. In this talk I will present AGM contraction and revision of rules using input/output logical theories. We replace propositional formulas in the AGM framework of theory change by pairs of propositional formulas, representing the rule based character of theories. In general, results from belief base dynamics can be transferred to rule base dynamics, but a similar transfer of AGM theory change to rule change turns out to be much more problematic. (Joint work with Leon van der Torre.)

Acknowledgements

Organizing a conference is team work. We are grateful to everyone who made this conference possible. Most of all, we thank our invited speakers, and all the authors of the presented papers. Special thanks go to the members of the Program Committee and the additional reviewers for their service in reviewing papers and advising us on the program. They were all forced to work on a very tight timescale to make this volume a reality. We also thank the Local Organizing Committee, especially Albert Anglberger, Huimin Dong, Franziska Poprawe and Nathan Wood, for taking care of all the countless details that a conference like this requires. We also thank Leon van der Torre and Jeff Horty, Chair and Vice Chair of the DEON Steering Committee, respectively, for their advice and continuing goodwill. Finally we are indebted to College Publications, and to Dov Gabbay and Jane Spurr in particular, for their support in getting these proceedings published.

June 2016 Olivier Roy, Allard Tamminga and Malte Willer

Organization

Program Chairs

Olivier Roy	University of Bayreuth
Allard Tamminga	University of Groningen and Utrecht University
Malte Willer	University of Chicago

Program Committee

Maria Aloni	University of Amsterdam
Albert Anglberger	University of Bayreuth
Mathieu Beirlaen	Ruhr University Bochum
Guido Boella	University of Torino
Jan Broersen	Utrecht University
Mark A. Brown	Syracuse University
Fabrizio Cariani	Northwestern University
Erik Carlson	Uppsala University
José Carmo	University of Madeira
Roberto Ciuni	University of Amsterdam, ILLC
Robert Demolombe	IRIT
Janice Dowell	Syracuse University
Stephen Finlay	University of Southern California
Lou Goble	Willamette University
Guido Governatori	Data61, CSIRO
Norbert Gratzl	Ludwig-Maximilians University Munich
Davide Grossi	University of Liverpool
Jörg Hansen	University of Leipzig
Andreas Herzig	IRIT-CNRS, Toulouse University
Jeff Horty	University of Maryland
Magdalena Kaufmann	University of Connecticut
Stefan Kaufmann	University of Connecticut
Gert-Jan Lokhorst	Delft University of Technology
Emiliano Lorini	IRIT-CNRS, Toulouse University
Paul McNamara	University of New Hampshire
Joke Meheus	Ghent University
John-Jules Meyer	Utrecht University
Eric Pacuit	University of Maryland
Xavier Parent	University of Luxembourg
Gabriella Pigozzi	Université Paris-Dauphine
Henry Prakken	Utrecht University and University of Groningen
Antonino Rotolo	University of Bologna
Olivier Roy	University of Bayreuth

Giovanni Sartor	European University Institute of Florence and University of Bologna
Audun Stolpe	Norwegian Defence Research Establishment (FFI)
Christian Straßer	Ruhr University Bochum
Allard Tamminga	University of Groningen and Utrecht University
Paolo Turrini	Imperial College London
Frederik Van De Putte	Ghent University
Leon van der Torre	University of Luxembourg
Peter Vranas	University of Wisconsin, Madison
Malte Willer	University of Chicago
Tomoyuki Yamada	Hokkaido University

Additional Reviewers

Aleks Knoks	University of Maryland
Johannes Korbmacher	Ludwig-Maximilians University Munich
Martin Rechenauer	Ludwig-Maximilians University Munich
Katsuhiko Sano	Japan Advanced Institute of Science and Technology
Igor Yanovich	University of Tübingen

Local Organizing Committee

Olivier Roy (Chair)	University of Bayreuth
Albert Anglberger	University of Bayreuth
Matthew Braham	University of Bayreuth
Huimin Dong	University of Bayreuth
Norbert Gratzl	Ludwig-Maximilians University Munich
Franziska Poprawe	University of Bayreuth
Nathan Wood	University of Bayreuth

Sponsoring Institutions

Deutsche Forschungsgemeinschaft (DFG), Germany
Department of Philosophy, University of Bayreuth, Germany

Table of Contents

Cumulative Aggregation

Diego Agustín Ambrossio Xavier Parent Leendert van der Torre

University of Luxembourg
6, rue Richard Coudenhove-Kalergi, Luxembourg
{diego.ambrossio,xavier.parent,leon.vandertorre}@uni.lu

Abstract

From any two conditional obligations "X if A" and "Y if B", cumulative aggregation derives the combined obligation "$X \cup Y$ if $A \cup (B \setminus X)$", whereas simple aggregation derives the obligation "$X \cup Y$ if $A \cup B$". We propose FC systems consisting of cumulative aggregation together with factual detachment, and we give a representation result for FC systems, as well as for FA systems consisting of simple aggregation together with factual detachment. We relate FC and FA systems to each other and to input/output logics recently introduced by Parent and van der Torre.

Keywords: cumulative aggregation, abstract normative systems, input/output logic.

1 Introduction

In this paper, we contrast and study two different principles of aggregation for norms in the context of the framework of Abstract Normative Systems (ANS) due to Tosatto et al. [9].

This one is intended as a general framework to compare logics for normative reasoning. Only fragments of the standard input/output logics [5] are covered by Tosatto et al., and so here we set ourselves the task of applying the framework to the input/output logic recently introduced by Parent and van der Torre [7]. (Cf. also [6].) Its most salient feature is the presence of a non-standard form of cumulative transitivity, called "aggregative" (ACT, for short). Such a rule is used in order to block the counter-examples usually given to the principle known as "deontic detachment": from the obligation of X and the obligation of Y if X, infer the obligation of Y.

Our contribution is first and foremost technical. We acknowledge that the benefits of using the theory of abstract normative systems may not be obvious to the reader. We will not discuss the question of whether it has a reasonable claim to be a general framework subsuming others, nor will we discuss the question of whether aggregative cumulative transitivity is, ultimately, the right form of transitivity.

A central feature of the Tosatto et al. account is that it abstracts away from the language of propositional logic. We recall that as initially conceived input/output logic is an attempt to generalize the study of conditional obligation from modal logic to the abstract study of conditional codes viewed as relations between Boolean formulas. The underlying language is taken from propositional logic. It contains truth-functional connectives, and is assumed to be closed under application of these connectives. It is natural to ask if one can extend the generality further, by working with an arbitrary language, viewed as a collection of items, and without requiring that the items under consideration be "given" or regimented in some special way. Similar programs have been run for propositional logic and modal logic. Koslow [4]'s structuralist approach to logic is perhaps one of the best-known examples of such a program. Unlike Koslow, we do not even assume that the items under consideration can enter into some special implication relations with each other. There are scholars who (rightly or wrongly) take the well-known Tarskian conditions for the consequence relation to be objectionable on the grounds that, for reasons of vagueness (or more), important consequence relations over natural languages (however formalized) are, for instance, not generally transitive. (See, e.g., [8].) The idea is just to investigate the possibility of a formal theory of normative reasoning that avoids such commitments (be they justified or not). [1]

Tosatto et al.'s account has no apparatus for handling conjunction of outputs, and our main purpose in this paper is to develop it to do so. We follow the ideas of so-called "multiple-conclusion logic", and treat normative consequence as a relation between sets, whose elements are understood conjunctively. No assumption about the inner structure of these elements is made.

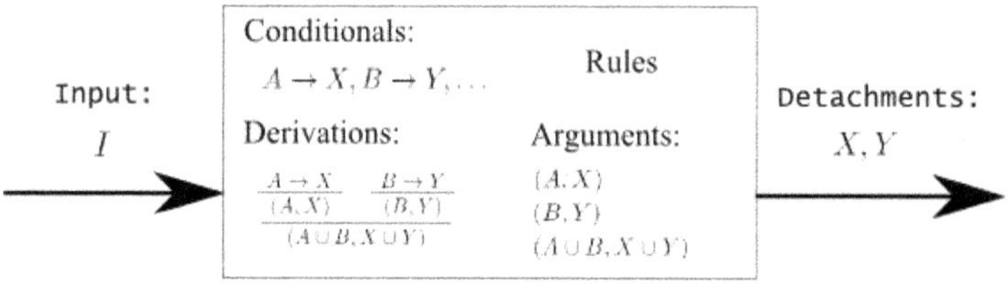

Fig. 1. An Abstract Normative System

An example of an abstract normative system studied in this paper is given in Figure 1. It should be read as follows. Conditionals $A \to X, B \to Y, \ldots$ are the norms of the normative system. Each of A, X, B and Y is a set of language elements (whose inner structure remains unanalyzed). Sets are understood conjunctively on both sides of $\to$. The input I is a collection of language elements representing the context. Rules are used to generate derivations and arguments based on I. The set of detachments $\{X, Y, \ldots\}$ is the output consisting of all detached obligations. The elements of Figure 1 are explained in more detail in the next two sections.

The prime focus in [7] was the contrast between two forms of transitivity, called "cumulative transitivity" and "aggregative cumulative transitivity".

[1] This motivation for using ANS is ours.

This paper shifts the emphasis on the contrast between the following two forms of aggregation.

Simple aggregation If X is obligatory in context A, and Y is obligatory in context B, then $X \cup Y$ is obligatory in context $A \cup B$. In other words, simple aggregation derives the obligation "$X \cup Y$ if $A \cup B$" from any two conditional obligations "X if A" and "Y if B". [2]

Cumulative aggregation If X is obligatory in context A, and Y is obligatory in context B, then $X \cup Y$ is obligatory in context $A \cup (B \setminus X)$. In other words, cumulative aggregation derives the combined obligation "$X \cup Y$ if $A \cup (B \setminus X)$" from the same two conditional obligations.

The rule of simple aggregation gives the most straightforward way of collecting items as detachments are performed. When $A = B$, simple aggregation gives the rule "If X is obligatory given A, and Y is obligatory given A, then $X \cup Y$ is obligatory given A." A drawback of simple aggregation is that it does not capture transitive reasoning. Given the two conditional obligations "$\{x\}$ if $\{\}$" and "$\{y\}$ if $\{x\}$", simple aggregation only yields "$\{x, y\}$ if $\{x\}$". This motivates the rule of cumulative aggregation. In the particular case where $B = A \cup X$, cumulative aggregation yields the form of transitivity introduced by Parent and van der Torre [7] under the name ACT. This is the rule $(A, X), (A \cup X, Y)/(A, X \cup Y)$. In our example, one gets "$\{x, y\}$ if $\{\}$." [3]

To summarize, we adress the following issues:

- How to develop the theory of abstract normative systems to handle conjunction of outputs and the form of cumulative transitivity described in [7]?
- How to define the proof theory of the system? What are the most significant properties of the framework?
- How to provide a semantical characterisation, along with a representation result linking it with the proof theory?

The layout of this paper is as follows. In Section 2, we introduce FA systems for simple aggregation. In Section 3, we introduce FC systems for cumulative aggregation. We give representation results for both systems. In Section 4, we show how FA and FC systems relate to one another, and we discuss some properties of the systems. In Section 5 we show how FA and FC systems relate with the input/output logics introduced by Parent and van der Torre [7].

Due to space limitation, we focus on the logical framework and the results, and leave the proofs of the representation theorems to a technical report [1]. We would like to stress that these are not just a re-run of the proofs given by Parent and van der Torre [7] in a classical logic setting. The two settings

---

[2] Note that intersection as used in abstract normative systems does not correspond to disjunction in propositional logic. Take $(\{p\}, \{x\})$ and $(\{q\}, \{x\})$. The intersection of the two contexts yields $(\{\}, \{x\})$. Reasoning by cases would yield $(\{p \vee q\}, \{x\})$ instead.

[3] As mentioned, it is not our purpose to discuss this rule in any greater depth. For more details on it, see Parent and van der Torre [7].

are very different. The question of whether the proofs of our representation results can be adapted to yield a completeness result in a classical logic setting remains an open problem.

2 FA systems for simple aggregation

In this section, we introduce abstract normative systems for simple aggregation, and we give a representation result. Though FA systems may be interesting in their own right, in this paper the main role of FA systems is to set the stage for FC systems for cumulative aggregation, introduced in the next section. Thus, although we talk about normative systems and use examples from normative system it must be kept in mind that FA systems are not appropriate for all kinds of normative reasoning.

In general, a system $\langle L, C, R\rangle$ consists of a language L, a set of conditionals C defined over this language, and a set of rules R. The input is a set of sentences from L. If $\langle L, C, R\rangle$ is a normative system, then a conditional $A \rightarrow X$ can be read as the norm "if A, then obligatory X". A normative system contains at least one set of norms, the regulative norms from which obligations and prohibitions can be detached. It may also contain permissive norms, from which explicit permissions can be detached, and constitutive norms, from which institutional facts can be detached. In this paper we do not consider permissive and constitutive norms. In the present setting, a system generates or produces an obligation set, a subset of the universe, reflecting the obligatory elements of the universe.

All abstract normative systems we consider satisfy at least factual detachment. To represent factual detachment, we write (A, X) for the argument for X in context A, in other words, for input A the output contains X. Factual detachment is the rule $A \rightarrow X/(A, X)$, and says that if there is a rule with the context as antecedent, then the output contains the consequent.

Besides factual detachment, FA systems have the rule of so-called simple aggregation. This one is usually given the form $(A, X), (A, Y)/(A, X \cup Y)$. In this paper aggregation is given the more general form $(A, X), (B, Y)/(A \cup B, X \cup Y)$. This more general form allows for the inputs not to be the same. Given strengthening of the input, $(A, X)/(A \cup B, X)$, the two rules are equivalent. Since we do not assume strengthening of the input, our rule is strictly stronger.

Definition 2.1 [FA system with input] A FA *system* is a triple $\langle L, C, R\rangle$ with L a language, $C \subseteq 2^L \times 2^L$ a set of conditionals written as $A \rightarrow X$, and R a set of rules. For every conditional $A \rightarrow X \in C$, A and X are finite sets. A FA system is a system $\langle L, C, R\rangle$ where R consists of the rule of *factual detachment* (FD) and the rule of *aggregation* (AND):

$$\text{FD}\ \frac{A \rightarrow X}{(A, X)} \qquad \text{AND}\ \frac{(A, X) \quad (B, Y)}{(A \cup B, X \cup Y)}$$

An *input* $I \subseteq L$ for *system* $\langle L, C, R\rangle$ is a subset of the language.

Let $FA = \{FD, AND\}$. We write $a(A \rightarrow X) = A$ for the *antecedent* of a conditional, and $c(A \rightarrow X) = X$ for the *consequent* of a conditional. We write

$a(C) = \cup\{a(A \to X) \mid A \to X \in C\}$ for the union of the antecedents of all the conditionals in C. We write $c(C) = \cup\{c(A \to X) \mid A \to X \in C\}$ for the union of the consequents of all the conditionals in C.[4]

The following example is meant to exercise the notation. We build a language, and introduce a set of conditionals and an input. The language L is the domain (or universe) of discourse. For the purpose of the example, L is a set of literals. Following Tosatto et al., we also introduce a complement function $\overline{e}$ for the elements e of the language L.

Example 2.2 [Sing and dance, adapted from Goble [3]] Given a language L_0 which does not contain formulas of the form $\sim a$, the language L is $L_0 \cup \{\sim a \mid a \in L_0\}$. For $a \in L$, if $a \in L_0$ then $\overline{a} = \sim a$, and otherwise $\overline{a} = b$ for the $b \in L_0$ such that $a = \sim b$.

Let L_0 be $\{x, y, d, s\}$. Intuitively: "it is Spring" (x); "it is Sunday" (y); "a dance is performed" (d); and "a song is performed" (s). The language L adds classical negation to the language, $L = L_0 \cup \{\sim y, \sim x, \sim d, \sim s\}$. The complement function says $\bar{x} = \sim x$, $\overline{\sim x} = x$, and so on.

Suppose the conditionals $C_1 = \{y \to d, x \to s\}$ apply to a wedding party. This says that on Sundays one ought to dance, and in Spring one ought to sing. The antecedents of the conditionals are: $a(y \to d) = y$; $a(x \to s) = x$; $a(C_1) = \{x, y\}$. Their consequents are: $c(y \to d) = d$; $c(x \to s) = s$; $c(C_1) = \{s, d\}$.

We distinguish three related kinds of output from a system and an input, called derivations, arguments and detachments, respectively. A *derivation* is a finite tree, whose leaves are elements from the set of conditionals and whose root is a pair (A, X) obtained by successive applications of the rules, with the further constraint that $A \subseteq I$.[5] An *argument* is a pair (A, X) for which such a derivation exists, and X is a *detachment* for which such an argument (A, X) exists.[6]

Definition 2.3 [Derivations der, Arguments *arg*, and Detachments *det*]
Given a system $\langle L, C, R\rangle$ and an input I,

- a *derivation* of (A, X) on the basis of I in system $\langle L, C\rangle$ is a finite tree[7] using the rules R, with as leaves elements of C, and as root the pair (A, X) where $A \subseteq I$ and $X \subseteq L$.

---

[4] To ease readability we will omit curly braces when referring to singleton sets, and we write $a \to x$ for $\{a\} \to \{x\}$.

[5] Alternatively, we could add the condition $A \subseteq I$ only to the definitions of arguments and detachments, or only to the definition of detachments. There are pros and cons to both choices. For example, the advantage of our definition is that the set of derivations is smaller, but the disadvantage is that the set of derivations is not closed under sub-derivations, which complicates the proofs of the formal results.

[6] Note the special feature of our formal framework that weakening of the output can be added in different ways. For example, one can add a rule $(A, X \cup Y)/(A, X)$, or one can adapt the definition of detachment such that X is detached for input I if there is an argument (A, Y) such that $A \subseteq I$ and $X \subseteq Y$. The same holds for other properties added to the formal system. We leave the formal analysis of such kinds of extensions to further research.

[7] By a finite tree, we mean one with finitely many nodes.

- an *argument* is a pair (A, X), such that there exists a derivation d with $root(d) = (A, X)$.
- a *detachment* is a set X such that there is an argument (A, X).

We write $\mathtt{der}(L, C, I, R)$ for the set of all the derivations which can be constructed in this way, we write $arg(L, C, I, R)$ for the set of all such arguments, and we write $det(L, C, I, R)$ for the set of all such detachments.

We write $leaves(d)$ for the set of all the leaves of derivation d, $i((A, X)) = A$ for the input of a pair (A, X) and $o((A, X)) = X$ for the output of a pair (A, X). Also we write $i(D) = \cup\{i((A, X)) \mid (A, X) \in D\}$ and $o(D) = \cup\{o((A, X)) \mid (A, X) \in D\}$ for the inputs and outputs of sets of such pairs.

The derivation rules take one datatype, norms, and outputs another, arguments. Nonetheless, the main idea is that derivations are always based on an input. This is reflected by the constraint $i(root(d)) \subseteq I$. But we stress that such a constraint is put on the root of the derivation only, and that all the other nodes need not verify this constraint. Otherwise we would not be able to chain conditionals together. Because of this, the property of closure under sub-derivations does not always hold. It depends on the rules being used. We will see an example of this phenomenon with system FC in Section 3. This also makes the proof of the representation theorem for FC trickier. The standard method of induction over the length of derivations is not available any more.

A derivation is a relative notion, since it is meant to represent the inner structure of an argument. As argued before derivations are tied to the context giving a justification for the argument put forward based on what is, or is not, the case. In the literature, the notion of argument is defined in two ways. Either an argument is viewed as either a pair whose first element is a set of formulas (the support) and second element a formula (the conclusion), or as a derivation in a logical proof system, i.e. a sequence, tree or graph of logical formulas. Here we choose the first definition. In the context of this study, the pair itself denotes a norm. However, it could represent any conditional statement. We use the term argument rather than norm, just to emphasize that we are interested in the relationship between a set of premises and its set of conclusions.

We now can briefly explain the notion of abstraction at stake in the theory of abstract normative systems. Intuitively, the detachment system treats the elements of L as atomic, in the sense that detachments have no relation with the logical structure of language L. Formally, we can replace one language L by another one L', define a one-to-one function f between elements of L and L', and extend f to subsets of L and C. Then we have $f(det(L, C, I, R)) = det(f(L), f(C), f(I), R)$. In this sense, it is an abstract theory.

We continue Example 2.2 to illustrate factual detachment and aggregation, as well as the distinction between derivations, arguments and detachments. In the absence of the rule of strengthening of the antecedent, one cannot derive that X is obligatory in context $A \cup B$ from the fact that X is obligatory in context A. This reflects the idea that arguments are minimal, in the sense that one cannot add irrelevant elements like B to their support. For example, if the input is $\{A, B\}$ and the sole conditional is $A \to X$, then there is no argument $(A \cup B, X)$. But X will be detached, since the input set triggers the conditional in question. The absence of the rule of strengthening of the antecedent does *not* reflect the fact that rules may leave room for exceptions.

Example 2.4 [Example 2.2 - Continued] Given $L = L_0 \cup \{\sim a \mid a \in L_0\}$, we say that an element $a \in I$ is a violation if there is a detachment containing $\overline{a}$, and this detachment is called a violated obligation. Moreover, we say that a detachment is a cue for action if it is not a violated obligation.

The derivations for $C_1 = \{y \to d, x \to s\}$ and $I_1 = \{x, y\}$ are $\mathtt{der}(L, C_1, I_1, FA) =$

$$\left\{ d_1 = \frac{y \to d}{(y,d)}\ \mathrm{FD}\ ,\ d_2 = \frac{x \to s}{(x,s)}\ \mathrm{FD}\ ,\ d_3 = \frac{\dfrac{x \to d}{(x,d)}\ \mathrm{FD} \quad \dfrac{y \to s}{(y,s)}\ \mathrm{FD}}{(\{x,y\},\{s,d\})}\ \mathrm{AND} \right\},$$

the arguments are $arg(L, C_1, I_1, FA) = \{(y, d), (x, s), (\{x, y\}, \{s, d\})\}$ and the detachments are $det(L, C_1, I_1, FA) = \{\{d\}, \{s\}, \{s, d\}\}$, which are all cues for action. Thus I_1 does not contain violations. Factual detachment derives d and s, and aggregation combines them to $\{d, s\}$. First, note that some strengthening of the input is built in the aggregation inference rule AND, as we derive the conditional norm $(\{x, y\}, \{s, d\})$ whose antecedent is stronger than the antecedent of the conditional norms in C_1. Second, note that, for the context where there is no singing $I_2 = \{x, y, \bar{s}\}$, we obtain exactly the same derivations, arguments and detachments. However, now $\bar{s}$ is a violation, and the detachments $\{s\}$ and $\{s, d\}$ are violated obligations, and only $\{d\}$ is a cue for action.

Now consider $C_2 = \{\{x, y\} \to \{s, d\}\}$ and, e.g., I_2. The derivation is

$$\mathtt{der}(L, C_2, I_2, FA) = \left\{ d_4 = \frac{\{x,y\} \to \{s,d\}}{(\{x,y\},\{s,d\})}\ \mathrm{FD} \right\},$$

the arguments are $arg(L, C_2, I_2, FA) = \{(\{x, y\}, \{s, d\})\}$ and the detachments are $det(L, C_2, I_2, FA) = \{\{s, d\}\}$.

It should not come as a surprise that the set of detachments is syntax-dependent. This follows at once from letting the rule of weakening of the output go. This phenomenon is familiar from the literature on belief revision. [8]

Theorem 2.5 gives a representation result for FA systems. The left-hand side of the bi-conditional pertains to the proof theory, while the right-hand side of it provides a semantic characterization in terms of subset selection. For X to be derivable from a set of conditionals C on the basis of input I, X must be the union of the consequents of finitely many conditionals in C, which are all 'triggered' by the input set I. [9]

---

[8] For more on the rule of weakening of the output, and the reason why it may be considered counter-intuitive, we refer the reader to the discussion in Goble [3] (see also Parent and van der Torre [7].)

[9] In FA systems, we call 'triggered' those conditionals whose antecedents are in I.

Theorem 2.5 (Representation result, FA) *$X \in det(L, C, I, FA)$ if and only if there is some non-empty and finite $C' \subseteq C$ such that $a(C') \subseteq I$ and $X = c(C')$.*

Proof. See [1]. □

Corollary 2.6 (Monotonicity of *det*) *$det(L, C, I, FA) \subseteq det(L, C', I, FA)$ whenever $C \subseteq C'$.*

The following example illustrates how to calculate the detachments using the semantic characterization described in the statement of Theorem 2.5.

Example 2.7 [Example 2.2 - Continued] We calculate $det(L, C_1, I_1, FA)$, now using Theorem 2.5. The set of conditionals C_1 has three non-empty subsets: $C_{1.1} = \{y \rightarrow d\}$, $C_{1.2} = \{x \rightarrow s\}$, and $C_{1.3} = \{y \rightarrow d, x \rightarrow s\}$. Here $a(C_{1.1}) \subseteq I_1$, $a(C_{1.2}) \subseteq I_1$ and $a(C_{1.3}) \subseteq I_1$. Also $c(C_{1.1}) = \{d\}$, $c(C_{1.2}) = \{s\}$ and $c(C_{1.3}) = \{s, d\}$. So $det(L, C_1, I_1, FA) = \{c(C_{1.1}), c(C_{1.2}), c(C_{1.3})\} = \{\{d\}, \{s\}, \{s, d\}\}$.

3 FC systems for cumulative aggregation

In this section we introduce FC systems for cumulative aggregation. FC is much alike FA except that the rule of aggregation AND is replaced with that of cumulative aggregation CAND.

Definition 3.1 [FC system with input] A FC system is a triple $\langle L, C, R\rangle$ where R consists of the following rule of *factual detachment* (FD), and the rule of *cumulative aggregation* (CAND). We write $FC = \{FD, CAND\}$.

$$\text{FD} = \frac{A \rightarrow X}{(A, X)} \quad \text{CAND} = \frac{(A, X) \quad (B, Y)}{(A \cup (B \setminus X), X \cup Y)}$$

To illustrate the difference between FA and FC systems, we use the same example as the one that Parent and van der Torre [7] use in order to motivate their rule ACT. We reckon that, compared to the framework described in [7], the present framework does not yield any new insights into the analysis of the example itself.

Example 3.2 [Exercise, from Broome [2]] C contains two conditionals. One says that you ought to exercise hard everyday: $\{\} \rightarrow x$. The other says that, if you exercise hard everyday, you ought to eat heartily: $x \rightarrow h$. Intuitively, in context $\{\}$, we would like to be able to derive $\{x, h\}$, but not $\{h\}$.

FA systems do not allow us to do it.

Let $I = \{\}$. With simple aggregation the set of derivations is $\mathsf{der}(L, C, I, FA) = \left\{d_1 = \frac{\{\} \rightarrow x}{(\{\}, x)}\ \text{FD}\right\}$, the set of arguments is $arg(L, C, I, FA) = \{(\{\}, x)\}$ and the set of detachments is $det(L, C, I, FA) = \{\{x\}\}$. Thus the desired obligation is not detached. Norms can be chained together only in so far as the input set contains their antecedent. Let $I' = \{x\}$.

Then the set of derivations is $\mathtt{der}(L, C, I', FA) =$

$$\left\{ d_1 = \frac{\{\} \to x}{(\{\}, x)}\text{FD} \;,\; d_2 = \frac{x \to h}{(x,h)}\text{FD} \;,\; d_3 = \frac{\dfrac{\{\} \to x}{(\{\}, x)}\text{FD} \quad \dfrac{x \to h}{(x,h)}\text{FD}}{(x, \{x,h\})}\text{AND} \right\},$$

the set of arguments is $arg(L, C, I', FA) = \{(\{\}, x), (x, h), (x, \{x, h\})\}$ and the detachments are $det(L, C, I', FA) = \{\{x\}, \{h\}, \{x, h\}\}$.

With cumulative aggregation, the derivations for C and $I = \{\}$ are $\mathtt{der}(L, C, I, FC) =$

$$\left\{ d_1 = \frac{\{\} \to x}{(\{\}, x)}\text{FD} \;,\; d_2 = \frac{\dfrac{\{\} \to x}{(\{\}, x)}\text{FD} \quad \dfrac{x \to h}{(x,h)}\text{FD}}{(\{\}, \{x,h\})}\text{CAND} \right\}$$

The arguments are $arg(L, C, I, FC) = \{(\{\}, x), (\{\}, \{x, h\})\}$ and the detachments are $det(L, C, I, FC) = \{\{x\}, \{x, h\}\}$. Factual detachment allows us to detach $\{x\}$, and cumulative aggregation allows us to detach $\{x, h\}$ in addition. Like in [7], h cannot be derived without x. Intuitively, the obligation to eat heartily no longer holds, if you take no exercise.

Definition 3.3 introduces the functions f and g, to be used later on in the semantic characterization of cumulative aggregation. Intuitively, given a set $D \subseteq L$, the function $f(C, D)$ gathers all the consequents of the conditionals in C that are triggered by D. The function $g(C, I)$ gathers all the sets D that extend the input set I and are closed under $f(C, D)$.

Definition 3.3 [f and g] We define

$$f(C, D) = \bigcup\{X \mid A \to X \in C; A \subseteq D\}$$

$$g(C, I) = \{D \mid I \subseteq D \supseteq f(C, D)\}$$

We illustrate the calculation of functions f and g continuing Example 3.2.

Example 3.4 [Example 3.2 - Continued] Consider the following table. The left-most column shows the relevant subsets C' of C. The middle column shows what consequents can be detached depending on what set D is used as input. The right-most column shows the sets D extending I and closed under $f(C', D)$, for each subset C'.

C'	$f(C', D)$	$g(C', \{\})$
$\{\} \to x$	$\{x\}$	$\{D \mid x \in D\}$
$x \to h$	$\{\}$ if $x \notin D$, $\{h\}$ if $x \in D$	$\{D \mid x \notin D$ or $\{x, h\} \subseteq D\}$
$\{\} \to x$, $x \to h$	$\{x\}$ if $x \notin D$, $\{x, h\}$ if $x \in D$	$\{D \mid \{x, h\} \subseteq D\}$

Theorem 3.5 gives a representation result for FC systems. For X to be derivable from a set of conditionals C on the basis of input I, X must be the union of the consequents of finitely many conditionals in C, which are either

directly triggered by the input set I (in the sense of Footnote 9), or indirectly triggered by the input set I (via a chain of norms).

Theorem 3.5 (Representation result, FC) *$X \in det(L,C,I,FC)$ if and only if there is some non-empty and finite $C' \subseteq C$ such that, for all $D \in g(C',I)$, we have $a(C') \subseteq D$ and $X = f(C',D)$.*

Proof. See [1]. □

We show with an example how to calculate the detachments using the semantic characterization given in the statement of Theorem 3.5.

Example 3.6 [Example 3.4 - Continued] We again calculate $det(L,C,I,FC)$, now using Theorem 3.5. We use the Table shown in Example 3.4.

The top row tells us that, $\{x\} \in det(L,C,I,FC)$. This is because, for all D in $g(C',\{\})$, $f(C',D) = \{x\}$.

The bottom row tells us that, $\{x,h\} \in det(L,C,I,FC)$. This is because, for all D in $g(C',\{\})$, $f(C',D) = \{x,h\}$.

We can also conclude that, $\{h\} \notin det(L,C,I,FC)$ because, for all C', there is a D in $g(C',\{\})$ such that $f(C',D) \neq \{h\}$.

Finally, the set of detachments is $det(L,C,I,FC) = \{\{x\},\{x,h\}\}$.

4 Some properties of FA systems and FC systems

We start by showing how FA systems and FC systems relate to each other.

Definition 4.1 [Argument subsumption] Argument (A,X) *subsumes* argument (B,Y) if $A \subseteq B$ and $X = Y$. Given two sets of arguments S and T, we say that T subsumes S (notation: $S \sqsubseteq T$), if for all $(B,Y) \in S$ there is an argument $(A,X) \in T$ such that (A,X) subsumes (B,Y).

Example 4.2 Consider the following derivation.

$$d = \frac{(A,X) \quad (A \cup B \cup X, X \cup Y)}{(A \cup B, X \cup Y)} \text{CAND}$$

The argument $(A \cup B, X \cup Y)$ subsumes the argument $(A \cup B \cup X, X \cup Y)$.

Proposition 4.3 *$arg(L,C,I,FA) \sqsubseteq arg(L,C,I,FC)$.*

Proof. Let $(A,X) \in arg(L,C,I,FA)$, where $A \subseteq I$. Let d be the derivation of (A,X) on the basis of I using the rules FD and AND. Let $leaves(d) = \{A_1 \to X_1,\dots,A_n \to X_n\}$. We have $A = \bigcup_{i=1}^n A_i$ and $X = \bigcup_{i=1}^n X_i$.[10] That is,

$$(A,X) = (\bigcup_{i=1}^{n} A_i, \bigcup_{i=1}^{n} X_i)$$

One may transform d into a derivation d' of (A',X) on the basis of I using the rules FD and CAND. Keep the leaves and their parent nodes (obtained using

[10] Strictly speaking, this follows from a lemma used in the proof of the representation result for FA systems, Lemma 1 in [1].

FD) as they are in d, and replace any application of AND by an application of CAND. The result will be a tree whose root is

$$(A', X) = (A_1 \cup \bigcup_{i=2}^{n}(A_i \setminus \bigcup_{j=1}^{i-1} X_j), \bigcup_{i=1}^{n} X_i)$$

We have

$$A_1 \cup \bigcup_{i=2}^{n}(A_i \setminus \bigcup_{j=1}^{i-1} X_j) \subseteq \bigcup_{i=1}^{n} A_i \subseteq I \text{ and } \bigcup_{i=1}^{n} X_i = \bigcup_{i=1}^{n} X_i$$

On the one hand, $(A', X) \in arg(L, C, I, FC)$. On the other hand, (A', X) subsumes (A, X). □

Corollary 4.4 $det(L, C, I, FA) \subseteq det(L, C, I, FC)$

Proof. This follows at once from Proposition 4.3. □

We now point out a number of other properties of FA and FC systems.

Proposition 4.5 (Applicability) *The rules AND and CAND can be applied to any arguments (A, X) and (B, Y).*

Proof. Trivial. Assume arguments (A, X) and (B, Y). By definition of an argument, $A \subseteq I$, $B \subseteq I$, $X \subseteq L$ and $Y \subseteq L$. Thus, $A \cup B \subseteq I$, $A \cup (B \setminus X) \subseteq I$ and $X \cup Y \subseteq L$. □

Proposition 4.6 (Premises permutation, *FA*) *AND can be applied to two arguments (A, X) and (B, Y) in any order.*

Proof. Straightforward. □

It is noteworthy that Proposition 4.6 fails for CAND, as shown by the following counterexample, where $A \neq B$:

$$\frac{(A, B) \qquad (B, A)}{(A, A \cup B)}\text{ CAND} \quad \not\Leftrightarrow \quad \frac{(B, A) \qquad (A, B)}{(B, A \cup B)}\text{ CAND}$$

The arguments $(A, A \cup B)$ and $(B, A \cup B)$ are distinct.

Proposition 4.7 considers two successive applications of AND, or of CAND.

Proposition 4.7 (Associativity) *Each of AND and CAND is associative, in the sense of being independent of the grouping of the pairs to which it is applied.*

Proof. The argument for AND is straightforward, and is omitted. For CAND, it suffices to show that the pairs appearing at the bottom of the following two derivations are equal:

$$\dfrac{\dfrac{\vdots}{(A, X)} \qquad \dfrac{\dfrac{\vdots}{(B, Y)} \qquad \dfrac{\vdots}{(C, Z)}}{(B \cup (C \setminus Y), Y \cup Z)}}{(A \cup ((B \cup (C \setminus Y)) \setminus X), X \cup Y \cup Z)} \qquad \dfrac{\dfrac{\dfrac{\vdots}{(A, X)} \qquad \dfrac{\vdots}{(B, Y)}}{(A \cup (B \setminus X), X \cup Y)} \qquad \dfrac{\vdots}{(C, Z)}}{(A \cup (B \setminus X) \cup (C \setminus (X \cup Y)), X \cup Y \cup Z)}$$

The fact that the two pairs in question are equal follows at once from the following two laws from set-theory:

$$(A \cup B) \setminus X = (A \setminus X) \cup (B \setminus X) \tag{1}$$
$$B \setminus (X \cup Y) = (B \setminus X) \setminus Y \tag{2}$$

We have:

$$\begin{aligned} A \cup ((B \cup (C \setminus Y)) \setminus X) &= A \cup (B \setminus X) \cup ((C \setminus Y) \setminus X) && \text{[by law (1)]} \\ &= A \cup (B \setminus X) \cup (C \setminus (X \cup Y)) && \text{[by law (2)]} \end{aligned}$$

□

Proposition 4.8 *FA systems are closed under sub-derivations in the following sense: given a derivation $d \in \mathtt{der}(L, C, I, FA)$, for all sub-derivations d' of d, $d' \in \mathtt{der}(L, C, I, FA)$–that is, $i(root(d')) \subseteq I$.*

Proof. Let $d \in \mathtt{der}(L, C, I, FA)$ with $root(d) = (A, X)$ and $A = A_1 \cup \ldots \cup A_n \subseteq I$ and $X = X_1 \cup \ldots \cup X_n$. Without loss of generality, we can assume that $n > 1$. By Proposition 4.7, d can be given the form:

$$\dfrac{\dfrac{\dfrac{\dfrac{A_1 \to X_1}{(A_1, X_1)}\,\text{FD} \quad \dfrac{A_2 \to X_2}{(A_2, X_2)}\,\text{FD}}{(A_1 \cup A_2, X_1 \cup X_2)}\,\text{AND} \quad \dfrac{\vdots}{(A_3, X_3)}\,\text{FD}}{(A_1 \cup A_2 \cup A_3, X_1 \cup X_2 \cup X_3)}\,\text{AND} \quad \vdots}{\vdots}$$

$$\text{AND}\,\dfrac{(A_1 \cup \ldots \cup A_{n-1}, X_1 \cup \ldots \cup X_{n-1}) \qquad \text{FD}\,\dfrac{A_n \to X_n}{(A_n, X_n)}}{(A_1 \cup \ldots \cup A_n, X_1 \cup \ldots \cup X_n)}$$

Let d' be a sub-derivation of d with root (A', X'). Clearly, $A' \subseteq A$, and so $A' \subseteq I$, since $A \subseteq I$. □

Proposition 4.9 *FC systems are not closed under sub-derivations.*

Proof. We prove this proposition by giving a counterexample. Let C be the set of conditionals $\{A \to X, X \to Y\}$ and let $I = \{A\}$. Consider the following derivation:

$$d = \dfrac{d_1 = \dfrac{A \to X}{(A, X)} \quad \dfrac{X \to Y}{(X, Y)} = d_2}{(A, X \cup Y)}$$

We have $i(root(d)) \subseteq I$, so that $d \in \mathtt{der}(L, C, I, FC)$. Since $i(root(d_2)) = X$ and $X \not\subseteq I$, $d_2 \notin \mathtt{der}(L, C, I, FC)$. □

Proposition 4.10 (Non-repetition) *For every $d \in \mathtt{der}(L, C, I, FA)$ with root (A, X) and leaves $leaves(d)$, there exists a derivation $d' \in \mathtt{der}(L, C, I, FA)$ with the same root and the same set of leaves, such that each leaf in $leaves(d')$ is used at most once. The same holds for every derivation $d \in \mathtt{der}(L, C, I, FC)$.*

Proof. We only consider the case of FC systems (the argument for FA systems is similar). Assume we have a derivation d with $root(d) = (A, X)$ and $leaves(d) = \{A_1 \to X_1, \dots A_n \to X_n\}$. By Proposition 4.7, one can transform d into a derivation d' of the form

$$\dfrac{\dfrac{\dfrac{\dfrac{\dfrac{A_1 \to X_1}{(A_1,X_1)}\,\text{FD} \qquad \dfrac{A_2 \to X_2}{(A_2,X_2)}\,\text{FD}}{\vdots}\,\text{AND} \qquad \dfrac{A_3 \to X_3}{(A_3,X_3)}\,\text{FD}}{\vdots \qquad \qquad \vdots}\,\text{AND}}{\vdots} \qquad \text{FD}\,\dfrac{A_n \to X_n}{(A_n,X_n)}}{(A,X)}\,\text{AND}$$

Suppose that in d' some $A_l \to X_l$ decorates at least two distinct leaves. We show that we can eliminate the second one. To aid comprehension, let B be mnemonic for the following union, where $l \leq j$:

$$A_1 \cup (A_2 \setminus X_1) \cup (A_3 \setminus (X_1 \cup X_2)) \cup ... \cup (A_j \setminus (X_1 \cup ... \cup X_{j-1}))$$

Suppose we have the step:

$$\dfrac{\dfrac{\begin{array}{c}\dfrac{\dfrac{\dfrac{A_1 \to X_1}{(A_1,X_1)} \quad \dfrac{A_2 \to X_2}{(A_2,X_2)}}{(A_1 \cup (A_2 \setminus X_1), X_1 \cup X_2)} \quad \dfrac{A_3 \to X_3}{(A_3,X_3)}}{(A_1 \cup (A_2 \setminus X_1) \cup (A_3 \setminus (X_1 \cup X_2)), X_1 \cup X_2 \cup X_3)} \\ \vdots \qquad \qquad \vdots \quad \dfrac{A_j \to X_j}{(A_j,X_j)}\end{array}}{(B, \bigcup_{i=1}^{j} X_i)} \qquad \dfrac{A_l \to X_l}{(A_l,X_l)}}{(B \cup (A_l \setminus \bigcup_{i=1}^{j} X_i), \bigcup_{i=1}^{j} X_i \cup X_l)}$$

where the sub-derivation with root $(B, \bigcup_{i=1}^{j} X_i)$ contains a leaf carrying $A_l \to X_l$. That is, $A_l \to X_l$ is one of $A_1 \to X_1$, ... and $A_j \to X_j$, and it is re-used immediately after $A_j \to X_j$. Since X_l is one of X_1, ... and X_j, $\bigcup_{i=1}^{j} X_i \cup X_l = \bigcup_{i=1}^{j} X_i$. On the other hand, $(A_l \setminus \bigcup_{i=1}^{j} X_i) \subseteq (A_l \setminus \bigcup_{i=1}^{l-1} X_i) \subseteq B$, so that $B \cup (A_l \setminus \bigcup_{i=1}^{j} X_i) = B$. Thus, we can remove from d' all the re-occurrences of the leaves as required.

□

5 Related research

As mentioned in Section 1, the present paper extends the framework described by Tosatto et al. [9] in order to handle conjunction of outputs along with the form of cumulative transitivity introduced by Parent and van der Torre [7].

At the time of writing this paper, we are not able to report any formal result showing how the Tosatto et al. framework relates with the present one. Care should be taken here. On the one hand, the present account does not validate the rule of strengthening of the input, while the Tosatto et al. one does in the following restricted form: from $(\top, x)$, infer (y, x). On the other hand, in order to relate the proof-theory with the semantics, the authors make a detour through the notion of deontic redundancy [10]. A more detailed comparison between the two accounts is left as a topic for future research.

There are close similarities between the systems described in this paper and the systems of I/O logic introduced by Parent and van der Torre [7]. As explained in the introductory section, our rule CAND is the set-theoretical counterpart of their rule ACT. In both systems, weakening of the output goes away. At the same time there are also important differences between the two settings. First, the present setting remains neutral about the specific language to be used. This one need not be the language of propositional logic. Second, the present account does not validate the rule of strengthening of the input.

Tosatto et al. explain how to instantiate the ANS with propositional logic to obtain fragments of the standard input/output logics [5]. In this section we rerun the same exercise for the systems studied in [7]. Unlike Tosatto et al., we argue semantically, and not proof-theoretically, because of the problem alluded to above: derivations in FC are not closed under sub-derivations.

For the reader's convenience, we first briefly recall the definitions of $\mathcal{O}_1$ and $\mathcal{O}_3$ given by Parent and van der Torre [7]. Given a set X of formulas, and a set N of norms (viewed as pairs of formulas), $N(X)$ denotes the image of N under X, i.e., $N(X) = \{x : (a, x) \in N \text{ for some } a \in X\}$. $Cn(X)$ is the consequence set of X in classical propositional logic. And $x \dashv\vdash y$ is short for $x \vdash y$ and $y \vdash x$. We have $x \in \mathcal{O}_1(N, I)$ whenever there is some finite $M \subseteq N$ such that $M(Cn(I)) \neq \{\}$ and $x \dashv\vdash \wedge M(Cn(I))$. We have $x \in \mathcal{O}_3(N, I)$ if and only if there is some finite $M \subseteq N$ such that $M(Cn(I)) \neq \{\}$ and for all B, if $I \subseteq B = Cn(B) \supseteq M(B)$, then $x \dashv\vdash \wedge M(B)$. [11]

Theorem 5.1 (Instantiation) *Let $\langle L, C, R\rangle$ be a FA system, or a FC system, with L the language of propositional logic (without $\top$) and C a set of conditionals whose antecedents and consequents are singleton sets. Define $N = \{(a, x) \mid \{a\} \to \{x\} \in C\}$. The following applies:*

i) If $X \in det(L, C, I, FA)$, then $\wedge X \in \mathcal{O}_1(N, I)$, where $\wedge X$ is the conjunction of all the elements of X;

ii) If $X \in det(L, C, I, FC)$, then $\wedge X \in \mathcal{O}_3(N, I)$.

Proof. See [1]. □

6 Summary and future work

We have extended the Tosatto et al. framework of abstract normative systems in order to handle conjunction of outputs along with the aggregative form of cumulative transitivity introduced by the last two co-authors of the present

[11] The proof-system corresponding to $\mathcal{O}_1$ has three rules: from (a, x) and $b \vdash a$, infer (b, x) (SI); from (a, x) and (a, y), infer $(a, x \wedge y)$ (AND); from (a, x) and $b \dashv\vdash a$, infer (b, x) (EQ). The proof-system corresponding to $\mathcal{O}_3$ may be obtained by replacing (AND) with (ACT). This is the rule: from (a, x) and $(a \wedge x, y)$, infer $(a, x \wedge y)$.

paper. We have introduced two abstract normative systems, the FA and FC systems. We have illustrated these two systems with examples from literature, and presented two representation theorems for these systems. We have also shown how they relate to the original I/O systems.

FA systems. They supplement factual detachment with the rule of simple aggregation, taking unions of inputs and outputs. The representation theorem shows that the sets of formulas that can be detached in FA precisely correspond to sets of conditionals that generate this output.

FC systems. They supplement factual detachment with the rule of cumulative aggregation, a subtle kind of transitivity or reuse of the output, as introduced in [7]. The representation theorem shows how the cumulative aggregation rule corresponds to the reuse of the detached formulas.

Besides the issues mentioned in the previous section, we are currently investigating the question of how to use FA and FC systems as a basis for a Dung-style argumentation framework.

Acknowledgments We thank three anonymous referees for valuable comments. Leendert van der Torre has received funding from the European Union's H2020 research and innovation programme under the Marie Skodowska-Curie grant agreement No. 690974 for the project "MIREL: MIning and REasoning with Legal texts".

References

[1] Ambrossio, D. A., X. Parent and L. van der Torre, *A representation theorem for abstract cumulative aggregation*, Technical report, University of Luxembourg (2016). URL `http://hdl.handle.net/10993/27364`

[2] Broome, J., "Rationality Through Reasoning," Wiley-Blackwell, 2013.

[3] Goble, L., *A logic of good, should, and would. Part I*, Journal of Philosophical Logic **19** (1990), pp. 169–199.

[4] Koslow, A., "A Structuralist Theory of Logic," Cambridge University Press, Cambridge, 2001.

[5] Makinson, D. and L. van der Torre, *Input-output logics*, Journal of Philosophical Logic **29** (2000), pp. 383–408.

[6] Parent, X. and L. van der Torre, *Aggregative deontic detachment for normative reasoning*, in: *Principles of Knowledge Representation and Reasoning: Proceedings of KR 2014, Vienna, Austria, July 20-24, 2014*.

[7] Parent, X. and L. van der Torre, *"Sing and dance!"*, in: F. Cariani, D. Grossi, J. Meheus and X. Parent, editors, *Deontic Logic and Normative Systems, DEON 2014*, Lecture Notes in Computer Science, pp. 149–165.

[8] Ripley, D., *Paradoxes and failures of cut*, Australasian Journal of Philosophy **91** (2013), pp. 139–164.

[9] Tosatto, S., G. Boella, L. van der Torre and S. Villata, *Abstract normative systems*, in: *Principles of Knowledge Representation and Reasoning: Proceedings of KR 2012*, pp. 358–368.

[10] van der Torre, L., *Deontic redundancy*, in: G. Governatori and G. Sartor, editors, *Deontic Logic in Computer Science: Proceedings of DEON 2010* (2010), pp. 11–32.

An Exact Truthmaker Semantics for Permission and Obligation

Albert J.J. Anglberger [1]

University of Bayreuth

Federico L.G. Faroldi [2]

University of Pisa & University of Florence

Johannes Korbmacher [3]

Munich Center for Mathematical Philosophy

Abstract

We develop an exact truthmaker semantics for permission and obligation. The idea is that with every singular act, we associate a *sphere of permissions* and a *sphere of requirements*: the acts that are rendered permissible and the acts that are rendered required by the act. We propose the following clauses for permissions and obligations:

- a singular act is an exact truthmaker of $P\varphi$ iff every exact truthmaker of φ is in the sphere of permissibility of the act, and
- a singular act is an exact truthmaker of $O\varphi$ iff some exact truthmaker of φ is in the sphere of requirements of the act.

We show that this semantics is *hyperintensional*, and that it can deal with some of the so-called *paradoxes of deontic logic* in a natural way. Finally, we give a sound and complete axiomatization of the semantics.

Keywords: strong permission, exact truthmaker semantics, free choice permission, Good Samaritan paradox

1 Introduction

The aim of this paper is to develop an exact truthmaker semantics for permission and obligation. The basic idea of exact truthmaker semantics is that we can give the semantic content of a statement by saying what precisely in the world makes the statement true: by giving its exact truthmakers. Intuitively,

[1] ✉ albert@anglberger.org

[2] ✉ faroldi@nyu.edu

[3] ✉ jkorbmacher@gmail.com

an exact truthmaker of a statement is a state (of affairs) such that whenever the state obtains it is directly and wholly responsible for the truth of the statement. In particular, an exact truthmaker of a statement will not contain as a part any other state that is not wholly responsible for the truth of the statement. So, for example, the state of the pen being black is an exact truthmaker of the statement "the pen is black." But the complex state of the pen being black and full of ink is *not* an exact truthmaker of the statement, since it contains as a part the state of the pen being full of ink, which is irrelevant to the truth of "the pen is black." This idea traces back to a paper by Bas van Fraassen [11]. But in recent work, Kit Fine uses it to give truth-conditions for: counterfactual conditionals [3], metaphysical ground [4], permission [5], and partial content and analytic equivalence [6]. [4]

It turns out that the framework of exact truthmaker semantics has a natural action-theoretic interpretation: we can take an exact truthmaker of a sentence to be any concrete singular act, such that the performance of the act is directly and wholly responsible for the truth of the sentence. For example, on this interpretation, President Obama's act of refilling the pen at his desk in the oval office on Monday morning at 7 a.m. would be an exact truthmaker of the statement "Obama refills the pen." In contrast, Obama's act of refilling the pen and spilling his coffee would *not* be an exact truthmaker of the statement, because it has as a part the irrelevant act of Obama spilling his coffee. In this paper, we will use this interpretation to provide a natural semantics for permissions and obligations, which are the direct result of normative acts.

Once we interpret the exact truthmaker framework in this way, there is a natural way to obtain truth-conditions for permissions and obligations. For this purpose, let's assume that we're given a set of normatively admissible and a set of normatively required acts. Then we can say:

- a statement of the form $P\varphi$ is true iff every act that is an exact truthmaker of φ is admissible, [5] and
- a statement of the form $O\varphi$ is true iff some act that is required is an exact truthmaker of φ.

But this only gives us the *truth-conditions* for permissions and obligations, and not their exact truthmakers. And from the perspective of exact truthmaker semantics, this means that these clauses don't give us the *content* of permissions and obligations. To make things worse, the clauses cannot be applied to *iterated* permissions and obligations, where a permission or obligation occurs in the context of another permission or obligation. To see this, consider a statement of the form $OP\varphi$, for example. According to the above truth-conditions, we get:

- a statement of the form $OP\varphi$ is true iff some act that is required *is an exact*

[4] Note that Fine only gives *truth-conditions* for the concepts in question and not their exact truthmakers.

[5] Such a clause is essentially proposed by Fine [5].

truthmaker of $P\varphi$.

But since we don't know what an exact truthmaker of $P\varphi$ is, we can't ascertain the truth-value of $OP\varphi$. In this paper, we shall propose recursive clauses for the exact truthmakers of permissions and obligations, which can deal with these issues.

We propose that with every act, a set of acts is associated that are admissible as a result of the act being performed, and a set of acts that are required as as result of the act being performed: we associate with every act a *sphere of permissions* and a *sphere of requirements*. For example, consider John's act of checking in at the airport. This act permits him to proceed to his gate, but it obligates him to keep his luggage with him at all times. Thus, the act of John going to the gate is in the sphere of permissions of him checking in, and the act of John keeping his luggage with him is in the act's sphere of obligations. We can then give the following clauses for the exact truthmakers of permissions and obligations:

- an act is an exact truthmaker of $P\varphi$ iff every exact truthmaker of φ is in the sphere of permission of the act, and
- an act is an exact truthmaker of $O\varphi$ iff some exact truthmaker of φ is in the sphere of requirements of the act.

In the following, we shall develop this informal idea in formal detail.

2 The Semantics

In the following we shall work in the context of a standard propositional deontic language with: a countable stock $p_1, p_2, \ldots$ of propositional variables, the truth-functional operators $\neg, \wedge, \vee$ and the deontic operators P, O. We write $\varphi \rightarrow \psi$ for $\neg\varphi \vee \psi$ and $\varphi \leftrightarrow \psi$ for $(\varphi \rightarrow \psi) \wedge (\psi \rightarrow \varphi)$.

To develop our semantics, we assume that we're given a non-empty set A of *atomic singular acts.*[6] These acts correspond to the concrete atomic acts an agent might perform, like the aforementioned concrete act by Obama of refilling the pen, for example. We then say that a *complex singular act* (over A) is a set of atomic acts:

$$X \text{ is a complex singular act iff } X \subseteq A.$$

Complex acts are "aggregates" of atomic acts, which we think of as being performed together, like the concrete act of Obama refilling the pen *and* spilling

[6] Two short comments are in order here. First, whenever we talk about concrete singular acts (atomic or complex), we do not presuppose that they are actually executed. By "concrete singular acts" we rather mean "(possible) concrete singular acts", and we will introduce executed singular acts later. Second, how to distinguish between atomic and complex singular acts certainly is an interesting philosophical question. Here we do not deal with this question though, and rather assume that this distinction is useful. However, nothing hinges on that. To get the theory off the ground, all we need is that we can individuate concrete acts $a_1 \ldots, a_n$ to construct the set $A = \{a_1, \ldots, a_n\}$ of concrete (atomic) acts. Anyone who deems the distinction between atomic and complex singular acts to be meaningless, may just take the singletons of A to be "complex" generic acts, which, in a sense, eliminates the distinction.

the coffee. We denote the set of complex acts (over A) by $\mathbf{A}$, i.e. $\mathbf{A} = \wp(A)$. A *generic action* over A is a set of complex singular acts over A:

$$\mathcal{X} \text{ is a generic action iff } \mathcal{X} \subseteq \mathbf{A}.$$

A generic action is a collection of complex acts, which we think of as the different ways of performing the generic action. For example, there are various concrete ways in which Obama can refill the pen, e.g. he may refill it with blue ink, black ink, green ink etc. All these concrete acts are realizations of the same generic action of refilling the pen. A similar phenomenon can be found in metaphysics: various (concrete) objects can be concrete instances of one and the same (abstract) type. Obviously, the same holds for singular acts and generic actions: there are numerous (concrete) ways in which Obama can refill the pen, all of which are instances of the (abstract) type *Obama-refills-the-pen.* Hence and in line with the usual terminology, we will occasionally use 'action token' to talk about a singular act (atomic or complex), and 'action type' to talk about a generic action.

We denote the set of generic actions over A by $\mathbf{T}$, i.e. $\mathbf{T} = \wp(\mathbf{A})$. Finally, we assume that some subset $Ex \subseteq A$ of atomic singular acts are *executed.* We say that a complex singular act $X \in \mathbf{A}$ is executed iff *all* the members of X are executed:

$$X \text{ is executed iff } X \subseteq Ex.$$

We denote the sets of executed complex singular acts by $\mathbf{Ex}$, i.e. $\mathbf{Ex} = \wp(Ex)$. And a generic action $\mathcal{X} \in \mathbf{T}$ is realized iff *some* member $X \in \mathcal{X}$ is executed:

$$\mathcal{X} \text{ is realized iff } \mathcal{X} \cap \mathbf{Ex} \neq \emptyset$$

Thus, we can think of a generic action as a disjunctive list of conjunctive complex acts. To realize a generic action means to execute (at least) one such complex act. We will call a structure of the form (A, Ex) an *action frame.* Structures of this form are the action theoretic backdrop to our semantics.

If (A, Ex) is an action frame, then we'll assume that we're given for every atomic singular act $x \in A$, both a *sphere of permissions* $Ok_x \subseteq \mathbf{A}$ and a *sphere of obligations* $Req_x \subseteq \mathbf{A}$. Intuitively, the members of Ok_x for a (singular) act $x \in A$ are exactly those (complex) acts that are rendered normatively admissible by x: it is a normative consequence of x being executed that all members of Ok_x are admissible. Similarly, the members of Req_x are the acts that are rendered required by the performance of x: it is a normative consequence of x being executed that all members of Req_x are required.[7]

Let us consider an example.[8] Suppose that Johannes executes the following, concrete act: he buys a day ticket on March 7, 2016 at 8am for the public transport in Munich (a_1). This renders quite a number of other concrete acts

[7] Note that not all acts have to be *normatively significant*, i.e. Ok_x and Req_x can also be empty.

[8] For reasons of simplicity, but without loss of generality, we take all the singular acts in the example to be atomic.

admissible: He may take the U3 at 8:04am and go to Moosach (a_2). He may take the U6 at 8:08am to go to Marienplatz (a_3). Since Johannes bought a day ticket, he is also entitled to take the S3 after work at 7pm from Marienplatz to go to Haidhausen (a_4). And so on. In our formal framework, this is expressed by $Ok_{a1} = \{\{a_2\}, \{a_3\}, \{a_4\}, \ldots\}$.

For a complex act $X \in \mathbf{A}$, we define the set Ok_X to be $\bigcup_{x \in X} Ok_x$ and Req_X to be $\bigcup_{x \in X} Req_x$. Thus, intuitively the members of Ok_X for an act $X \in \mathbf{A}$ are the acts that are rendered admissible by the performance of the members of X and the members of Req_X are the acts that are rendered required by the performance of the members of X. We call a structure of the form $(A, Ex, (Ok_x)_{x \in A}, (Req_x)_{x \in A})$, where (A, Ex) is an action frame and $((Ok_x)_{x \in A}, (Req_x)_{x \in A})$ are spheres of permissions and obligations for every act $x \in A$ a *deontic action frame*. Thus, a deontic action frame consists of a basic action theoretic structure together with a normative framework on top, which determines the normative consequences of actions.

Following von Wright [12], we take formulas of our language to represent action types. More formally, if $(A, Ex, (Ok_x)_{x \in A}, (Req_x)_{x \in A})$ is a deontic action frame, then we assign to every atomic formula p an action type $V(p) \in \mathbf{T}$, where we think of the members of $V(p)$ as all concrete acts that exactly realize what's expressed by p under V. We furthermore assign to every atomic formula p an action type $F(p) \in \mathbf{T}$, where we think of the members of $F(p)$ as all those acts that exactly prevent what's expressed by p under F.

To illustrate, think of the example with Obama again. Let us suppose that $a_1, a_2, \ldots, a_5$ are all atomic singular acts Obama can execute. Let $\{\{a_1, a_2\}, \{a_1, a_3\}\}$ be the set of all the (complex) singular acts that exactly realize the action type of Obama refilling the pen, i.e. $V(Obama - refills - the - pen) = \{\{a_1, a_2\}, \{a_1, a_3\}\}$. For example, $\{a_1, a_2\}$ might be the complex act of Obama filling the pen by opening the pen (a_1) and inserting a blue cartridge (a_2), while $\{a_1, a_3\}$ is the complex act of him opening the pen and putting in a black cartridge (a_3). Some of the other (complex) concrete singular acts, let's say $\{a_1, a_4\}$ and $\{a_1, a_5\}$, exactly realize Obama signing the document. Since Obama cannot refill the pen and sign the document (at the same time, of course), both $\{a_1, a_4\} \in F(Obama - refills - the - pen)$, and $\{a_1, a_5\} \in F(Obama - refills - the - pen)$.

We now extend the exact truthmakers and exact falsemakers to arbitrary propositional formulas by a simultaneous recursion on the construction of formulas using van Fraassen's clauses [11]:

- $V(\neg\varphi) = F(\varphi)$
- $F(\neg\varphi) = V(\varphi)$
- $V(\varphi \vee \psi) = V(\varphi) \cup V(\psi)$
- $F(\varphi \vee \psi) = \{X \cup Y \mid X \in F(\varphi), Y \in F(\psi)\}$
- $V(\varphi \wedge \psi) = \{X \cup Y \mid X \in V(\varphi), Y \in V(\psi)\}$
- $F(\varphi \wedge \psi) = F(\varphi) \cup F(\psi)$

If $\mathcal{F} = (A, Ex, (Ok_x)_{x \in A}, (Req_x)_{x \in A})$ is a deontic action frame and V and F are truthmaker (falsemaker) assignments of the sort just described, then $(\mathcal{F}, V, F)$ is a *deontic action model.*

Since the underlying action frame tells us which actions are executed, we can define what it means for a formula to be true (false) under an interpretation of the sort just described: it is true iff the action type it expresses (prevents what it expresses) is executed. More precisely, if $\mathcal{M} = (\mathcal{F}, V, F)$ is a deontic action model, then:

- $\mathcal{M} \vDash \varphi$ iff $V(\varphi)$ is realized, i.e. $V(\varphi) \cap \mathbf{Ex} \neq \emptyset$

φ is true in a model $\mathcal{M} = (\mathcal{F}, V, F)$ iff there is an exact realization of φ that is executed according to the deontic action frame $\mathcal{F} = (A, Ex, (Ok_x)_{x \in A}, (Req_x)_{x \in A})$. This simply means that for at least one exact realization of φ, all atomic acts that constitute an exact realization of φ are in Ex.

- $\mathcal{M} \dashv \varphi$ iff $F(\varphi)$ is realized, i.e. $F(\varphi) \cap \mathbf{Ex} \neq \emptyset$

φ is false in a model $\mathcal{M} = (\mathcal{F}, V, F)$ iff there is an executed act according to the deontic action frame $\mathcal{F} = (A, Ex, (Ok_x)_{x \in A}, (Req_x)_{x \in A})$, such that an execution of φ is prevented. This simply means that for at least one exact exact falsemaker of φ, all atomic acts that constitute such an exact exact falsemaker of φ are in Ex.

As usual, validity ($\vDash$) is defined as truth in all deontic action models. We can then show the following lemma:

Lemma 2.1 *If $\mathcal{M}$ is deontic action model, then:*

i) a) $\mathcal{M} \vDash \neg\varphi$ *iff* $\mathcal{M} \dashv \varphi$
b) $\mathcal{M} \dashv \neg\varphi$ *iff* $\mathcal{M} \vDash \varphi$

ii) a) $\mathcal{M} \vDash \varphi \wedge \psi$ *iff* $\mathcal{M} \vDash \varphi$ *and* $\mathcal{M} \vDash \psi$
b) $\mathcal{M} \dashv \varphi \wedge \psi$ *iff* $\mathcal{M} \dashv \varphi$ *or* $\mathcal{M} \dashv \psi$

iii) a) $\mathcal{M} \vDash \varphi \vee \psi$ *iff* $\mathcal{M} \vDash \varphi$ *or* $\mathcal{M} \vDash \psi$
b) $\mathcal{M} \dashv \varphi \vee \psi$ *iff* $\mathcal{M} \dashv \varphi$ *and* $\mathcal{M} \dashv \psi$

We might want to put conditions on the exact truthmakers and exact falsemakers of formulas. If V and F are exact truthmaker and exact falsemaker assignments in a deontic action frame, we say that:

- (V, F) is complete iff for all p, $V(p) \cap \mathbf{Ex} \neq \emptyset$ or $F(p) \cap \mathbf{Ex} \neq \emptyset$
  (i.e. p is either realized or prevented)
- (V, F) is consistent iff for no p, $V(p) \cap \mathbf{Ex} \neq \emptyset$ and $F(p) \cap \mathbf{Ex} \neq \emptyset$
  (i.e. p is not realized and prevented)
- (V, F) is classical iff for all p, either $V(p) \cap \mathbf{Ex} \neq \emptyset$ or $F(p) \cap \mathbf{Ex} \neq \emptyset$
  (i.e. p is either realized or prevented, but not both)

It it easily shown, that these conditions extend to all formulas:

Lemma 2.2 *If $\mathcal{M} = (\mathcal{F}, V, F)$ is a deontic action model, then for all φ without P or O:*

i) if (V, F) is complete, then for all φ, $\mathcal{M} \vDash \varphi$ or $\mathcal{M} \dashv \varphi$

ii) if (V, F) is consistent, then for all φ, not both $\mathcal{M} \vDash \varphi$ and $\mathcal{M} \dashv \varphi$

iii) if (V, F) is classical, then for all φ, either $\mathcal{M} \vDash \varphi$ or $\mathcal{M} \dashv \varphi$

In particular, this means that by imposing conditions on the assignments, we can ensure that we obtain a certain background logic: [9]

Lemma 2.3 *For all Γ and φ without P or O,*

i) $\Gamma \vDash_{\mathbf{FDE}} \varphi$ iff for all deontic action models $\mathcal{M}$, if $\mathcal{M} \vDash \Gamma$, then $\mathcal{M} \vDash \varphi$

ii) $\Gamma \vDash_{\mathbf{K3}} \varphi$ iff for all deontic action models $\mathcal{M}$ such that (V, F) is consistent, if $\mathcal{M} \vDash \Gamma$, then $\mathcal{M} \vDash \varphi$

iii) $\Gamma \vDash_{\mathbf{LP}} \varphi$ iff for all deontic action models $\mathcal{M}$ such that (V, F) is complete, if $\mathcal{M} \vDash \Gamma$, then $\mathcal{M} \vDash \varphi$

iv) $\Gamma \vDash_{\mathbf{CL}} \varphi$ iff for all deontic action models $\mathcal{M}$ such that (V, F) is classical, if $\mathcal{M} \vDash \Gamma$, then $\mathcal{M} \vDash \varphi$

Proof. This follows from the previous two lemmas. □

We could therefore, in principle, use different background logics, but in the following we shall restrict ourselves to classical logic: we shall assume that all exact truthmaker and exact falsemaker assignments are classical. We shall call a deontic action model $(A, Ex, (Ok_x)_{x \in A}, (Req_x)_{x \in A}, V, F)$ classical iff (V, F) is classical.

It is now high time to introduce our clauses for the exact truthmakers and exact falsemakers of permissions and obligations. The case for the exact truthmakers is relatively straightforward. If $(A, Ex, (Ok_x)_{x \in A}, (Req_x)_{x \in A}, V, F)$ is a classical deontic action model, we say that:

- $V(P\varphi) = \{X \mid V(\varphi) \subseteq Ok_X\}$

  A complex act exactly realizes that φ is permitted iff the execution of that act renders all exact realizations of φ admissible.

- $V(O\varphi) = \{X \mid V(\varphi) \cap Req_X \neq \emptyset\}$

  A complex act exactly realizes that φ is obligatory iff the execution of that act renders at least one exact realization of φ required.

In other words, a complex act is a exact truthmaker of an permission $P\varphi$ iff every exact truthmaker of φ is in the sphere of permissions of the act, and an act is a exact truthmaker of an obligation $O\varphi$ iff some exact truthmaker of φ is in the sphere of obligations of the act.

---

[9] Here we assume that the reader is familiar with the many valued semantics for the logic of first-degree entailment (**FDE**), strong Kleene logic (**K3**), the logic of paradox (**LP**), and (of course) classical logic (**CL**). For the details of these semantics, see e.g. [9, §§ 7–8].

But when it comes to the exact falsemakers of permissions and obligations, the issue becomes a bit more complicated. Intuitively, what makes an permission or obligation false is that no corresponding normative acts have been executed. But what is an act that makes this the case then? We propose that if indeed no corresponding normative act has been executed, then it is the totality of the executed acts that jointly makes it the case that something is not permitted or obligatory:

- $F(P\varphi) = \begin{cases} \{Ex\} & if\ V(P\varphi) \cap \mathbf{Ex} = \emptyset \\ \emptyset & otherwise \end{cases}$
- $F(O\varphi) = \begin{cases} \{Ex\} & if\ V(O\varphi) \cap \mathbf{Ex} = \emptyset \\ \emptyset & otherwise \end{cases}$

Remember that we confined our semantics to classical deontic action models. The classicality of deontic action models and the definition of $F(P\varphi)$ and $F(O\varphi)$ result in a very natural reading of what prevents a permission (an obligation) to hold in that model. On the one hand, classicality implies completeness: given a classical deontic action model $\mathcal{M}$, every φ is either realized or prevented (given the set of executed singular acts $\mathbf{Ex}$ of the model $\mathcal{M}$). As a consequence, either $P\varphi$ is realized or $P\varphi$ is prevented in a classical deontic action model $\mathcal{M}$. Now suppose that there is no executed act that allows φ, i.e. $V(P\varphi) \cap \mathbf{Ex} = \emptyset$. Since the model is maximal, there is no further executable act, and the totality of all executed atomic acts (Ex) is responsible for $P\varphi$ ($O\varphi$) being prevented. On the other hand, classicality also implies consistency: if there is an executed act that allows φ, i.e. $V(P\varphi) \cap \mathbf{Ex} \neq \emptyset$, then there cannot be an act that exactly prevents it from being permitted, i.e. $F(P\varphi) = \emptyset$ (same for $O\varphi$).

We shall conclude this section with an observation about how our semantics relates to the truth-conditions that we sketched in the introduction to this paper. Remember that we said that once we've identified what states are admissible and required, natural truth conditions for P and O are as follows:

- a statement of the form $P\varphi$ is true iff every act that is an exact truthmaker of φ is admissible, and
- a statement of the form $O\varphi$ is true iff some act that is required is an exact truthmaker of φ.

Indeed, in our semantics above, we can recover these truth-conditions in the following lemma:

Lemma 2.4 *If $\mathcal{M}$ is a classical deontic action model, then:*

i) $\mathcal{M} \vDash \neg\varphi$ *iff* $\mathcal{M} \nvDash \varphi$

ii) $\mathcal{M} \vDash \varphi \wedge \psi$ *iff* $\mathcal{M} \vDash \varphi$ *and* $\mathcal{M} \vDash \psi$

iii) $\mathcal{M} \vDash \varphi \vee \psi$ *iff* $\mathcal{M} \vDash \varphi$ *or* $\mathcal{M} \vDash \psi$

iv) $\mathcal{M} \vDash P\varphi$ *iff* $V(\varphi) \subseteq \bigcup_{x \in Ex} Ok_x$

v) $\mathcal{M} \models O\varphi$ *iff* $V(\varphi) \cap \bigcup_{x \in Ex} Req_x \neq \emptyset$

In other words, in a given classical deontic action model, we can identify the admissible acts in the model with the acts that are rendered admissible by the executed acts ($\bigcup_{x \in Ex} Ok_x$) and the required acts with the acts rendered required by the executed acts ($\bigcup_{x \in Ex} Req_x$).

Note that the semantics for P and O is *hyperintensional*: there is a model $\mathcal{M}$ and formulas φ, ψ such that $\models \varphi \leftrightarrow \psi$ ('φ is logically equivalent to ψ') and $\mathcal{M} \models P\varphi$ but $\mathcal{M} \not\models P\psi$ (and similarly for O). In other words, on our semantics permission and obligation (in a model) are not closed under logical equivalence. For example, it's easy to find a model $\mathcal{M}$ such that $\mathcal{M} \models P(p \vee \neg p)$ but $\mathcal{M} \not\models P(q \vee \neg q)$. But this is a feature rather than a bug: for example we might want it to be permitted to go home or not to go home without it being permitted to kill the cat or not kill the cat.

According to our semantics, whether something is permitted (obligatory) depends entirely on the executed actions, on the members of Ex, and their spheres of permissibility (requirements). In particular, if *no* action is executed ($Ex = \emptyset$), then *nothing* is permitted (obligatory).[10] At first glance, this seems to rule out the possibility categorical permissions (obligations), which are independent of what actions are executed. But there are ways we can allow for categorical permissions and obligations by making slight changes to our framework. First note that there are at least two ways in which we can think of a permission (obligation) being independent of the actions. We can understand this as meaning that every possible act renders a generic action permissible (obligatory) or that there is some necessarily executed act which renders it permissible (obligatory). To allow for categorical permissions in the first sense, we would simply have to require that the set of executed actions is always non-empty, i.e. $Ex \neq \emptyset$. Then it would follow that categorical permissions (obligations) in fact imply permissions (obligations) in the present sense. To allow for categorical permissions in the second sense, we might introduce a special atomic singular act $a_\top \in A$, and interpret $a_\top$ as the empty action. $a_\top$ is further always executed ($a_\top \in Ex$). Categorical permissions (obligations) can then be modelled as permissions (obligations) which result from the execution of $a_\top$.

3 The Paradoxes

In this section, we shall show that our semantics deals in a natural way with some well-known paradoxes of deontic logic.

3.1 The Paradox of Free Choice Permission

Suppose Johannes issues the following permission "Albert, you may have tiramisu or zabaglione for dessert." Albert (naturally) concludes that he is free to choose: that he may have zabaglione, and that he may have tiramisu. In everyday discourse, the permission of a disjunction seems to imply the permission

[10] We are particularly grateful to one reviewer for raising this point.

of both disjuncts (cf. [8]):

$$(FCP) \qquad P(\varphi \vee \psi) \rightarrow P\varphi \wedge P\psi$$

Put differently, permitting Albert to have tiramisu or zabaglione, but not permitting him to have tiramisu seems to be inconsistent. It is well-known that FCP is recipe for disaster: already very weak principles, if augmented with FCP, lead to unacceptable consequences. Take, for instance, the rule RE, that warrants substitution of logically equivalent formulas:

$$(RE) \qquad \frac{\vdash \varphi \leftrightarrow \psi}{\vdash P\varphi \leftrightarrow P\psi}$$

According to classical logic, we have $\vdash \varphi \leftrightarrow (\varphi \wedge \psi) \vee (\varphi \wedge \neg\psi)$. This equivalence and RE+FCP already leads to a disastrous result, i.e. if φ is permitted, then φ together with any ψ is permitted, in formal terms:

$$(IC) \qquad P\varphi \rightarrow P(\varphi \wedge \psi)$$

is a theorem of CL+FCP+RE, and it seems to be completely unacceptable as a theorem of any useful deontic logic. This suggests that it is generally very hard to find a logic which contains FCP but also avoids problematic consequences like IC. As Sven Ove Hansson puts it: "It [i.e. the derivation of IC] indicates that the free choice permission postulate may be faulty in itself, even if not combined with other deontic principles such as those of SDL." [7, p.208] This is *the problem of free choice permission.*

It probably doesn't come as a surprise that FCP is highly controversial and regarded to be implausible by most deontic logicians. Given certain interpretations of permission, FCP turns out to be valid though. Take, for instance, the open reading of permission (cf. [2],[1]) where $P\varphi$ is interpreted as "every way to ensure φ is admissible".[11] Now, given that $\varphi \rightarrow \varphi \vee \psi$ is a theorem (cf. [7]), this interpretation validates FCP. However, this reading (intuitively and formally) also validates IC: Since every way to ensure $\varphi \wedge \psi$ is a way to ensure φ, the permission of φ implies the permission of $\varphi \wedge \psi$. However, accepting the (intuitively) unacceptable consequence IC in order to make sense of the (intuitively) acceptable principle FCP is far from an ideal solution to the problem of free choice permission. This approach just replaces one evil with another.

In our opinion, the semantics developed in the previous section offers a *real* solution to the problem of free choice permission. First, note that according to our reading of permission, FCP turns out to be valid. In this respect, our semantics is similar to the open reading of permission. In more formal terms:

Lemma 3.1 $\models P(\varphi \vee \psi) \rightarrow P\varphi \wedge P\psi$.

Proof. Let $\mathcal{M} = (\mathcal{F}, V, F)$ be deontic action model and suppose $\mathcal{M} \models P(\varphi \vee \psi)$ i.e. (by Lemma 2.4) $V(\varphi \vee \psi) \subseteq \bigcup_{x \in Ex} Ok_x$. Hence, $V(\varphi) \cup V(\psi) \subseteq \bigcup_{x \in Ex} Ok_x$,

[11] Or "every execution of φ leads to an Ok-state", depending on your preferred framework, cf. [2]

by the construction of exact realizations of disjunctive generic actions. Basic set theory now gives us $V(\varphi) \subseteq \bigcup_{x \in Ex} Ok_x$ and $V(\psi) \subseteq \bigcup_{x \in Ex} Ok_x$, which according to Lemma 2.4 means that $\mathcal{M} \models P\varphi \wedge P\psi$. □

But how do we now avoid the seemingly unavoidable consequence IC? The solution to this is quite simple: RE is not a sound rule in our semantics. The semantics we developed in the previous section is hyperintensional: logical (even necessary) equivalences may not generally be substituted for one another. In order to see why RE is not a plausible rule in exact truthmaker semantics, take, for example, the problematic equivalence statement $\varphi \leftrightarrow (\varphi \wedge \psi) \vee (\varphi \wedge \neg\psi)$ again. Although classically equivalent, φ and $(\varphi \wedge \psi) \vee (\varphi \wedge \neg\psi)$ may have completely different exact realizations. An exact realization of $(\varphi \wedge \psi)$ must consist of an exact realization of both φ and ψ, and exact realization of $(\varphi \wedge \neg\psi)$ of an exact realization of φ and an exact prevention of ψ. An exact realization of φ does not have to be either, just take an exact realization of φ that is neither an exact realization of ψ nor an exact prevention of ψ. This idea shows us how to find a countermodel for IC:

Lemma 3.2 $\not\models P\varphi \rightarrow P(\varphi \wedge \psi)$.

Proof. Let $\mathcal{F} = (A, Ex, (Ok_x)_{x \in A}, (Req_x)_{x \in A})$ be a deontic action frame with $A = \{a_1, a_2\}$, $Ex = \{a_1\}$, $Ok_{a_1} = \{\{a_1\}\}$. Let $\mathcal{M} = (\mathcal{F}, V, F)$ based on $\mathcal{F}$ s.t. $V(\varphi) = \{\{a_1\}\}$ and $V(\psi) = \{\{a_2\}\}$. This gives us $\mathcal{M} \models P\varphi$ (since $V(\varphi) \subseteq Ok_{a_1}$ and $a_1 \in Ex$). We also have $V(\varphi \wedge \psi) = \{\{a_1, a_2\}\}$, but since there is no x with $\{\{a_1, a_2\}\} \not\subseteq Ok_x$ and $x \in Ex$, we get $\mathcal{M} \not\models P(\varphi \wedge \psi)$. □

To conclude this section, let us now consider the converse of FCP

$$(CFCP) \qquad (P\varphi \wedge P\psi) \rightarrow P(\varphi \vee \psi).$$

CFCP does not seem to have sparked much controversy, and just as in many other deontic logics, it is also valid in our semantics:

Lemma 3.3 $\models (P\varphi \wedge P\psi) \rightarrow P(\varphi \vee \psi)$.

Proof. Left to the reader. □

Note that $P\varphi \wedge P\psi$ and $P(\varphi \vee \psi)$ are not just logically equivalent, but they even have the same exact truthmakers. Hence, offering a choice between φ or ψ (i.e. $P(\varphi \vee \psi)$) is permitting both φ and ψ (i.e. $P\varphi \wedge P\psi$), and permitting both is offering a choice. We take this to be very plausible property of our semantics.

However, as we will now show, the reasons why our logic validates the converse of FCP are quite different from the ones usually brought forward: Many deontic logics (e.g. SDL, DDeL, etc.) contain closure of permission under logical consequence

$$(CLP) \qquad \frac{\vdash \varphi \rightarrow \psi}{\vdash P\varphi \rightarrow P\psi},$$

which immediately gives you CFCP. But of course, closure usually gives you *more* than just that. Whereas CFCP might seem plausible, some formulas used in its standard derivation

(i) $\varphi \to (\varphi \vee \psi)$ (Tautology)

(ii) $P\varphi \to P(\varphi \vee \psi)$ (CLP)

(iii) $(P\varphi \wedge P\psi) \to P(\varphi \vee \psi)$ (Monotonicity)

have been considered to be intuitively problematic. In particular, formula (ii) has the permission variant of the Ross' Paradox

"If it is permitted to post the letter, then it is permitted to post the letter or burn it."

as one of its many counterintuitive instances.[12] (ii) is logically stronger than CFCP, so it is obviously is more prone to counterexamples. Where in CFCP both φ and ψ have to be permitted to result in the permission of $\varphi \vee \psi$, it is according to (ii) sufficient that only φ is. What makes this intuitively problematic is that we might also add an (intuitively) forbidden action type ψ ("burn the letter", in the example). In our semantics, permission is *not* closed under logical consequence (it's not even closed under logical equivalence, see above) so we avoid the validity of problematic formula (ii), while keeping the desired property CFCP.
The lesson to be drawn from all of this is that material equivalence does not adequately express identity of exact realizations. This is how we solve the problem of free choice permission in our semantics.

3.2 The Good Samaritan Paradox

Another paradox of deontic logic that has a natural solution in our semantics is Prior's Good Samaritan paradox [10]. This paradox arises in systems where obligation is closed under logical consequence, i.e. systems which validate the following rule:

$$(CL) \quad \frac{\vdash \varphi \to \psi}{\vdash O\varphi \to O\psi}$$

This rule is validated by many systems of deontic logic, such as the system SDL of standard deontic logic, but it leads to counterintuitive results in certain cases. Consider the case of Smith who has been robbed. Intuitively, it is obligatory that Jones helps smith. Thus, it is obligatory that John helps Smith who has been robbed. According to Prior, we can formalize this by the formula $O(p \wedge q)$, where p stands for *John helps Smith* and q stands for *Smith has been robbed.* But since in classical logic we have $\vdash p \wedge q \to p$, it follows by CL that Oq, which

[12] One might also have general worries about the classical action theoretic background logic (and thereby (i) and (iii)) or about monotonicity (iii). We want to keep the action theoretic background as classical as possible though, and rather focus on what is special about deontic contexts.

means that its obligatory that Smith has been robbed and is absurd.

Prior's concrete example may be more or less convincing, but there are many examples of the same logical structure that lead to the same result: CL is intuitively flawed. For example, it is intuitively obligatory for the nurse to give his patient the medicine A and medicine B, if together they heal him, but medicine A alone might kill the patient, so it is not obligatory for the nurse to give his patient medicine A. Intuitively, the problem is that certain acts, such as the nurse giving the patient medicine A and the nurse giving him medicine B, are only required in conjunction and not by themselves. And in our semantics, we can faithfully represent this intuitive claim.

To see this, let's model this situation in our semantics. Consider an action frame (A, Ex) with two atomic acts $A = \{a, b, c\}$ and one executed action $\{a\}$. Intuitively, a is the act of the doctor telling the nurse that he should give medicine A and B to the patient, b is the act of the nurse giving medicine A to the patient, and c is the act of the nurse giving medicine B to the patient. Since a is the act of the doctor telling the nurse that he should give medicine A and B to the patient, we can plausibly assume that $Req_a = Req_{\{a\}} = \{\{b, c\}\}$, and for simplicity we can assume that the spheres of permissions and obligations for all the other acts are empty. Let $\mathcal{M}$ be the corresponding deontic action frame. Now let p stand for *the nurse gives the patient medicine* A and q for *the nurse gives the patient medicine* B. We will have $V(p) = \{\{b\}\}$, $V(q) = \{\{c\}\}$, and thus $V(p \wedge q) = \{\{b, c\}\}$. Moreover, we'll have $V(O(p \wedge q)) = \{\{a\}\}$ and hence that $\mathcal{M} \models O(p \wedge q)$. But we'll neither have $\mathcal{M} \models Op$ nor $\mathcal{M} \models Oq$, exactly as we want. More generally, this model shows that CL is not sound with respect to our semantics:

Lemma 3.4 $\not\models O(\varphi \wedge \psi) \rightarrow O\varphi$

In this consists our solution to the Good Samaritan paradox.[13]

4 Axioms

In this section, we give a sound and complete axiomatization of our semantics. However, we shall use a slightly non-standard technique to obtain such an axiomatization, which is nevertheless adequate to the hyperintensional spirit of our semantics.

To formulate our proof-theory, we shall extend our language with the binary operator $\varphi \leftrightharpoons \psi$, which we give the intended reading that φ and ψ have the same truthmakers in all models. This allows us to significantly simplify the proof-theory for our logic, which will simply be the restriction of the proof-theory we develop here to the language without $\leftrightharpoons$.

[13] It's well known that CL can also be made invalid by moving to a non-normal modal logic, e.g. by using a neighborhood semantics without upward closure of the obligation neighborhood (of a world). What makes our approach different, though, is that we do not just get rid of it (formally), but that we can give a natural explanation of its invalidity in terms of exact truthmakers.

In a recent paper, Fine sketches how to obtain an axiomatization of sameness of exact truthmakers according to van Fraassen's clauses, which we've used in our above semantics [6]. The axiomatization consists of the following axioms and rules:

$$\varphi \leftrightharpoons \varphi \qquad \varphi \leftrightharpoons \neg\neg\varphi$$
$$\varphi \wedge \psi \leftrightharpoons \psi \wedge \varphi \qquad \varphi \vee \psi \leftrightharpoons \psi \vee \varphi$$
$$\varphi \vee \varphi \leftrightharpoons \varphi \qquad \varphi \wedge (\psi \vee \theta) \leftrightharpoons (\varphi \wedge \psi) \vee (\varphi \vee \theta)$$
$$\varphi \vee (\psi \vee \theta) \leftrightharpoons (\varphi \vee \psi) \vee \theta \qquad \varphi \wedge (\psi \wedge \theta) \leftrightharpoons (\varphi \wedge \psi) \wedge \theta$$
$$\neg(\varphi \vee \psi) \leftrightharpoons \neg\varphi \wedge \neg\psi \qquad \neg(\varphi \wedge \psi) \leftrightharpoons \neg\varphi \vee \neg\psi$$

(Replacement) $\quad \theta(\varphi), \varphi \leftrightharpoons \psi / \theta(\psi)$

Let's denote derivability in this system by $\vdash_E$. Then we get the following theorem:

Theorem 4.1 *For all φ and ψ without $P, O, \leftrightharpoons$, we have: $\vdash_E \varphi \leftrightharpoons \psi$ iff for all deontic action models $(\mathcal{F}, V, F)$, we have $V(\varphi) = V(\psi)$.*

Our goal is to use this system to obtain an axiomatization for our semantics of permission and obligation. The first step along the way is to get a grip of the truthmakers of permissions and obligations. We get this in the following lemma:

Lemma 4.2 *For all deontic action models $(\mathcal{F}, V, F)$, we have for all φ and ψ:*

i) $V(P(\varphi \vee \psi)) = V(P\varphi \wedge P\psi)$.

ii) $V(O(\varphi \vee \psi)) = V(O\varphi \vee O\psi)$.

Proof. Note that since on our semantics we have that $Ok_X = \bigcup_{x \in X} Ok_x$ and $Req_X = \bigcup_{x \in X} Req_x$, we get:

- $Ok_{\bigcup_i X_i} = \bigcup_i Ok_{X_i}$
- $Req_{\bigcup_i X_i} = \bigcup_i Req_{X_i}$

Using these identities, we get:

i) $\underbrace{\{X \mid V(\varphi) \cup V(\psi) \subseteq Ok_X\}}_{=V(P(\varphi \vee \psi))} = \underbrace{\{X \cup Y \mid V(\varphi) \subseteq Ok_X, V(\psi) \subseteq Ok_Y\}}_{=V(P\varphi \wedge P\psi)}$

ii) $\underbrace{\{X \mid (V(\varphi) \cup V(\psi)) \cap Req_X \neq \emptyset\}}_{=V(O(\varphi \vee \psi))} = \underbrace{\{X \mid V(\varphi) \cap Req_X \neq \emptyset\} \cup \{X \mid V(\psi) \cap Req_X \neq \emptyset\}}_{=V(O\varphi \vee O\psi)}$

□

It turns out that these two identities are enough to obtain a sound and complete axiomatization of our semantics. The system consists of the above axioms and rules plus all axioms (over the full language including P and O) and rules of classical propositional logic and:

$$P(\varphi \vee \psi) \leftrightharpoons P\varphi \wedge P\psi$$
$$O(\varphi \vee \psi) \leftrightharpoons O\varphi \vee O\psi$$

We shall denote derivability in this system by $\vdash_{EDL}$.

Theorem 4.3 *For all* φ *and* Γ *without* $\leftrightharpoons$, $\Gamma \vdash_{EDL} \varphi$ *iff* $\Gamma \vDash \varphi$.

Let's conclude with a few sample derivations to show how the system works:

(i) $\vdash_{EDL} P(\varphi \vee \psi) \leftrightarrow P\varphi \wedge P\psi$

(a)	$P(\varphi \vee \psi) \leftrightarrow P(\varphi \vee \psi)$	(Tautology)
(b)	$P(\varphi \vee \psi) \leftrightharpoons P\varphi \wedge P\psi$	(Axiom)
(c)	$P(\varphi \vee \psi) \leftrightarrow P\varphi \wedge P\psi$	(a,b, Replacement)

(ii) $\vdash_{EDL} O\neg(\varphi \wedge \psi) \leftrightarrow O\neg\varphi \vee O\neg\psi$

(a)	$O\neg(\varphi \wedge \psi) \leftrightarrow O\neg(\varphi \wedge \psi)$	(Tautology)
(b)	$\neg(\varphi \wedge \psi) \leftrightharpoons \neg\varphi \vee \neg\psi$	(Axiom)
(c)	$O\neg(\varphi \wedge \psi) \leftrightarrow O(\neg\varphi \vee \neg\psi)$	(a,b, Replacement)
(d)	$O(\neg\varphi \vee \neg\psi) \leftrightharpoons O\neg\varphi \vee O\neg\psi$	(Axiom)
(e)	$\neg(\varphi \wedge \psi) \leftrightarrow O\neg\varphi \vee O\neg\psi$	(c,d, Replacement)

(iii) $P\neg\neg(\varphi \vee \psi) \vdash_{EDL} P\varphi$

(a)	$P\neg\neg(\varphi \vee \psi)$	(Assumption)
(b)	$\neg\neg(\varphi \vee \psi) \leftrightharpoons \varphi \vee \psi$	(Axiom)
(c)	$P(\varphi \vee \psi)$	(a,b,Replacement)
(d)	$P(\varphi \vee \psi) \leftrightarrow P\varphi \wedge P\psi$	(1.)
(e)	$P\varphi \wedge P\psi$	(c,d, Logic)
(f)	$P\varphi$	(e, Logic)

5 Summary and Future Research

We've developed a new, and so we believe exciting semantics for permission and obligation in terms of truthmakers and falsemakers. We've argued that the semantics is quite natural on intuitive grounds and we've shown that it solves the Problem of Free Choice Permission and the Good Samaritan Paradox in intuitively plausible ways.

But the work doesn't end here. Note, for example, that many standardly held principles in deontic logic fail on our semantics. Here are just a few:

(Obligation Aggregation)	$\vDash O\varphi \wedge O\psi \rightarrow O(\varphi \wedge \psi)$
(Obligation Weakening)	$\vDash O(\varphi \wedge \psi) \rightarrow O\varphi \wedge O\psi$
(Obligation Implies Permission)	$\vDash O\varphi \rightarrow P\varphi$
(No Conflicting Obligations)	$\vDash \neg(O\varphi \wedge O\neg\varphi)$

The fact that our semantics doesn't validate (Obligation Aggregation) and (Obligation Weakening) is integral to our solution of the Good Samaritan Paradox. But it might be interesting to investigate what is possible with respect to (Obligation Implies Permission) and (No Conflicting Obligations). But we postpone this work to another day.[14]

[14] Acknowledgements: We would like thank the following people for very useful comments on earlier versions of this paper: the three anonymous DEON referees, J. Broersen, A. Tamminga, O. Foisch, the Bayreuth logic group, attendees of the Deontic Logic and Formal

References

[1] Albert J.J. Anglberger, Norbert Gratzl, and Olivier Roy. Obligation, free choice, and the logic of weakest permissions. *The Review of Symbolic Logic*, 8:807–827, 12 2015.

[2] Jan Broersen. Action negation and alternative reductions for dynamic deontic logics. *Journal of Applied Logic*, 2:153–168, 2004.

[3] Kit Fine. Counterfactuals without possible worlds. *Journal of Philosophy*, 109(3):221–46, 2012.

[4] Kit Fine. Guide to ground. In Fabrice Correia and Benjamin Schnieder, editors, *Metaphysical Grounding: Understanding the Structure of Reality*, pages 37–80. Cambridge University Press, Cambridge, UK, 2012.

[5] Kit Fine. Permission and possible worlds. *Dialectica*, 68(3):317–336, 2014.

[6] Kit Fine. Angellic content. *Journal of Philosophical Logic*, pages 1–28, 2015.

[7] Sven Ove Hansson. The varieties of permission. In Dov Gabbay, John Horty, Xavier Parent, Ron van der Meyden, and Leendert van der Torre, editors, *Handbook of Deontic Logic and Normative Systems*, pages 195–240. College Publications, 2013.

[8] Hans Kamp. Free Choice Permission. *Proceedings of the Aristotelian Society*, 74:57–74, 1973.

[9] Graham Priest. *An Introduction to Non-Classical Logic: From If to Is*. Cambridge University Press, 2008.

[10] A. N. Prior. Escapism: The logical basis of ethics. In A. I. Melden, editor, *Journal of Symbolic Logic*, pages 610–611. University of Washington Press, 1958.

[11] Bas C. van Fraassen. Facts and tautological entailment. *The Journal of Philosophy*, 66(15):477–87, 1969.

[12] G. H. von Wright. Deontic Logic. *Mind*, 60:1–15, 1951.

Ethics Workshop (Venice, November 2015), the people at TiLPS, and the logic group of the Wahrheitsseminare in Salzburg.

A Structured Argumentation Framework for Detaching Conditional Obligations

Mathieu Beirlaen and Christian Straßer [1]

Research Group for Non-Monotonic Logics and Formal Argumentation
Institute for Philosophy II
Ruhr University Bochum

Abstract

We present a general formal argumentation system for dealing with the detachment of conditional obligations. Given a set of facts, constraints, and conditional obligations, we answer the question whether an unconditional obligation is detachable by considering reasons for and against its detachment. For the evaluation of arguments in favor of detaching obligations we use a Dung-style argumentation-theoretical semantics. We illustrate the modularity of the general framework by considering some extensions, and we compare the framework to some related approaches from the literature.

Keywords: formal argumentation, ASPIC$^+$, conditional norms, conflicting norms, prioritized norms, factual detachment, deontic detachment.

1 Introduction

We take an argumentative perspective on the problem of detaching conditional obligations relative to a set of facts and constraints. We allow for the construction of arguments the deontic conclusions of which are candidates for detachment. Next, we define a number of ways in which these arguments may attack one another, as when the conclusions of two arguments are conflicting. We borrow Dung's semantics [6] for evaluating arguments relative to the attack relations that hold between them. Conclusions of arguments which are evaluated positively are safely detachable in our framework. They can be interpreted as all-things-considered obligations – following Ross [28] – or output obligations – following Makinson & van der Torre [18,19].

The argumentative approach defended in this paper is both natural and precise. Norms which guide reasoning are naturally construed as conclusions of proof sequences. Objections raised against the derivation of certain obligations are naturally construed as argumentative attacks. Arguments are naturally evaluated in terms of the objections raised against them.

---

[1] Email: mathieu.beirlaen@rub.de, christian.strasser@rub.de. The research of both authors was supported by a Sofja Kovalevskaja award of the Alexander von Humboldt-Foundation, funded by the German Ministry for Education and Research.

In Section 2 we introduce a basic argumentation system for evaluating arguments the conclusions of which can be interpreted as all-things-considered obligations. This generic, modular framework can be extended in various ways, as we illustrate in Section 3. We show how various mechanisms for conflict-resolution can be implemented (Section 3.1), and how we can rule out obligations committing us to further violations or conflicts (Section 3.2). In Section 4 we compare our approach to related systems from the literature. We end by pointing to some further expansions of our framework, which we aim to present in a follow-up paper (Section 5).

Due to space limitations we had to omit the Appendix with meta-proofs in this manuscript. They are included in the online version of this article available at `http://arxiv.org/abs/1606.00339`.

2 The basic framework

We start by reviewing the basic concepts needed from Dung's semantics (Section 2.1). Next we turn to the construction of deontic arguments (Section 2.2) and attack definitions (Section 2.3). We define a consequence relation for detaching all-things-considered obligations in deontic argumentation frameworks (Section 2.4), and present some of its meta-theoretical properties (Section 2.5).

2.1 Abstract argumentation

A Dung-style *abstract argumentation framework* (AF) is a pair $(\mathcal{A}, \mathsf{Att})$ where $\mathcal{A}$ is a set of arguments and $\mathsf{Att} \subseteq \mathcal{A} \times \mathcal{A}$ is a binary relation of attack. Relative to an AF, Dung defines a number of extensions – subsets of $\mathcal{A}$ – on the basis of which we can evaluate the arguments in $\mathcal{A}$.

Definition 1 (Complete and grounded extension). *Let $(\mathcal{A}, \mathsf{Att})$ be an AF. For any $a \in \mathcal{A}$, a is acceptable w.r.t. some $\mathcal{S} \subseteq \mathcal{A}$ (or, $\mathcal{S}$ defends a) iff for all b such that $(b, a) \in \mathsf{Att}$ there is a $c \in \mathcal{S}$ for which $(c, b) \in \mathsf{Att}$.*
If $\mathcal{S} \subseteq \mathcal{A}$ is conflict-free, i.e. there are no $a, b \in \mathcal{S}$ for which $(a, b) \in \mathsf{Att}$, then:

- *$\mathcal{S}$ is a complete extension iff $a \in \mathcal{S}$ whenever a is acceptable w.r.t. $\mathcal{S}$;*
- *$\mathcal{S}$ is the grounded extension iff it is the set inclusion minimal complete extension.*

Dung [6] showed that for every AF there is a grounded extension, it is *unique*, and it can be constructed as follows.

Definition 2 (Defense). *A set of arguments $\mathcal{X}$ defends an argument a iff every attacker of a is attacked by some $b \in \mathcal{X}$.*

Definition 3 (Construction of the grounded extension). *The grounded extension $\mathcal{G}$ relative to an AF $(\mathcal{A}, \mathsf{Att})$ is defined as follows (where $\mathcal{A}$ is countable):*

- *$\mathcal{G}_0$: the set of all arguments in $\mathcal{A}$ without attackers;*
- *$\mathcal{G}_{i+1}$: all arguments defended by $\mathcal{G}_i$;*
- $\mathcal{G} = \bigcup_{i \geq 0} \mathcal{G}_i$

Besides the grounded extension, a number of further extensions (preferred, (semi-)stable, ideal etc.) have been defined in the literature. Due to space limitations, we focus exclusively on grounded extensions in the remainder.

On Dung's abstract approach [6], arguments are basic units of analysis the internal structure of which is not represented. But nothing prevents us from *instantiating* such abstract arguments by conceptualizing them as proof trees for deriving a conclusion based on a set of premises and inference rules. Frameworks with instantiated arguments are called *structured argumentation frameworks* (for examples, see e.g. [1]).[2] In the remainder of Section 2 we show how questions regarding obligation detachment in deontic logic can be addressed and answered within structured *deontic* argumentation frameworks.

2.2 Instantiating deontic arguments

Our formal language $\mathcal{L}$ is defined as follows:

$$\mathcal{P} := \{p, q, r, \ldots\} \qquad \mathcal{L}^{\Rightarrow} := \langle \mathcal{L}^P \rangle \Rightarrow \langle \mathcal{L}^P \rangle$$
$$\mathcal{L}^P := \mathcal{P} \mid \top \mid \bot \mid \neg\langle \mathcal{L}^P \rangle \mid \langle \mathcal{L}^P \rangle \vee \langle \mathcal{L}^P \rangle \qquad \mathcal{L}^{\mathsf{O}} := \mathsf{O}\langle \mathcal{L}^P \rangle$$
$$\mathcal{L}^{\Box} := \Box\langle \mathcal{L}^P \rangle \mid \langle \mathcal{L}^P \rangle \mid \neg\langle \mathcal{L}^{\Box} \rangle \mid \langle \mathcal{L}^{\Box} \rangle \vee \langle \mathcal{L}^{\Box} \rangle \qquad \mathcal{L} := \mathcal{L}^P \mid \mathcal{L}^{\Box} \mid \mathcal{L}^{\Rightarrow} \mid \mathcal{L}^{\mathsf{O}}$$

The classical connectives $\wedge, \supset, \equiv$ are defined in terms of $\neg$ and $\vee$. We represent *facts* as members of $\mathcal{L}^P$. Where $A, B \in \mathcal{L}^P$, *conditional obligations* are formulas of the form $A \Rightarrow B$, read 'If A, then it – prima facie – ought to be that B' or 'If A, then B is prima facie obligatory'.[3] Where $A \in \mathcal{L}^P$, a *constraint* $\Box A$ abbreviates that A is settled, i.e. that A holds unalterably.[4] Formulas of the form $\mathsf{O}A$ (where $A \in \mathcal{L}^P$) represent *all-things-considered* obligations.

Unless specified otherwise, upper case letters $A, B, \ldots$ denote members of $\mathcal{L}^P$ and upper case Greek letters $\Gamma, \Delta, \ldots$ denote subsets of $\mathcal{L}^P \cup \mathcal{L}^{\Box} \cup \mathcal{L}^{\Rightarrow}$. Where $\Gamma \subseteq \mathcal{L}$ and $\dagger \in \{P, \Box, \Rightarrow, \mathsf{O}\}$, $\Gamma^{\dagger} = \Gamma \cap \mathcal{L}^{\dagger}$.

$\mathrm{Cn}_{\mathbf{CL}}(\Gamma)$ denotes the closure of $\Gamma \subseteq \mathcal{L}^P$ under propositional classical logic, **CL**. $\mathrm{Cn}_{\mathbf{L}^{\Box}}(\Gamma)$ denotes the closure of $\Gamma \subseteq \mathcal{L}^{\Box}$ under $\mathbf{L}^{\Box}$, which we use as a generic name for a modal logic for representing background constraints, e.g. **T**, **S4**, **S5**, etc. In our examples below, we will assume that $\mathbf{L}^{\Box}$ is normal and validates the axiom $\Box A \supset A$.[5]

Arguments are ordered pairs $\langle A : \mathsf{s} \rangle$ in which A is called the *conclusion*, and s a *proof sequence* for deriving A. We use lower case letters $a, b, c, \ldots$ as

---

[2] Our approach is similar in spirit to the $ASPIC^+$ framework for structured argumentation from e.g. [20]. We return to this point in Section 4.2.

[3] Depending on the context of application, the following alternative readings are also fine: 'If A is the case, then B is pro tanto obligatory', 'If A, then the agent ought (prima facie, pro tanto) to bring about B'. On the latter, agentive reading, we can think of '$\Rightarrow$' as implicitly indexed by an agent.

[4] If $\Box A$ holds, then the fact that A is deemed fixed, necessary, and unalterable. Obligations which contradict these facts are unalterably violated. Carmo & Jones cite three factors giving rise to such unalterable violations. The first is time, e.g. when you did not return a book you ought to have returned by its due date. The second is causal necessity, e.g. when you killed a person you ought not to have killed. The third is practical impossibility, e.g. when a dog owner stubbornly refuses to keep her dog against the house regulations, and nobody else dares to try and convince her to remove it [4, pp. 283-284].

[5] Moreover, where $\Delta^{\Rightarrow} \subseteq \mathcal{L}^{\Rightarrow}$, we assume that $\Gamma \vdash_{\mathbf{L}^{\Box}} \Box A$ iff $\Gamma \cup \Delta^{\Rightarrow} \vdash_{\mathbf{L}^{\Box}} \Box A$.

placeholders for arguments.

Definition 4. *Given a premise set Γ, we allow the following rules for constructing arguments:*

(i) If $\Box A \in \mathrm{Cn}_{\mathbf{L}^\Box}(\Gamma)$, then $\langle \Box A : -- \rangle$ is an argument; (where $--$ denotes the empty proof sequence)

(ii) If $A \Rightarrow B \in \Gamma^{\Rightarrow}$ and $A \in \mathrm{Cn}_{\mathbf{L}^\Box}(\Gamma)$, then $\langle \mathsf{O}B : A, A \Rightarrow B \rangle$ is an argument;

(iii) If $A \Rightarrow B \in \Gamma^{\Rightarrow}$ and $a = \langle \mathsf{O}A : \ldots \rangle$ is an argument, then $\langle \mathsf{O}B : a, A \Rightarrow B \rangle$ is an argument;

(iv) If $a = \langle \mathsf{O}A : \ldots \rangle$ and $b = \langle \mathsf{O}B : \ldots \rangle$ are arguments, then $\langle \mathsf{O}(A \wedge B) : a, b \rangle$ is an argument.

(v) If $a = \langle \mathsf{O}A : \ldots \rangle$ is an argument and $\Box(A \supset B) \in \mathrm{Cn}_{\mathbf{L}^\Box}(\Gamma)$, then $\langle \mathsf{O}B : a, \Box(A \supset B) \rangle$ is an argument.

Argument a is a deontic argument *if a is of the form $\langle \mathsf{O}A : \ldots \rangle$. We use $\mathsf{C}(a)$ to denote the set of all formulas in $\mathcal{L}$ used in the construction of a, including its conclusion. E.g. where $a = \langle \mathsf{O}q : p, p \Rightarrow q \rangle$ and $b = \langle \mathsf{O}r : a, q \Rightarrow r \rangle$, $\mathsf{C}(a) = \{p, p \Rightarrow q, \mathsf{O}q\}$ and $\mathsf{C}(b) = \{p, p \Rightarrow q, \mathsf{O}q, q \Rightarrow r, \mathsf{O}r\}$. Argument a is a* sub-argument *of argument b if $\mathsf{C}(a) \subseteq \mathsf{C}(b)$; a is a* proper *sub-argument of argument b if $\mathsf{C}(a) \subset \mathsf{C}(b)$; and b is a* super-argument *of argument a if a is a proper sub-argument of b.*

(ii)-(v) correspond to inference rules well-known from deontic logic. (ii) allows for the factual detachment of an all-things-considered obligation $\mathsf{O}B$ from a conditional prima facie obligation $A \Rightarrow B$ and a fact A. (iii) is a deontic detachment principle. (iv) and (v) allow for obligation aggregation (or agglomeration), resp. inheritance (or weakening).

Example 1 (Constructing arguments). *Let $\Gamma_1 = \{\Box p, \top \Rightarrow \neg p, \neg p \Rightarrow \neg q, p \Rightarrow q\}$. By Definition 4 we can construct – amongst others – the following arguments from Γ_1:*

a_1:	$\langle \Box p : -- \rangle$	a_4:	$\langle \mathsf{O}q : p, p \Rightarrow q \rangle$
a_2:	$\langle \mathsf{O}\neg p : \top, \top \Rightarrow \neg p \rangle$	a_5:	$\langle \mathsf{O}(\neg q \wedge q) : a_3, a_4 \rangle$
a_3:	$\langle \mathsf{O}\neg q : a_2, \neg p \Rightarrow \neg q \rangle$	a_6:	$\langle \mathsf{O}(q \vee r) : a_4, \Box(q \supset (q \vee r)) \rangle$

Argument a_1 is constructed from $\Box p \in \Gamma_1$ in view of (i). Arguments a_2 and a_4 are constructed by means of (ii)[6]*; a_3 is constructed from a_2 by means of (iii); a_5 is constructed from a_3 and a_4 by (iv); and a_6 is constructed from a_4 by (v).*

We can interpret Γ_1 as representing a classic *contrary-to-duty* (CTD) scenario (for the sake of readability, we omit the qualifier 'prima facie' in our reading of conditional obligations):[7]

---

[6] Note that, in the construction of argument a_4, the formula p follows from Γ_1 by $\Box p$ and since $\vdash_{\mathbf{L}^\Box} \Box p \supset p$.

[7] The example is adapted from [27].

$\top \Rightarrow \neg p$	There ought not be a dog.
$\neg p \Rightarrow \neg q$	If there is no dog, there ought not be a warning sign.
$p \Rightarrow q$	If there is a dog, there ought to be a warning sign.
$\Box p$	It is settled that there is a dog.

Of course, not all of the conclusions of arguments a_2-a_6 qualify as all-things-considered obligations. Argument a_5, for instance, is internally incoherent and should be filtered out when evaluating the arguments constructed from Γ_1. Arguments are evaluated in terms of the attack relations which hold amongst them. Before we turn to the definition of these relations, we point out that rules (i)-(v) in Definition 4 allow for a version of the necessitation rule whenever $\mathbf{L}^{\Box}$ is a normal modal logic. For instance, given a premise set $\{\Box p, \top \Rightarrow q\}$, we can construct the argument $a_1 = \langle \mathsf{O}q : \top, \top \Rightarrow q \rangle$ by (ii). Since $\Box p \vdash_{\mathbf{L}^{\Box}} \Box(q \supset p)$, we can construct the argument $a_2 = \langle \mathsf{O}p : a_1, \Box(q \supset p)\rangle$ by (v). If desired, the construction of a_2 can be prevented by defining – in addition to '$\supset$' – a weaker (non-material) implication connective in $\mathbf{L}^{\Box}$ on the basis of which to construct arguments in line with clause (v) in Definition 4.

2.3 Attacking deontic arguments

In our basic framework, we define two ways in which arguments may attack one another. First, we take care that unalterably violated obligations are attacked by the constraints which violate them. (We write $A = -B$ in case $A = \neg B$ or $B = \neg A$.)

Definition 5 (Fact attack). *Where $a = \langle \mathsf{O}A : \ldots \rangle$ is an argument, let $\mathsf{UO}(a) = \{B \mid \mathsf{O}B \in \mathsf{C}(a)\}$. Where $\emptyset \neq \Theta \subseteq \mathsf{UO}(a)$, $\langle \Box - \bigwedge \Theta : -- \rangle$ attacks a.*

In Example 1 the obligation $\mathsf{O}\neg p$ cannot guide the agent's actions, since it cannot be acted upon in view of the constraint $\Box p$. Definition 5 takes care that a_1 attacks a_2, since $\mathsf{UO}(b) = \{\neg p\}$. Note that, as soon as $A \in \mathsf{UO}(a)$ for some argument a and formula A, $A \in \mathsf{UO}(b)$ for any super-argument b of a. Consequently, if an argument c attacks a in view of Definition 5, then c also attacks all super-arguments b of a. So in Example 1 the argument a_1 attacks a_2 as well as its super-arguments a_3 and a_5.

Since we assume that $\mathbf{L}^{\Box}$ is a normal modal logic, we know that $\Box(\neg(\neg q \wedge q)) \in \mathrm{Cn}_{\mathbf{L}^{\Box}}(\Gamma_1)$. Hence, by Definition 5 again, argument $a_7 = \langle \Box(\neg(\neg q \wedge q)) : -- \rangle$ attacks argument a_5 from Example 1.

Example 2 (Attacks on incoherent arguments). *Let $\Gamma_2 = \{\top \Rightarrow p, \top \Rightarrow \neg p, \top \Rightarrow q\}$. We construct the following arguments on the basis of Γ_2:*

a_1:	$\langle \mathsf{O}p : \top, \top \Rightarrow p \rangle$	a_4:	$\langle \mathsf{O}(p \vee \neg q) : a_1, \Box(p \supset (p \vee \neg q))\rangle$
a_2:	$\langle \mathsf{O}\neg p : \top, \top \Rightarrow \neg p \rangle$	a_5:	$\langle \mathsf{O}(\neg p \wedge (p \vee \neg q) : a_2, a_4 \rangle$
a_3:	$\langle \mathsf{O}q : \top, \top \Rightarrow q \rangle$	a_6:	$\langle \mathsf{O}\neg q : a_5, \Box((\neg p \wedge (p \vee \neg q)) \supset \neg q)\rangle$

By Definition 5:

$\mathsf{UO}(a_5) = \{p, p \vee \neg q, \neg p, \neg p \wedge (p \vee \neg q)\}$
$\mathsf{UO}(a_6) = \{p, p \vee \neg q, \neg p, \neg p \wedge (p \vee \neg q), \neg q\}$

Hence, both a_5 and a_6 are attacked by a_7:

$a_7 = \langle \Box\neg(p \wedge \neg p) : --\rangle$

Arguments a_5 and a_6 are incoherent in the sense that in constructing them we relied on arguments the conclusions of which are conflicting (namely a_1 and a_2). It is vital that we are able to filter out such incoherent arguments. Definition 5 takes care of that. By attacking a_6, argument a_7 protects (defends) the unproblematic a_3, which is attacked by a_6 in view of Definition 6 below. We return to this point in footnote 9, after we explained how arguments are evaluated.

The second type of attack relation ensures that mutually incompatible obligations attack each other:

Definition 6 (Conflict attack). *$a = \langle \mathsf{O}{-}A : \ldots\rangle$ attacks $b = \langle \mathsf{O}A : \ldots\rangle$, and a attacks all of b's super-arguments.*

In Example 1, arguments a_3 and a_4 attack each other according to Definition 6. Moreover, a_3 attacks a_5 and a_6; and a_4 attacks a_5. Likewise, in Example 2, a_1 and a_2 attack each other, and so do a_3 and a_6. Moreover, a_1 attacks a_5 and a_6; and a_2 attacks a_4, a_5, and a_6.

Example 3 (Conflict attack). *Let $\Gamma_3 = \{p, q, p \Rightarrow r, (p \wedge q) \Rightarrow s, \Box\neg(r \wedge s)\}$. We construct the following arguments on the basis of Γ_3:*

a_1:	$\langle \mathsf{O}r : p, p \Rightarrow r\rangle$	a_4:	$\langle \Box\neg(r \wedge s) : --\rangle$
a_2:	$\langle \mathsf{O}s : p \wedge q, (p \wedge q) \Rightarrow s\rangle$	a_5:	$\langle \mathsf{O}\neg r : a_2, \Box(s \supset \neg r)\rangle$
a_3:	$\langle \mathsf{O}(r \wedge s) : a_1, a_2\rangle$	a_6:	$\langle \mathsf{O}\neg s : a_1, \Box(r \supset \neg s)\rangle$

a_4 attacks a_3 by Definition 5. By Definition 6 a_1 attacks a_5; a_5 attacks a_1, a_3, and a_6; a_2 attacks a_6; and a_6 attacks a_2, a_3, and a_5.

2.4 Evaluating deontic arguments

For the evaluation of deontic arguments relative to a premise set, we extend Dung-style AFs to deontic argumentation frameworks, and we borrow Dung's argument evaluation mechanism from Definitions 1-3:

Definition 7 (DAF). *The* deontic argumentation framework *(DAF) for $\Gamma \subseteq \mathcal{L}^P \cup \mathcal{L}^\Box \cup \mathcal{L}^\Rightarrow$ is an ordered pair $\langle \mathcal{A}(\Gamma), \mathsf{Att}(\Gamma)\rangle$ where*

- *$\mathcal{A}(\Gamma)$ is the set of arguments constructed from Γ in line with Definition 4; and*
- *where $a, b \in \mathcal{A}(\Gamma)$: $(a, b) \in \mathsf{Att}(\Gamma)$ iff a attacks b according to Definition 5 or Definition 6.*

Like AFs, DAFs can be represented as directed graphs. Here, for instance, is a graph depicting the arguments we constructed on the basis of Γ_1:[8]

Nodes in the graph represent the arguments constructed on the basis of Γ_1 in Example 1. Below the arguments' names, we stated their conclusions. Arrows represent attacks. Dotted lines represent sub-argument relations.

We evaluate arguments in a DAF using Dung's grounded semantics from

[8] Due to space limitations, we leave it to the reader to construct similar graphs for the other examples in this paper.

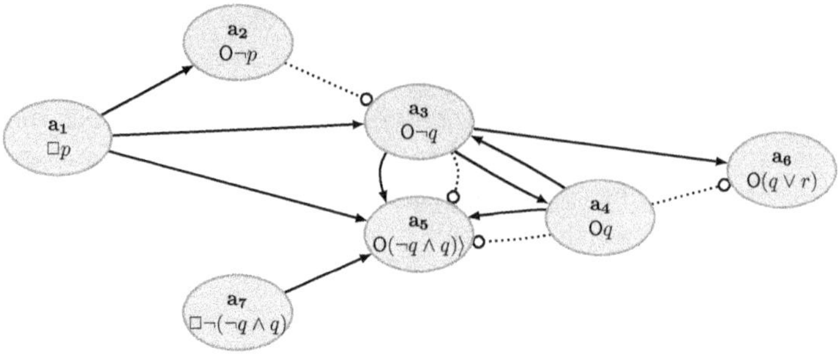

Fig. 1. Arguments and attack relations for Γ_1.

Section 2.1: In Definition 1, replace $\mathcal{A}$ (resp. Att) with $\mathcal{A}(\Gamma)$ (resp. $\mathsf{Att}(\Gamma)$). Similarly for Definition 3, where we also replace occurrences of $\mathcal{G}$ and $\mathcal{G}_i$ with $\mathcal{G}(\Gamma)$ and $\mathcal{G}_i(\Gamma)$ respectively.

Let us now apply Definition 3 to Example 1. Clearly, $a_1, a_7 \in \mathcal{G}_0(\Gamma_1)$, since Definitions 5 and 6 provide us with no means to attack arguments the conclusions of which are members of $\Gamma_1^{\Box}$. In the next step of our construction, $a_4, a_6 \in \mathcal{G}_1(\Gamma_1)$, since they are defended by $a_1 \in \mathcal{G}_0(\Gamma_1)$. $a_2, a_3, a_5 \notin \mathcal{G}_1(\Gamma_1)$, since each of these arguments is attacked by a_1 (hence undefended).

We cannot construct any further arguments which attack a_4 or a_6 and which do not contain any of the undefended arguments a_2 or a_3 as sub-arguments. Moreover, we show in the Appendix (Lemma 2) that, for any premise set Γ, if $a \in \mathcal{G}(\Gamma)$, then $a \in \mathcal{G}_1(\Gamma)$. By the Definition 3, $a_1, a_4, a_6, a_7 \in \mathcal{G}(\Gamma_1)$ while $a_2, a_3, a_5 \notin \mathcal{G}(\Gamma_1)$.

Definition 8 (DAF-consequence)**.** *Where* $\Gamma \subseteq \mathcal{L}^P \cup \mathcal{L}^{\Box} \cup \mathcal{L}^{\Rightarrow}$ *and* $A \in \mathcal{L}^P$, $\Gamma \vdash_{\mathbf{DAF}} \mathsf{O}A$ *iff there is an argument* $a \in \mathcal{G}(\Gamma)$ *with conclusion* $\mathsf{O}A$.

By Definition 8, $\Gamma_1 \vdash_{\mathbf{DAF}} \mathsf{O}q$ and $\Gamma_1 \vdash_{\mathbf{DAF}} \mathsf{O}(q \vee r)$, while $\Gamma_1 \nvdash_{\mathbf{DAF}} \mathsf{O}\neg p$ and $\Gamma_1 \nvdash_{\mathbf{DAF}} \mathsf{O}\neg q$.

In Example 2, $\Gamma_2 \vdash_{\mathbf{DAF}} \mathsf{O}q$.[9] We leave it to the reader to check that none of $\mathsf{O}p, \mathsf{O}\neg p, \mathsf{O}(p \vee \neg q)$, or $\mathsf{O}\neg q$ is a **DAF**-consequence of Γ_2, and that none of $\mathsf{O}r, \mathsf{O}s, \mathsf{O}(r \wedge s), \mathsf{O}\neg r$, or $\mathsf{O}\neg s$ is a **DAF**-consequence of Γ_3.

2.5 Rationality postulates

In [3, Sec. 4] the properties of output closure and output consistency were proposed as desiderata for well-behaved argumentation systems. Where $Output(\Gamma) = \{A \mid \Gamma \vdash_{\mathbf{DAF}} \mathsf{O}A\}$:

Property 1 (Closure)**.** $Output(\Gamma) = \mathrm{Cn}_{\mathbf{CL}}(Output(\Gamma))$.

Property 2 (Consistency)**.** $\mathrm{Cn}_{\mathbf{CL}}(Output(\Gamma))$ *is consistent.*

Properties 1 and 2 follow for **DAF** in view of resp. Theorems 1 and 2 in

9 The conclusion $\mathsf{O}q$ of argument a_3 in Example 2 is accepted despite its being attacked by a_6. The reason is that a_6 is in turn attacked by a_7, so that a_7 defends a_3 from the attack by a_6.

the Appendix. Property 3 is proven in Theorem 3 in the Appendix:

Property 3 (Cautious cut/cumulative transitivity). *Let* $\Delta_{\Rightarrow} = \{\top \Rightarrow A \mid A \in \Delta\}$. *If* $\Gamma \vdash_{\mathbf{DAF}} \mathsf{O}A$ *for all* $A \in \Delta$ *and* $\Gamma \cup \Delta_{\Rightarrow} \vdash_{\mathbf{DAF}} \mathsf{O}B$, *then* $\Gamma \vdash_{\mathbf{DAF}} \mathsf{O}B$.

Properties 4 and 5 fail for **DAF**:

Property 4 (Cautious monotonicity). *If* $\Gamma \vdash_{\mathbf{DAF}} \mathsf{O}A$ *and* $\Gamma \vdash_{\mathbf{DAF}} \mathsf{O}B$, *then* $\Gamma \cup \{\top \Rightarrow A\} \vdash_{\mathbf{DAF}} \mathsf{O}B$.

Property 5 (Rational monotonicity). *If* $\Gamma \vdash_{\mathbf{DAF}} \mathsf{O}A$ *and* $\Gamma \nvdash_{\mathbf{DAF}} \mathsf{O}\neg B$, *then* $\Gamma \cup \{\top \Rightarrow B\} \vdash_{\mathbf{DAF}} \mathsf{O}A$

Example 4 (Failure of properties 4 and 5, adapted from [2]). *Let* $\Gamma_4 = \{p, p \Rightarrow q, q \Rightarrow r, r \Rightarrow \neg q, \neg q \Rightarrow s, \top \Rightarrow \neg s\}$. *We construct the following arguments on the basis of* Γ_4*:*

a_1:	$\langle \mathsf{O}q : p, p \Rightarrow q\rangle$	a_4:	$\langle \mathsf{O}s : a_3, \neg q \Rightarrow s\rangle$
a_2:	$\langle \mathsf{O}r : a_1, q \Rightarrow r\rangle$	a_5:	$\langle \mathsf{O}\neg s : \top, \top \Rightarrow \neg s\rangle$
a_3:	$\langle \mathsf{O}\neg q : a_2, r \Rightarrow \neg q\rangle$	a_6:	$\langle \Box\neg(q \wedge \neg q) : --\rangle$

By Definition 6: a_1 *attacks* a_3 *and* a_4*;* a_3 *attacks all of* a_1*-*a_4 *(including itself); and* a_4 *and* a_5 *attack each other. By Definition 5,* a_6 *attacks* a_3 *and* a_4*, since both* q *and* $\neg q$ *are members of* $\mathsf{UO}(a_3)$ *and* $\mathsf{UO}(a_4)$*. As a result,* $\mathsf{O}q, \mathsf{O}r$*, and* $\mathsf{O}\neg s$ *are* **DAF***-consequences of* Γ_4*, while* $\mathsf{O}\neg q$ *and* $\mathsf{O}s$ *are not.*

Now add the new conditional obligation $\top \Rightarrow r$ *to* Γ_4*, so that we obtain the new arguments*

a_7:	$\langle \mathsf{O}r : \top, \top \Rightarrow r\rangle$	a_9:	$\langle \mathsf{O}s : a_8, \neg q \Rightarrow s\rangle$
a_8:	$\langle \mathsf{O}\neg q : a_7, r \Rightarrow \neg q\rangle$		

None of these new arguments is attacked by a_6*, which defends* a_1 *and* a_5 *from the attacks by* a_3 *and* a_4 *respectively. By Definition 6,* a_8 *and* a_1 *attack each other. So do* a_9 *and* a_5*. As a result, none of* a_1, a_5, a_8*, and* a_9 *is in the grounded extension of* $\Gamma_4 \cup \{\top \Rightarrow r\}$*. So we have a counter-example to Property 4:* $\Gamma_4 \vdash_{\mathbf{DAF}} \mathsf{O}r$ *and* $\Gamma_4 \vdash_{\mathbf{DAF}} \mathsf{O}\neg s$*, while* $\Gamma_4 \cup \{\top \Rightarrow r\} \nvdash_{\mathbf{DAF}} \mathsf{O}\neg s$*.*

This example also serves to illustrate the failure of Property 5 for **DAF**. Arguments with conclusion $\mathsf{O}\neg r$ can be constructed on the basis of Γ_4 only on the basis of incoherent arguments. Let, for instance:

a_{10}: $\langle \mathsf{O}(q \wedge \neg q) : a_1, a_3\rangle$ a_{11}: $\langle \mathsf{O}\neg r : a_{10}, \Box((q \wedge \neg q) \supset \neg r)\rangle$

In view of Definition 5, arguments constructed on an incoherent basis are attacked by an otherwise unattacked argument. For instance, a_{11} is attacked by the unattacked argument a_6. Because of this, $\Gamma_4 \nvdash_{\mathbf{DAF}} \mathsf{O}\neg r$. But then, since $\Gamma_4 \vdash_{\mathbf{DAF}} \mathsf{O}\neg s$ and $\Gamma_4 \cup \{\top \Rightarrow r\} \nvdash_{\mathbf{DAF}} \mathsf{O}\neg s$, Property 5 fails for **DAF**.

3 Beyond the basics

3.1 Conflict-resolution

3.1.1 Resolving conflicts via logical analysis

It has been argued that, in cases of conflict, more specific obligations should be given precedence over less specific ones. [10] Consider the following example:

Example 5 (Specificity). *Let $\Gamma_5 = \{q, r, q \Rightarrow p, (q \wedge r) \Rightarrow \neg p\}$. We can interpret Γ_5 as representing a scenario in which an agent is making carrot soup. Let p, q, and, respectively, r abbreviate 'there is fennel', 'there are carrots', and 'there is celery'. If there are carrots in the garden still, our agent should take care that he buys fennel in order to make the soup ($q \Rightarrow p$). However, if both carrots and celery are in the garden, he should not get fennel ($(q \wedge r) \Rightarrow \neg p$), because celery can be used instead of fennel. As it turns out, both carrots and celery are in his garden (q, r). The desirable outcome in this case is that the agent ought not go out and buy fennel.*

A principled way of obtaining outcomes in which more specific obligations are preferred over less specific ones, is to define specificity in terms of logical strength, and to define a new attack relation for letting more specific arguments attack less specific ones. Let the *factual support* of a deontic argument a be the set $\mathsf{S}(a) = \{B \mid B \in (\mathsf{C}(a) \cap \mathcal{L}^P)\}$.

We write $\mathsf{S}(a) \sqsubseteq \mathsf{S}(b)$ iff for all $A \in \mathsf{S}(a)$ there is a $B \in \mathsf{S}(b)$ such that $A \vdash B$ and for all $B \in \mathsf{S}(b)$ there is an $A \in \mathsf{S}(a)$ such that $A \vdash B$. $\mathsf{S}(a) \sqsubset \mathsf{S}(b)$ (a is *more specific* than b) iff $\mathsf{S}(a) \sqsubseteq \mathsf{S}(b)$ and $\mathsf{S}(b) \not\sqsubseteq \mathsf{S}(a)$.

We replace Definition 6 with Definition 9:

Definition 9 (Conflict attack w/specificity). *Let $a = \langle \mathsf{O}{-}A : \ldots \rangle$ and $b = \langle \mathsf{O}A : \ldots \rangle$.*

(i) If $\mathsf{S}(a) \sqsubset \mathsf{S}(b)$, then a attacks b and all of b's super-arguments,

(ii) b attacks a and all of a's super-arguments, unless a attacks b in view of clause (i).

Let $\mathbf{DAF_s}$ (with subscript 's' for specificity) be the logic resulting from constructing the attack relation Att on the basis of Definitions 5 and 9.

In Example 5, we construct the following arguments from Γ_5:

a_1: $\langle \mathsf{O}p : q, q \Rightarrow p \rangle$
a_2: $\langle \mathsf{O}\neg p : q \wedge r, (q \wedge r) \Rightarrow \neg p \rangle$

Since $\mathsf{S}(a_2) \sqsubset \mathsf{S}(a_1)$, a_2 attacks a_1 by Definition 9, but not vice versa. As a result, only a_2 is in Γ_5's grounded extension, and $\Gamma_5 \vdash_{\mathbf{DAF_s}} \mathsf{O}\neg p$, while $\Gamma_5 \nvdash_{\mathbf{DAF_s}} \mathsf{O}p$.

In Example 3, the factual support of the arguments constructed from Γ_3 is such that $\mathsf{S}(a_2) = \mathsf{S}(a_5) \sqsubset \mathsf{S}(a_1) = \mathsf{S}(a_6)$. By Definition 9, a_5 attacks a_1 and a_2 attacks a_6. As a result, the more specific arguments a_2 and a_5 defeat

[10] Understood in this way, specificity cases have been studied extensively in the fields of non-monotonic logic (see e.g. [7,5]) and deontic logic (see e.g. [4,27,30,31]).

the less specific a_1 and a_6, so that $\Gamma_3 \vdash_{\mathbf{DAF_s}} \mathsf{O}s$ and $\Gamma_3 \vdash_{\mathbf{DAF_s}} \mathsf{O}\neg r$, while $\Gamma_3 \nvdash_{\mathbf{DAF_s}} \mathsf{O}r$ and $\Gamma_3 \nvdash_{\mathbf{DAF_s}} \mathsf{O}\neg s$. As before, $\Gamma_3 \nvdash_{\mathbf{DAF_s}} \mathsf{O}(r \wedge s)$.

In dealing with conflict-resolution via logical analysis, we have chosen for a cautious notion of specificity. For instance, $\{p\} \not\sqsubset \{p, q\}$ and $\{p\} \not\sqsubset \{p \wedge q, r\}$. In certain contexts it may be sensible to opt for a stronger characterization of '$\sqsubset$'. A detailed discussion of such issues would lead us too far astray given our present purposes. Instead, we point out that our framework readily accommodates alternative characterizations of '$\sqsubset$' to be used in Definition 9.

3.1.2 Resolving conflicts via priorities

Instead of (or in combination with) conflict-resolution via logical analysis, a priority ordering $\leq$ can be introduced over conditional norms, and our formal language can be adjusted accordingly. Conditional norms then come with an associated degree of priority $\alpha \in \mathbb{Z}^+$, written $A \Rightarrow_\alpha B$ (higher numbers denote higher priorities).

We lift $\leq$ to a priority ordering $\preceq$ over arguments via the *weakest link* principle: an argument is only as strong as the weakest priority conditional used in its construction [25]. Let $\mathsf{Pr}(\Delta) = \{\alpha \mid A \Rightarrow_\alpha B \in \Delta\}$ and let $min(\mathsf{Pr}(\Delta))$ be the lowest $\alpha \in \mathsf{Pr}(\Delta)$. Then $\Delta \preceq \Delta'$ iff $min(\mathsf{Pr}(\Delta)) \leq min(\mathsf{Pr}(\Delta'))$. Relative to a premise set Γ, we write $a \preceq b$ iff $\mathsf{C}(a) \cap \Gamma^{\Rightarrow} \preceq \mathsf{C}(b) \cap \Gamma^{\Rightarrow}$. $a \prec b$ iff $a \preceq b$ and $b \npreceq a$.

We replace Definition 6 with the following definition:

Definition 10 (Prioritized conflict attack). *If $a \nprec b$, then $a = \langle \mathsf{O}{-}A : \ldots \rangle$ attacks $b = \langle \mathsf{O}A : \ldots \rangle$ and all of b's super-arguments.*

Let $\mathbf{DAF}_\leq$ be the logic resulting from constructing the attack relation Att on the basis of Definitions 5 and 10.

Example 6 (Prioritized conflict attack). *Let $\Gamma_6 = \{p, q, r, \Box\neg(s \wedge t \wedge u), p \Rightarrow_1 s, q \Rightarrow_2 t, r \Rightarrow_3 u\}$. We construct the following arguments on the basis of Γ_6:*

a_1:	$\langle \Box\neg(s \wedge t \wedge u) : -- \rangle$	a_8:	$\langle \mathsf{O}(s \wedge t \wedge u) : a_4, a_5 \rangle$
a_2:	$\langle \mathsf{O}s : p, p \Rightarrow_1 s \rangle$	a_9:	$\langle \mathsf{O}\neg(t \wedge u) : a_2, \Box(s \supset \neg(t \wedge u)) \rangle$
a_3:	$\langle \mathsf{O}t : q, q \Rightarrow_2 t \rangle$	a_{10}:	$\langle \mathsf{O}\neg(s \wedge u) : a_3, \Box(t \supset \neg(s \wedge u)) \rangle$
a_4:	$\langle \mathsf{O}u : r, r \Rightarrow_3 u \rangle$	a_{11}:	$\langle \mathsf{O}\neg(s \wedge t) : a_4, \Box(u \supset \neg(s \wedge t)) \rangle$
a_5:	$\langle \mathsf{O}(s \wedge t) : a_2, a_3 \rangle$	a_{12}:	$\langle \mathsf{O}\neg u : a_5, \Box((s \wedge t) \supset \neg u) \rangle$
a_6:	$\langle \mathsf{O}(s \wedge u) : a_2, a_4 \rangle$	a_{13}:	$\langle \mathsf{O}\neg t : a_6, \Box((s \wedge u) \supset \neg t) \rangle$
a_7:	$\langle \mathsf{O}(t \wedge u) : a_3, a_4 \rangle$	a_{14}:	$\langle \mathsf{O}\neg s : a_7, \Box((t \wedge u) \supset \neg s) \rangle$

The order of arguments is such that $a_2, a_5, a_6, a_8, a_9, a_{12}, a_{13} \prec a_3, a_7, a_{10}, a_{14} \prec a_4, a_{11}$. By Definition 10, a_{14} attacks a_2, a_5, a_6, a_8, a_9, a_{12}, and a_{13}; a_3 attacks a_{13}; a_4 attacks a_{12}; a_{11} attacks a_5, a_8, and a_{12}; a_{10} attacks a_6 and a_{13}; and a_7 attacks a_9. By Definition 5, a_1 attacks a_8. As a result, $a_1, a_3, a_4, a_7, a_{10}, a_{11}, a_{14} \in \mathcal{G}(\Gamma_6)$, while $a_2, a_5, a_6, a_8, a_9, a_{12}, a_{13} \notin \mathcal{G}(\Gamma_6)$. The following obligations are $\mathbf{DAF}_\leq$-consequences of Γ_6 : $\mathsf{O}t, \mathsf{O}u, \mathsf{O}(t \wedge u), \mathsf{O}\neg(s \wedge u), \mathsf{O}\neg(s \wedge t), \mathsf{O}\neg s$. The following obligations are not $\mathbf{DAF}_\leq$-derivable from Γ_6 : $\mathsf{O}s, \mathsf{O}(s \wedge t), \mathsf{O}(s \wedge u), \mathsf{O}(s \wedge t \wedge u), \mathsf{O}\neg(t \wedge u), \mathsf{O}\neg u, \mathsf{O}\neg t$.

As with '$\sqsubset$' in Definition 9, there are other ways of characterizing '$\prec$' in

Definition 10. For instance, instead of lifting $\leq$ via the weakest link principle, we could lift it via the *strongest link* principle, according to which an argument is as strong as the strongest priority conditional used in its construction.[11] Depending on the way $\leq$ is lifted to $\preceq$, different outcomes are possible with respect to the priority puzzles studied in e.g. [9,14,15]. A thorough investigation of these puzzles within our framework is left for an extended version of this paper.

3.2 Anticipating violations and conflicts

Obligations which are violated or conflicted should not be detached. But what about obligations that *commit* us to violations or conflicts? Consider the following example, adapted from [16,19].

Example 7. *Let* $\Gamma_7 = \{p, p \Rightarrow q, q \Rightarrow r, r \Rightarrow \neg p\}$. *We construct the following arguments on the basis of* Γ_7*:*

a_1:	$\langle \Box p : -- \rangle$	a_3:	$\langle \mathsf{O}r : a_2, q \Rightarrow r \rangle$
a_2:	$\langle \mathsf{O}q : p, p \Rightarrow q \rangle$	a_4:	$\langle \mathsf{O}\neg p : a_3, r \Rightarrow \neg p \rangle$

Suppose you are throwing a party. Let p (resp. q, r) abbreviate 'Peggy (resp. Quincy, Ruth) is invited to the party'. If Peggy is invited, then Quincy should be invited as well (perhaps because they are good friends and we know both of them). Likewise, if Quincy is invited then Ruth should be invited as well. But if Ruth is invited, then Peggy should not be (perhaps because we know Ruth and Peggy do not get along well). It is settled that Peggy is invited. You already sent her the official invitation, and it would be too awkward to tell her she can't come. Should Quincy and/or Ruth be invited?

Arguments a_1, a_2, and a_3 are in Γ_7's grounded extension $\mathcal{G}(\Gamma_7)$. a_4 is not in $\mathcal{G}(\Gamma_7)$ since it is attacked by a_1 according to Definition 5; consequently, $\Gamma_7 \vdash_{\mathbf{DAF}} \mathsf{O}q$ and $\Gamma_7 \vdash_{\mathbf{DAF}} \mathsf{O}r$, while $\Gamma_7 \nvdash_{\mathbf{DAF}} \mathsf{O}\neg p$.

A more cautious reasoner may argue that $\mathsf{O}q$ and $\mathsf{O}r$ should not be detached, since they lead to a commitment to $\mathsf{O}\neg p$: they form part of the detachment chain of a_4. This commitment reflects very badly on arguments a_2 and a_3, since $\mathsf{O}\neg p$ is violated.

To model this behavior, we introduce the *deontic doubt operator* $\odot$. We will use this operator to construct new arguments, called *shadow arguments*, the conclusion of which is of the form $\odot A$. A shadow argument with conclusion $\odot A$ casts doubt on – and attacks – arguments with conclusion $\mathsf{O}A$. Shadow arguments cannot be used to support obligations, but only to attack other arguments. They can *only* rule out deontic arguments. They cannot generate new consequences.[12]

---

[11] If the strongest link principle is used, Definition 10 should no longer allow for attacks on super-arguments, since $a \nprec b$ no longer warrants that $a \nprec c$ where c is a super-argument of b. A further alternative is to use the *last link* principle, according to which an argument gets the priority of the conditional which occurs last in its proof sequence.

[12] Shadow arguments are similar in spirit to Caminada's *HY-arguments* from [2]. An HY-argument a is an incoherent argument constructed on the basis of the conclusion of another

In the resulting system $\mathbf{DAF}_{\odot}$, our language $\mathcal{L}$ is adjusted so as to include members of $\mathcal{P}$ within the scope of the new operator $\odot$. Arguments are constructed in line with Definition 11:

Definition 11. *Given a premise set* Γ, *we allow rules (i)-(vii) for constructing arguments, where (i)-(v) are the rules from Definition 4:*

(vi) If $a = \langle \Box A : -- \rangle$ *is an argument, then* $\langle \odot -A : a \rangle$ *is an argument;*
(vii) If $a = \langle \mathsf{O}A : \ldots \rangle$ *is an argument, then* $\langle \odot -A : a \rangle$ *is an argument.*

We say that an argument a has *minimal support* if there is no argument b with the same conclusion such that $\mathsf{C}(b) \subset \mathsf{C}(a)$. In $\mathbf{DAF}_{\odot}$ the attack relation is constructed on the basis of Definition 12:[13]

Definition 12 (Shadow attack). *Where* $a = \langle \mathsf{O}A : \ldots \rangle$ *has minimal support:*

(i) Where b *is a deontic sub-argument of* a, $\langle \odot A : \ldots \rangle$ *attacks* b *as well as all of* b*'s super-arguments,*

(ii) Where b *is a deontic sub-argument of* a *and* $\emptyset \neq \Theta \subseteq \mathsf{UO}(a)$, $\langle \odot \bigwedge \Theta : \ldots \rangle$ *attacks* b *as well as all of* b*'s super-arguments.*

Reconsider Γ_7 from Example 7. From a_1, we can construct the shadow argument $a_5 = \langle \odot \neg p : a_1 \rangle$. By clause (i) of Definition 12, a_5 attacks a_4, a_3, and a_2. As a result, a_2 and a_3 are no longer in $\mathcal{G}(\Gamma_7)$. $\Gamma_7 \nvdash_{\mathbf{DAF}_{\odot}} \mathsf{O}q$ and $\Gamma_7 \nvdash_{\mathbf{DAF}_{\odot}} \mathsf{O}r$.

Example 8. *Let* $\Gamma_8 = \{\Box s, \top \Rightarrow p, \top \Rightarrow q, (p \wedge q) \Rightarrow r, r \Rightarrow \neg s, q \Rightarrow t\}$. *We construct the following arguments on the basis of* Γ_8*:*

a_1:	$\langle \Box s : -- \rangle$	a_5:	$\langle \mathsf{O}r : a_4, (p \wedge q) \Rightarrow r \rangle$
a_2:	$\langle \mathsf{O}p : \top, \top \Rightarrow p \rangle$	a_6:	$\langle \mathsf{O}\neg s : a_5, r \Rightarrow \neg s \rangle$
a_3:	$\langle \mathsf{O}q : \top, \top \Rightarrow q \rangle$	a_7:	$\langle \mathsf{O}t : a_3, q \Rightarrow t \rangle$
a_4:	$\langle \mathsf{O}(p \wedge q) : a_2, a_3 \rangle$	a_8:	$\langle \odot \neg s : a_1 \rangle$

By Definition 12 the shadow argument a_8 *attacks* a_6 *as well as its sub-arguments* $a_2 - a_5$. *Moreover, it attacks* a_7, *which is a super-argument of* a_3. *As a result, none of the conclusions of arguments* a_2*-*a_7 *are* $\mathbf{DAF}_{\odot}$*-consequences of* Γ_8.

Example 4 no longer serves as a counter-example to properties 4 and 5 provided in Section 2.5. We can construct the shadow argument $a_{12} : \langle \odot s : a_5 \rangle$. By clause (i) of Definition 12, this argument attacks a_4 as well as its sub-arguments a_1-a_3. As a result of this attack, $\Gamma_4 \nvdash_{\mathbf{DAF}_{\odot}} \mathsf{O}q$ and $\Gamma_4 \nvdash_{\mathbf{DAF}_{\odot}} \mathsf{O}r$. More generally, we can show that the cautious monotonicity property (Property

---

argument b. Since a shows that b leads to incoherence, b's conclusion is attacked by the HY-argument a. Caminada shows how in the presence of HY-arguments, the property of cautious monotonicity may be restored for AFs. The same holds true for shadow arguments in our setting (cfr. infra). As Caminada's construction is defined within a framework consisting only of literals and (defeasible) rules relating (conjunctions of) literals, we cannot employ it in our setting.

[13] By the construction of Definition 12, Definitions 5 and 6 become redundant in $\mathbf{DAF}_{\odot}$. All cases covered by these definitions are covered already by Definition 12.

4 in Section 2.5) holds for $\mathbf{DAF}_{\odot}$. A proof is provided in Theorem 4 of the Appendix.

Instead of – and equivalently to – working with the $\odot$-operator and Definitions 11 and 12, we could have generalized Definitions 5 and 6 so as to include attacks on sub-arguments. Definitions 5 and 6 currently entail that if a attacks b, then a attacks all super-arguments of b. In the generalized form, these definitions would entail that if a attacks b, then a attacks all superarguments of all sub-arguments of b.

There are two additional reasons for working with the doubt operator $\odot$, however. First, this operator has a clear and intuitive meaning, and adds expressivity to our argumentation frameworks. Second, by characterizing shadow arguments via a separate operator we can think more transparently about (a) the implementation of additional logical properties of this operator, and (b) alternatives to Definition 12. Regarding (a), think about the strengthening rule ('If $\odot A$, then $\odot B$ whenever $B \vdash A$'), which carries some intuitive force. Regarding (b), reconsider Example 8, and suppose we add the premise $\top \Rightarrow \neg p$ to Γ_8. A not-so-skeptical reasoner may argue that in this case we should not be able to cast doubt on the arguments a_3 and a_7, since the doubt casted on argument a_4 arguably arises in view of the conflicted conditional obligation to see to it that p. [14]

4 Related work

Due to space limitations, we restrict our discussion of related formalisms to those of input/output logic (Section 4.1) and those based on formal argumentation frameworks (Section 4.2). A comparison with other related deontic systems, such as Nute's defeasible deontic logic [22,21] and Horty's default-based deontic logic [13,10,11,15] is left for an extended version of this article.

4.1 Input/output logic

Like the constrained input/output (I/O) logics from [19], the DAFs defined here are tools for detaching conditional obligations relative to a set of inputs and constraints. Unlike most I/O logics, none of these DAFs validates strengthening of the antecedent (SA) for conditional obligations – from $A \Rightarrow C$ to infer $(A \wedge B) \Rightarrow C$. Unrestricted (SA) is counter-intuitive if we allow for conflict-resolution via logical analysis as defined Section 3.1.1, since it allows the unrestricted derivation of more specific from less specific conditional obligations. [15]

Example 9 (**DAF** and I/O logic). *Let* $\Gamma_9 = \{p, p \Rightarrow q, p \Rightarrow \neg r, q \Rightarrow r\}$. *We construct the following arguments on the basis of* Γ_9*:*

a_1: $\langle \mathsf{O}q : p, p \Rightarrow q\rangle$ $\qquad$ a_3: $\langle \mathsf{O}r : a_1, q \Rightarrow r\rangle$

a_2 $\langle \mathsf{O}\neg r : p, p \Rightarrow \neg r\rangle$

Since a_2 *and* a_3 *attack each other in view of Definition 6,* $a_2, a_3 \notin \mathcal{G}(\Gamma_9)$,

[14] Caminada's *HY-arguments* from [2] are similar in spirit to this less skeptical proposal.

[15] In [29] an I/O system is presented which invalidates (SA) in the context of exempted permissions which are subject to conflict-resolution via logical analysis (specificity).

while $a_1 \in \mathcal{G}(\Gamma_9)$. *Consequently,* $\Gamma_9 \nvdash_{\mathbf{DAF}} \mathsf{O}r$ *and* $\Gamma_9 \nvdash_{\mathbf{DAF}} \mathsf{O}\neg r$ *while* $\Gamma_9 \vdash_{\mathbf{DAF}} \mathsf{O}q$.

In constrained I/O logic, triggered conditional obligations in the input are divided into maximally consistent subsets (MCSs). $\Gamma_9^{\Rightarrow}$ has three MCSs: $\{p \Rightarrow q, q \Rightarrow \neg r\}$, $\{p \Rightarrow q, p \Rightarrow r\}$, and $\{q \Rightarrow \neg r, p \Rightarrow r\}$. In [19] two ways are presented for dealing with conflicts and constraints: via a full meet operation on the generated MCSs, or via a full join operation on the generated MCSs. The first approach gives us none of q, r, and $\neg r$ for Γ_9. The second gives us all three.

Some of the I/O logics defined in e.g. [18,19,24] validate intuitively appealing rules which are not generally valid in our DAFs, such as the rule (OR) – from $A \Rightarrow C$ and $B \Rightarrow C$ to infer $(A \vee B) \Rightarrow C$. A detailed study of the appeal and implementation of (OR) and similar rules in the present argumentative setting is left for future investigation.

4.2 Formal argumentation

Several ways of modeling normative reasoning on the basis of formal argumentation have been proposed in the literature. For instance, the approach in [8] is based on bipolar abstract argumentation frameworks. Dung's abstract argumentation frameworks are enriched with a support relation that is defined over the set of abstract arguments. This device is used to express deontic conditionals. A similar idea is used in [23] where a relation for evidential support is introduced. Argumentation schemes of normative reasoning are there expressed by means of Prolog-like predicates and subsequently translated into an argumentation framework. Here, we follow the tradition of structured or instantiated argumentation in which no support relation between arguments is needed. In our approach conditional obligations are modeled by a dyadic operator $\Rightarrow$ that is part of the object language. Arguments consist of sequences of applications of factual and deontic detachment. As a consequence, for instance, evidential or factual support is an intrinsic feature of our arguments and is modeled via the factual detachment rule.

The general setting of our DAFs is close to ASPIC$^+$. For instance, in the dynamic legal argumentation systems (in short, DLAS) from [26], deontic conditionals are also modeled via a defeasible conditional $\leadsto$ in the object language. There are several differences to our approach. For instance, our conditionals are not restricted to conjunctions of literals as antecedents. As a consequence we needed to define a strong fact attack rule (Def. 5) that, in order to avoid contamination problems (see Ex. 2), warrants that arguments with inconsistent supports are defeated.[16] Our fact attack and our shadow attack rules do not conform to the standard attack types defined in ASPIC$^+$ (rebutting, undercutting, and undermining). Our conflict attacks can be seen as forms of ASPIC$^+$-type rebuttals where the contrary of $\mathsf{O}A$ is defined by $\mathsf{O}\neg A$.

Unlike DLAS or Horty's deontic default logics, we follow the tradition in de-

[16] Other solutions to this problem have been proposed, e.g., in [33].

ontic logic to have a dedicated operator O for unconditional obligations which, for instance, allows to formally distinguish between cases of deontic and cases of factual detachment.

Recently, van der Torre & Villata extended the DLAS approach with deontic modalities [32], adopting the input/output methodology from Section 4.1. The resulting systems, like **DAF**, allow for versions of the factual and deontic detachment rules. Moreover, they allow for the representation of permissive norms. Unlike **DAF**, and unlike the I/O logics from Section 4.1, these systems do not have inheritance (weakening) or aggregation rules.

Another approach in which formal argumentation is used for the analysis of traditional problems of deontic logic, such as contrary-to-duty and specificity cases is [31]. There, arguments are Gentzen-type sequents in the language of standard deontic logic and conditionals are expressed using material implication. One drawback which is avoided in our setting is that there conditionals are contrapositable and subject to strengthening of the antecedent.

5 Outlook

We presented a basic logic, **DAF**, for detaching conditional obligations based on Dung's grounded semantics for formal argumentation. We extended **DAF** with mechanisms for conflict-resolution and for the anticipation of conflicts and violations. For now, these mechanisms mainly serve to illustrate the modularity of our framework. A detailed study of e.g. different approaches to prioritized reasoning, or different conceptions of specificity-based conflict-resolution, is left for an extended companion paper.

We conclude by mentioning three challenges for future research. The first is to include permission statements. The second is to increase the 'logicality' of our framework by allowing for the nesting and for the truth-functional combination of formulas of the form $\mathsf{O}A$, $A \Rightarrow B$, or $\Box A$. The third is to extend our focus beyond grounded extensions, and to study how our framework behaves when subjected to different types of acceptability semantics for formal argumentation. Working with Dung's preferred semantics [6], for instance, allows for the derivation of so-called floating conclusions [12,17].

References

[1] Besnard, P., A. Garcia, A. Hunter, S. Modgil, H. Prakken, G. Simari and F. Toni (eds.), *Special issue: Tutorials on structured argumentation*, Argument and Computation **5(1)** (2014).

[2] Caminada, M., *Dialogues and HY-arguments*, in: J. Delgrande and T. Schaub, editors, *10th International Workshop on Non-Monotonic Reasoning (NMR 2004), Whistler, Canada, June 6-8, 2004, Proceedings*, 2004, pp. 94–99.

[3] Caminada, M. and L. Amgoud, *On the evaluation of argumentation formalisms*, Artificial Intelligence **171** (2007), pp. 286 – 310.

[4] Carmo, J. and A. Jones, *Deontic logic and contrary-to-duties*, in: D. Gabbay and F. Guenthner, editors, *Handbook of Philosophical Logic (2nd edition) Vol. 8*, Kluwer Academic Publishers, 2002 pp. 265–343.

[5] Delgrande, J. and T. Schaub, *Compiling specificity into approaches to nonmonotonic reasoning*, Artificial Intelligence **90** (1997), pp. 301–348.

[6] Dung, P., *On the acceptability of arguments and its fundamental role in nonmonotonic reasoning, logic programming and n-person games*, Artificial Intelligence **77** (1995), pp. 321–357.

[7] Dung, P. and T. Son, *An argument-based approach to reasoning with specifity*, Artificial Intelligence **133** (2001), pp. 35–85.

[8] Gabbay, D., *Bipolar argumentation frames and contrary to duty obligations, preliminary report*, in: M. Fisher, L. van der Torre, M. Dastani and G. Governatori, editors, *Computational Logic in Multi-Agent Systems: Proceedings of the 13th International Workshop, CLIMA XIII, Montpellier, France*, Springer, 2012 pp. 1–24.

[9] Hansen, J., *Prioritized conditional imperatives: problems and a new proposal*, Autonomous Agents and Multi-Agent Systems **17** (2008), pp. 11–35.

[10] Horty, J., *Deontic logic as founded on nonmonotonic logic*, Annals of Mathematics and Artificial Intelligence **9** (1993), pp. 69–91.

[11] Horty, J., *Nonmonotonic foundations for deontic logic*, in: D. Nute, editor, *Defeasible Deontic Logic: Essays in Nonmonotonic Normative Reasoning*, Kluwer Academic Publishers, 1997 pp. 17–44.

[12] Horty, J., *Skepticism and floating conclusions*, Artificial Intelligence **135** (2002), pp. 55–72.

[13] Horty, J., *Reasoning with moral conflicts*, Noûs **37** (2003), pp. 557–605.

[14] Horty, J., *Defaults with priorities*, Journal of Philosophical Logic **36** (2007), pp. 367–413.

[15] Horty, J., "Reasons as Defaults," Oxford University Press, 2012.

[16] Makinson, D., *General patterns in nonmonotonic reasoning*, in: D. M. Gabbay, C. J. Hogger and J. A. Robinson, editors, *Handbook of Logic in Artificial Intelligence and Logic Programming, Vol. 3*, Oxford University Press, 1994 pp. 35–110.

[17] Makinson, D. and K. Schlechta, *Floating conclusions and zombie paths: two deep difficulties in the "directly skeptical" approach to defeasible inheritance nets*, Artificial Intelligence **48** (1991), pp. 199–209.

[18] Makinson, D. and L. van der Torre, *Input/output logics*, Journal of Philosophical Logic **29** (2000), pp. 383–408.

[19] Makinson, D. and L. van der Torre, *Constraints for input/output logics*, Journal of Philosophical Logic **30** (2001), pp. 155–185.

[20] Modgil, S. and H. Prakken, *The ASPIC+ framework for structured argumentation: a tutorial*, Argument & Computation **5** (2014), pp. 31–62.

[21] Nute, D., *Apparent obligation*, in: D. Nute, editor, *Defeasible Deontic Logic: Essays in Nonmonotonic Normative Reasoning*, Kluwer Academic Publishers, 1997 pp. 287–315.

[22] Nute, D., editor, "Defeasible Deontic Logic: Essays in Nonmonotonic Normative Reasoning," Kluwer Academic Publishers, 1997.

[23] Oren, N., M. Luck, S. Miles and T. Norman, *An argumentation inspired heuristic for resolving normative conflict*, in: *Proceedings of the fifth workshop on coordination, organizations, institutionsm and norms in agent systems, AAMAS-08, Toronto*, 2008, pp. 41–56.

[24] Parent, X. and L. van der Torre, *"Sing and dance!" Input/output logics without weakening*, in: F. Cariani, D. Grossi, J. Meheus and X. Parent, editors, *DEON (12th International Conference on Deontic Logic in Computer Science)*, Lecture Notes in Artificial Intelligence **8554** (2014), pp. 149–165.

[25] Prakken, H., *An abstract framework for argumentation with structured arguments*, Argument and Computation **1** (2011), pp. 93–124.

[26] Prakken, H. and G. Sartor, *Formalising arguments about norms.*, in: *JURIX*, 2013, pp. 121–130.

[27] Prakken, H. and M. Sergot, *Contrary-to-duty obligations*, Studia Logica **57** (1996), pp. 91–115.

[28] Ross, D. W., "The Right and the Good," Oxford University Press, 1930.

[29] Stolpe, A., *A theory of permission based on the notion of derogation*, Journal of Applied Logic **8** (2010), pp. 97–113.

[30] Straßer, C., *A deontic logic framework allowing for factual detachment*, Journal of Applied Logic **9** (2011), pp. 61–80.

[31] Straßer, C. and O. Arieli, *Normative reasoning by sequent-based argumentation*, Journal of Logic and Computation (in print), doi:10.1093/logcom/exv050.

[32] van der Torre, L. and S. Villata, *An ASPIC-based legal argumentation framework for deontic reasoning*, in: S. Parsons, N. Oren, C. Reed and F. Cerutti, editors, *Computational Models of Argument - Proceedings of COMMA 2014, Atholl Palace Hotel, Scottish Highlands, UK, September 9-12, 2014*, 2014, pp. 421–432.

[33] Wu, Y., "Between Argument and Conclusion. Argument-based Approaches to Discussion, Inference and Uncertainty," Ph.D. thesis, Universite Du Luxembourg (2012).

Argumentation Frameworks with Justified Attacks

Sjur K. Dyrkolbotn [1]

Utrecht University, Department of Philosophy and Religious Studies

Truls Pedersen

University of Bergen, Department of Information Science and Media Studies

Abstract

After Dung's seminal paper on argumentation frameworks, the relation of attack between arguments has occupied pride of place in formal argumentation theory. Yet, very little attention has been devoted to modelling argumentation about attack relations. Argumentation of this kind is encountered in many situations, especially when arguers discuss the relevance of each other's arguments. To model it, we introduce argumentation frameworks with justified attacks, where an attack succeeds only if one of its justification arguments are accepted. The main technical result is a representation theorem, showing how to translate argumentation frameworks with justified attacks into standard argumentation frameworks with dummy arguments whose combinatorial properties encode the support-function of attack justifications.

1 Introduction

The computational theory of argumentation has become an important research topic in artificial intelligence [14]. At the core of the theory we find the notion of an argumentation framework (AF), introduced by Dung in the early 1990s [7]. By proposing to treat arguments as atoms that are connected to each other by an attack relation, Dung was able to better harness the power of graph-theoretic methods in the study of non-classical logic. Specifically, he was able to define important semantic concepts in terms of combinatorial properties of directed graphs. Many technical results have followed, and formal argumentation theory is a field in rapid growth within the field of artificial intelligence, see, e.g., [6].

However, not everyone agrees that representing arguments as nodes in directed graphs strikes the appropriate balance between abstraction and representational adequacy. Specifically, it has been argued that taking the attack relation as a primitive and ignoring the internal content of arguments threatens to render the theory overly abstract [13, p. 94-95]. A key intuition behind this

[1] sjur.dyrkolbotn@uu.nl.

criticism is that when an argument a attacks another argument b, then it does so *for a reason.* This reason, moreover, must depend on the internal structure of a and b. However, on Dung's account of argumentation, the internal structure is hidden from view; an argument is nothing but an atom in a network, and the network itself is often drawn up by the modeller in an *ad hoc* fashion, not according to formal rules. This, it may be argued, is where a high level of abstraction becomes a potential problem, since it leaves out something that seems crucial when it comes to *justifying* the attack relation used in a given model.[2]

To address this concern, much recent work in formal argumentation is based on formally representing also the content of arguments, to make sure that attack relations are *instantiated.* That is, one requires the attack relation to be drawn up in a way that is justified by the internal argument structure, according to some agreed-upon rules of non-monotonic reasoning [2]. However, as show by much recent work in structured argumentation, it is hard to settle on a canonical set of rules for drawing up attack relations across different domains. In practice, the modeller will always have significant room for making discretionary decisions in this regard, for instance by stipulating a preference relation over arguments to prune away some attacks that would otherwise arise from the rules in the system, e.g., as it is done in the $ASPIC^+$ framework for structured argumentation [13,12].

In light of the inherent defeasibility of attack relations, the present paper proposes to complement existing techniques by a framework that approaches justification of attacks in a different way, not by trying to pin them down according to general rules, but rather by treating them as a topic of argumentation in their own right. Specifically, we propose to model attack justification as a form of meta-argumentation, where the justification of an attack (a, b) is itself an argument, one that has (a, b) as its conclusion.

We are not the first to discuss meta-level argumentation about properties of argumentation frameworks, see, e.g., [11,3,8]. We would especially like to highlight Gabbay's work in [8], which contains many of the same intuitions and technical ideas that we develop in a systematic way in this article. What sets us apart from previous work is that we anchor our discussion in the notion of an attack justification, proposing to treat such justifications as arguments. This greatly simplifies Gabbay's approach, while also generalising it in an in-

2 Some might wonder why we are not concerned also about the meta-level question of how to justify the inclusion of a given argument in the model. The reason why we do not worry about this is that the justification question for arguments can be seen as a special case of the justification question for attacks. In a system with an argumentation-based semantics, an argument that does not feature in any attacks plays no role whatsoever in the evaluation of any other arguments. Such an argument can safely be removed from consideration; it is irrelevant. Conversely, it seems clear that a reason to doubt the inclusion of an argument in a model is also a reason to doubt all justifications for including any attack involving that argument. In light of these two observations, there does not appear to be any reason to formally distinguish between justifications for including attacks and justifications for including arguments in a given model.

teresting way, allowing for non-wellfounded attack relations. Like Gabbay, we stay very close to Dung's original proposal. This allows us to prove a representation theorem, showing how argumentation models with justified attacks can be faithfully modelled using standard AFs, provided we add new atomic arguments and make sure the combinatorial properties of these arguments encode meta-level information about attacks. This result implies that algorithms and techniques developed for AFs can be applied also to argumentation frameworks with justified attacks. More generally, our results shows that there is no need to abandon the graph-theoretic view in order to account more fully for the argumentative origins of attacks.

The structure of the paper is as follows. In the next section, we motivate our own formalism by formulating three design principles that existing formal frameworks fail to satisfy. In Section 3, we define and explain our proposed *argumentation frameworks with justified attacks* (AFJAs) and give definitions of AF-style semantics for these structures. In Section 4, we show how any AFJA can be reduced to a classical AF. Finally in Section 5, we conclude.

2 Background and motivation

An argumentation framework (AF) is defined as a directed graph $\mathcal{F} = (A, R)$. The intuition is that A is a set of atomic argument names while $R \subseteq A \times A$ is an indefeasible attack relation over those arguments. Importantly, the attack relation is not justified but given as a primitive by the modeller. This is where we want to generalise the formalism, by asking for attacks to be justified by arguments. [3] An obvious consequence of asking for this is that we must then be prepared to account for the fact that attack arguments sometimes fail. Specifically, any formal model of justified attacks need to encompass a *defeasible* notion of attack, where attacks are sometimes not taken into account because their justifications are rejected.

The literature on formal argumentation has already seen several proposals for frameworks with defeasible attack relations [10,1,8]. However, none of the existing formalisms give us what we want, because they all fail to satisfy at least one among the following list of three design principles.

Principle I: attack justifications should be represented as arguments In general, an attack (a, b) can be justified in many different ways, by many *distinct* justifications. To capture this, we will demand that all attack justifications must be modelled explicitly as arguments. If the aim is to come up with something like a general theory of arguments, this demand must surely be met; in this case, attack justifications would fall under the scope of the theory already because a justification is a kind of argument. Usually, however,

[3] We remain agnostic about what theories of justification such arguments might be based on. This is the benefit of a highly abstract model; justifications can be anything, what matters is not where they come from but how their argumentative functions can be analysed by studying an appropriate attack relation (e.g., one that has been generated by some system of instantiated argumentation).

one has a more specific argumentation theory in mind, for instance a theory of arguments based on non-monotonic inference rules. If this is the case, what we effectively require is that the reasoning used to justify attacks between arguments should be representable *within* the formal system that generates what we call arguments. This is a natural property for a system of non-monotonic reasoning to satisfy, ensuring that it can represent a key component of its own argumentation-theoretic meta-properties. In future work, we plan to make this aspect explicit by considering instantiations of justified attacks in a system of structured argumentation with meta-level inference rules. For now, we hope the reader will agree that while the principle we propose here is not axiomatic by any means, it makes a natural demand on the expressive power of argumentation theories.

In addition, we highlight a more localised reason to adopt our proposal. Specifically, once we introduce higher-order attacks into our framework, our principle ensures that attacks on attacks make sense. We are now in a position to say why they are there, to enable also the higher-order attacks to be justified in a systematic manner. By contrast, if attack justifications are not represented explicitly as arguments in the model, it is not clear what the basis is for saying that you can attack an attack in the first place. What exactly are you attacking, if not some argument in favour of the attack you don't want? This does not seem clear in previous work on higher-order attacks, serving as a reason why our principle should at least be observed in this special context. The next design principle clarifies this point further.

Principle II: attacks on attacks are attacks on attack justifications

Given Principle I, it becomes perfectly natural to consider attacks on attack justifications; such justifications, after all, are regarded as arguments in their own right. Intuitively, in order to make an attack on an attack you have to specify *which* of the justifications for the attack you are challenging. We include this as our second design principle, to stress that an attack on an attack will only be considered well-formed if it targets specific justification arguments. In previous work, attacks on attacks have been modelled as a relation that targets attacks directly. Hence, an attack on an attack will either have to be taken to defeat all conceivable justifications, or else it must be taken to defeat none of them. In our opinion, it is better to model attacks on attacks by explicitly accounting for the justification arguments that such an attack challenges. This makes it possible to account for the fact that an attack can defeat *some* of the justifications for another attack, while still failing to force the attack to be retracted.

Principle III: the acceptability of an attack justification should not depend on the acceptability of the source of the underlying attack

Since we model attack justifications as arguments, it also becomes natural to consider attacks made by such arguments. In effect, attacks can sometimes attack other arguments and attacks, not directly, but through their justification arguments. This is similar to an intuition pursued by Gabbay [8]. However,

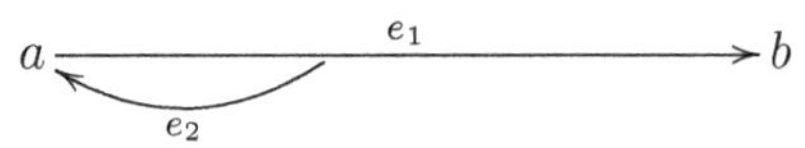

Fig. 1. A structure including an attack emanating from an attack justification.

attacks emanating from attacks become quite mysterious objects when no distinction is made between attacks and the justification arguments that support them. For this reason, Gabbay's definitions are both more complicated and less general than those presented here. Nevertheless, Gabbay's work contains a crucial semantic insight: the acceptability of an attack justification does not depend in any way on the acceptability of the argument from which the underlying attack emanates. If you have a convincing argument that a attacks b, then your argument is still convincing even if a is not accepted. This means that while there should be a dependence linking the acceptability of your argument and the acceptability of the attack (a, b), there should not necessarily be any such dependence between the acceptability of a and the acceptability of an argument justifying (a, b). Several previous theories of higher-order attacks violate this constraint, with Gabbay's work being a notable exception.

In the next section, we provide a formal framework that supports all the design principles stipulated above.

3 Formalising justified attacks

To satisfy Principles I and II, we need to be able to define attacks as relations over a set of arguments that includes at least one argument for every attack, corresponding to its justification. A simple formal structure that allows us to do this is presented below.

Definition 3.1 Given a set of propositions Π, an argumentation framework with justified attacks (AFJA) is a pair $\mathcal{A} = (N, E)$ where

- N is a non-empty set of argument names, and
- $E : N \to (N \times N) \cup \Pi$ maps argument names to conclusions.

If $E(n) \in \Pi$ we say that the argument n is *atomic*. Otherwise, we say it is a *meta-argument* (meaning its conclusion is that one argument attacks another).

To illustrate the definition, notice how the structure depicted in Figure 1 is an example of an AFJA with $N = \{a, b, e_1, e_2\}$ and $E(e_1) = (a, b), E(e_2) = (e_1, a)$.

AFJA models provide a representation formalism that is more general than any existing proposal of which we are aware. Specifically, our model supports representation of multiple attacks from a to b, as well as modelling of *non-wellfounded* attacks. For an example of the latter, consider the AFJA on the left in Figure 2. Arguably, the attack depicted here is the canonical example of an argument that successfully makes an attack that is not explicitly justified, meaning that the attack it makes must be (at least implicitly) *self-justifying*. The natural converse, depicted in the middle of Figure 2, is a new type of fallacious structure, namely the attack that is *self-defeating*. Finally, on the right, we see an example of a structure that is both self-defeating and self-justifying – the ultimate fallacy, namely the conceited claim that self-defeat is the only possible explanation for rejection.

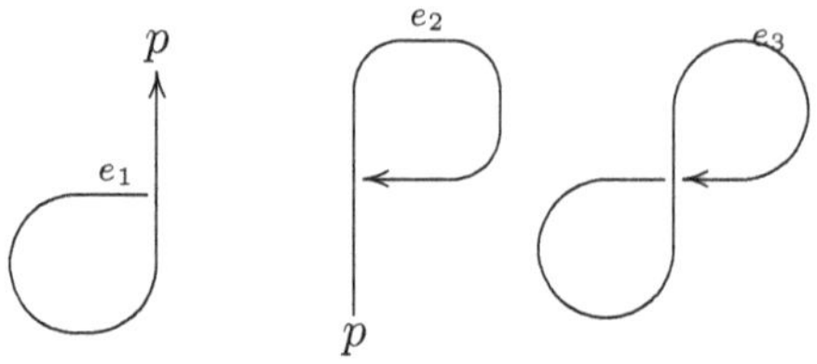

Fig. 2. Three kinds of non-wellfounded attack justifications. For instance, these can be used to model e_1: "not p, therefore e_1 attacks p" ($E(e_1) = (e_1, p)$), e_2: "not p, therefore p attacks e_2" ($E(e_2) = (p, e_2)$) and e_3: "e_3 is rejected, therefore e_3 attacks e_3" ($E(e_3) = (e_3, e_3)$).

The increased expressiveness of AFJAs compared to earlier proposals stems from the introduction of explicit attack arguments. Specifically, having such arguments available enables us to link attacks with their justifications using a map from N to $N \times N$, a simple representation of how attack claims can be the conclusions of arguments.[4] For uniformity, we have also endowed atomic arguments with explicit conclusions, taken from some arbitrary set of propositions Π. The perspective induced by this signature strikes us as the appropriate one, since a distinction should always be made between an argument and the proposition it is used to support. For the theoretic purposes of this paper, however, it is safe to simply conflate argument names with their conclusions, i.e., to assume that $\Pi \subseteq N$ and that $E(n) = id$ whenever n is not an attack justification.

Notice that AFJAs, like AFs, suppress the premises of arguments. We believe this is natural, since premises can take many forms; arguably, anything from proof trees to photographs can be used to build arguments that we might want to represent and reason about using formal tools. Therefore, we should rely on abstraction, to ensure that the computational theory does not become tied to a specific format for representing the content of arguments. Formalisms that attempt to derive attack relations from the internal structure of arguments tend to violate this constraint. For instance, in order to use the attack-generating features of $ASPIC^+$, one must first rewrite all arguments as derivations in some formal logic. Such a requirement might not always be appropriate, much less realistically fulfilled. Moreover, a computational theory

[4] By contrast, if one treats objects from $N \times N$ as arguments *and* attacks simultaneously, as in [8], giving well-defined attack relations becomes more complicated. Moreover, expressive power is lost, since an attack is then simply conflated with its justification.

of argument should be able to represent and process argumentative content sourced from a variety of domains, where premises can be represented using various kinds of data structures and may support their conclusions in less convincing ways than formal logicians tend to expect. In a concrete modelling context, one may then explore this aspect of instantiation further by providing a map $P : N \to I$, linking arguments with their premises (taken from some collection I of information). In general, however, we believe it is a strength if the theory of argumentation can be developed in general terms, without reference to any specific P or I.

Now, to demonstrate in more depth how AFJA models can be used, we will model a meta-argument based on a classical example in default logic. Recall that a (propositional) default rule has the form $a : \frac{p;\{j_1,\ldots,j_n\}}{q}$ where p, q, j_i are propositional formulas for all $1 \leq i \leq n$. Then we say that p is the prerequisite of a, all the j_is are justifications for it, while q is its conclusion. Default reasoning can be looked at as argumentation, a perspective that has been adopted often in previous work. Arguably, however, the most important arguments that can be sourced from the field of default logic are meta-arguments that do not correspond to chains of default rules, but instead pertain to the question of how to define a semantics for default theories. Here the AFJA formalism can help to formalise an aspect of default reasoning that is not commonly formalised at all, as demonstrated in the following example.

Example 3.2 Let $\Pi = \{p, q\}$ and assume we have the default rules $a : \frac{p:q}{q}$ $b : \frac{\top:\neg q}{\neg q}$ and $c : \frac{\top:p}{p}$. Moreover, assume the agent has a priority over these rules given by $a > b > c$. The example is important in default logic, since there is meta-level disagreement about it. According to some, it shows the inadequacy of early variants of prioritised default logic, which would tend to give $\{\neg q, p\}$ as the preferred extension of the theory [4]. Let us use an AFJA model to represent a simple argument about whether $\{\neg q, p\}$ is a reasonable outcome. First, we take a, b and c to be arguments with $E(a) = q, E(b) = \neg q$ and $E(c) = p$. Then we generate attack justifications on the basis of the following general principles: (1) we argue that symmetric attacks must be present between any two conflicting defaults, (2) we argue that an asymmetric attack must be present from a to b if a and b are in conflict and a has higher priority than b, and (3) we argue that when a attacks b via an asymmetric attack, then b must be rejected if the preconditions of a are added to the knowledge base. These principles appear justified by meta-logical properties of default logic. However, the arguments resulting from (3) are meta-arguments, since they talk about asymmetric attacks between default rules. Let us apply principles (1)-(3) to our example. Then by (1) we get e_1 with $E(e_1) = (a, b)$ and e_2 with $E(e_2) = (b, a)$. By (2) we get e_3 with $E(e_3) = (a, b)$. Finally, from principle (3) and the fact that p is necessarily added to the knowledge base, we get e_4 with $E(e_4) = (e_3, b)$. The resulting AFJA is depicted below.

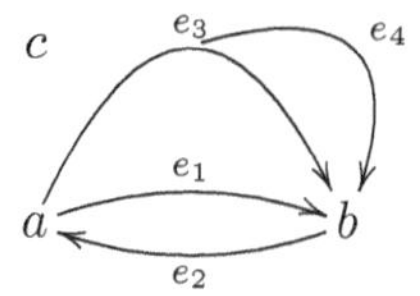

Intuitively, since there are no counterarguments against either e_3 or e_4, we should accept both of these arguments. Hence, we should reject b and accept a and c. In other words, $\{p, q\}$ should be the outcome of applying these rules to an empty knowledge base, not $\{p, \neg q\}$. If you disagree with this conclusion, of course you can try to add suitable attacks to the AFJA above. The question becomes: can you justify them? [5]

This example shows the potential usefulness of AFJA models when doing meta-reasoning in the context of default logic. Notice, in particular, that it would be incorrect, or at least counter-intuitive, to model the rejection of b as arising directly from the fact that a attacks b and is the preferred argument. Clearly, the rejection of b depends on holistic properties of the structure, properties that are more naturally modelled by meta-level attacks like e_4. Indeed, in examples like the one considered above, one of the intuitions we form is that we need to reject b *before* accepting a, since otherwise a can never be accepted. But if this is how we reason, a as such can hardly be the argument we rely on when rejecting b. Indeed, this is why the example is important, and why it leads to disagreement. It would be great to have an argumentation system that can systematically encode this sort of disagreement and produce default extensions in response to it (perhaps according to new principles, e.g., without relying on a fixed scheme to "lift" priorities to chains of defaults). We believe the example above shows that AFJA models can potentially play a useful role in the development of such kinds of argumentation systems. Of course, the true extent of the potential identified here will only become clear once we enhance AFJA models by an instantiation layer to systematically produce appropriate attack justifications, e.g., by a suitable modification of $ASPIC^+$. Further exploration of this will be left for future work.

In the next section, we develop a formal semantics for AFJAs that satisfies our design principles and also agrees with the intuitions we have presented in examples so far.

3.1 Semantics

A key assumption in formal argumentation is that arguments that are not attacked should be accepted. That is, the default position is to accept all arguments, unless there is a reason to do otherwise. This basic assumption is what permits the theory to make do with only an attack relation; an argument

[5] Perhaps you can; in what sense exactly is it correct to say that p *is added* to the knowledge base in this case, c.f, principle (3)?

is not in need of any support unless it is attacked, in which case the only relevant support is that offered by an attack on an attacker.

To arrive at a semantics for AFJA models we will rely on a generalised version of this intuition. Here we need to take into account that the effectiveness of an AFJA-attack depends on two separate semantic entities: (1) the attacking argument, and (2) the argument justifying the attack. On this basis, the idea behind the semantics we propose can be stated very simply, as the principle that the strength of an argument should be taken to equal the maximum possible acceptance value minus the strength of the weakest component of the strongest attacking argument. We mention that this idea is essentially present already in [8].[6] However, as noted earlier, our formalisation using explicit attack justifications is at once more general and easier to define.

To formalise things, we first define the auxiliary notions of source and target, provided in Equation 1.

$$\begin{aligned} src(n) &= \begin{cases} x & \text{if } E(n) = (x, y) \\ * & \text{otherwise} \end{cases} \\ trg(n) &= \begin{cases} y & \text{if } E(n) = (x, y) \\ * & \text{otherwise} \end{cases} \end{aligned} \tag{1}$$

In addition, for every $n \in N$, we define $in(n) = \{m \in N \mid n = trg(m)\}$. Moreover, if $f : N \to \{0, {}^1\!/_2, 1\}$, we use the notation $f^0 = \{n \in N \mid f(n) = 0\}$ and similarly for $f^{1/2}$ and f^1. Let us introduce the convention that by default

$$\max_{e \in \emptyset} (\min\{f(e), f(src(e))\}) = 0.$$

Then, for any $f : N \to \{0, {}^1\!/_2, 1\}$, we can define the argument evaluation $\overline{f} : N \to \{0, {}^1\!/_2, 1\}$ as follows, for all $n \in N$:

$$\overline{f}(n) = 1 - \max_{e \in in(n)} \min\{f(e), f(src(e))\} \tag{2}$$

This evaluation map allows us to generalise all the standard Dung-style argumentation semantics to AFJA models. Specifically, we are ready to define complete labellings as those that satisfy the following, for all $n \in N$:

$$f(n) = \overline{f}(n) \tag{3}$$

In the following, we will focus on the complete semantics. However, we note that semantics corresponding to the other classical semantics for AFs can also be defined, as listed in Figure 6. Notice that all the semantics defined here give the desired result in all examples considered previously. For instance, if we return to the AFJA in Figure 1, it is not hard to see that any f satisfying Equation 3 will have to provide $f(e_2) = f(e_1) = f(b) = 1$ (meaning these arguments are accepted) and $f(a) = 0$ (meaning a is rejected).

[6] Compare also with Gabbay's later work on equational semantics for argumentation [9].

Domain	*Condition*
Admissible, $adm(S)$:	
$f \in \{0, 1/2, 1\}^n$	$\forall n \in N : f(n) = 1 \Rightarrow f(n) = \overline{f}(n)$
Complete, $com(S)$:	
$f \in \{0, 1/2, 1\}^n$	$\forall n \in N : f(n) = \overline{f}(n)$
Grounded, $grd(S)$:	
$f \in com(S)$	$\forall f' \in com(S) : f^{1/2} \not\subset f'^{1/2}$
Preferred, $prf(S)$:	
$f \in adn(S)$	$\forall f' \in com(S) : f^1 \not\subset f'^1$
Semi-stable, $sem(S)$:	
$f \in adm(S)$	$\forall f' \in com(S) : f^{1/2} \not\supset f'^{1/2}$
Stable, $stb(S)$:	
$f \in adm(S)$	$f^{1/2} = \emptyset$

Fig. 3. Other semantics for AFJAs (the left column shows the domain of f, while the right column states the condition for membership in the semantics).

It is not *prima facie* clear that all AFJAs admit complete labellings. The remainder of this section is devoted to establishing an existence theorem. We begin by defining, for all $f : N \to \{0, 1/2, 1\}$, a corresponding sequence $F = \{f = f_1, f_2, \ldots, f_i, \ldots\}$ inductively as follows, for all $i > 1$, for all $x \in N$:

$$f_i(x) = 1 - \max_{e \in in(n)} \min\{f_{i-1}(e), f_{i-1}(src(e))\} \tag{4}$$

Clearly F contains a complete f_i just in case $f_i = f_{i+1}$, in which case $f_i = f_j$ for all $f_j \in F$ with $j \geq i$. This forms the basis for the proof of the following result.

Theorem 3.3 *For all AFJAs $\mathcal{A}$, there exists at least one complete labelling $f : N \to \{0, 1/2, 1\}$ (i.e., a labelling satisfying Equation 2).*

Proof. Let $f_0(n) = 1/2$ for every $n \in N$, and define the sequence $f_0, f_1, \ldots$ as previously discussed. Notice that since $f_0^{1/2} = N$, $f_0^0 = f_0^1 = \emptyset$. We show that $\forall i$, $f_i^0 \subseteq f_{i+1}^0$ and $f_i^0 \subseteq f_{i+1}^1$. Suppose towards a contradiction that this is not the case, and that c is the smallest number which violates this, i.e., either (i) $f_c^0 \not\subseteq f_{c+1}^0$ or (ii) $f_c^1 \not\subseteq f_{c+1}^1$ (or both). By definition of f_0 we know that $c = 0$ is impossible. To proceed, let us assume the monotonicity claim holds for all $0 \leq i < c$ and use this to derive a contradiction.

(i) Suppose $f_c^0 \not\subseteq f_{c+1}^0$. Let x be an element such that $x \in f_c^0$, but $x \notin f_{c+1}^0$. This means that

$$1 - \max_{e \in in(x)} \min\{f_{c-1}(e), f_{c-1}(src(e))\} = 0 \tag{5}$$

$$1 - \max_{e \in in(x)} \min\{f_c(e), f_c(src(e))\} \neq 0 \tag{6}$$

Equation 5 holds (by witness e) if, and only if,

$$\exists y \in in(x)\, [f_{c-1}(y) = f_{c-1}(src(y)) = 1]$$

and Equation 6 holds if, and only if,

$$\forall y \in in(x)\,[f_c(y) \neq 1 \text{ or } f_c(src(y)) \neq 1]$$

It follows that for the witness e we have $f_{c-1}(e) = f_{c-1}(src(y)) = 1$, but also that $\min\{f_c(e), f_c(src(y))\} \neq 1$. Suppose, without loss of generality, that $f_c(e) \neq 1$. Then $e \in f^1_{c-1}$, but $e \notin f^1_c$. This contradicts the assumption that c was the smallest index for which the monotonicity claim failed.

(ii) Suppose $f^1_c \not\subseteq f^1_{c+1}$. Let x be an element such that $x \in f^1_c$, but $x \notin f^1_{c+1}$. This means that

$$1 - \max_{e \in in(x)} \min\{f_{c-1}(e), f_{c-1}(src(e))\} = 1 \tag{7}$$

$$1 - \max_{e \in in(x)} \min\{f_c(e), f_c(src(e))\} \neq 1 \tag{8}$$

Equation 7 holds if, and only if,

$$\forall y \in in(x)\,[f_{c-1}(y) = 0 \text{ or } f_{c-1}(src(y)) = 0]$$

And Equation 8 holds, by witness e if, and only if,

$$\exists y \in in(x)\,[f_c(y) \neq 0 \text{ and } f_c(src(y)) \neq 0]$$

It follows that for the witness e we have $f_i(e) \neq 0$ and $f_c(src(e)) \neq 0$, but at least one of $f_{c-1}(e) = 0$ or $f_{c-1}(src(e)) = 0$, so $\{e, src(e)\} \cap f^0_{c-1} \setminus f^0_c \neq \emptyset$. But this contradicts the assumption that c was minimal.

We have shown that the assumption that monotonicity fails at some index $c > 0$ leads to a contradiction, hence the sequences of sets f^0_i and f^1_i grow (concurrently) monotonically with i. With an appeal to the Knaster-Tarski fixed-point theorem we have a fixed point. □

4 Reduction (AFJA $\rightsquigarrow$ AF)

In this section we show how to represent any AFJA model $\mathcal{A}$ as an AF $\tau(\mathcal{A})$. To do this, we need to be able to approach all arguments from $\mathcal{A}$ as though they were atomic. This risks losing important information, a concern we will address by adding auxiliary (atomic) arguments to ensure that all attack justifications can be represented implicitly, as combinatorial properties of those (atomic) arguments from $\tau(\mathcal{A})$ that correspond to meta-arguments in $\mathcal{A}$. Specifically, we provide the following definition, which is similar to a construction used in [8] to provide several AF-based semantics for higher-order attack relations.

Definition 4.1 For all AFJA models (N, E), we define $\tau((N, E)) = (A, R)$ as follows

- $A = N \cup \overline{N} \cup N_e$ where
  - $\overline{N} = \{\bar{x} \mid n \in N, src(E(n)) = x\}$

- $\cdot\ N_e = \{e_n, \bar{n} \mid E(n) \notin \Pi\}$
- $R = \{(e_n, y), (\bar{x}, e_n), (\bar{n}, e_n) \mid E(n) = (x, y)\} \cup \overline{R}$ where
  $\cdot\ \overline{R} = \{(n, \bar{n}) \mid \bar{n} \in \overline{N}\}$

Intuitively, each named argumentative element becomes an element $N \subseteq A$. Furthermore, every edge n has an "edge-node" e_n (from N_e), Finally, all edges and every node which is the source for some edge have an argument claiming that they must be rejected: $\bar{n}$. Notice that $\overline{N}$ and N_e may overlap. For an example, consider the AFJA and the corresponding AF-reduction depicted in Figure 4.

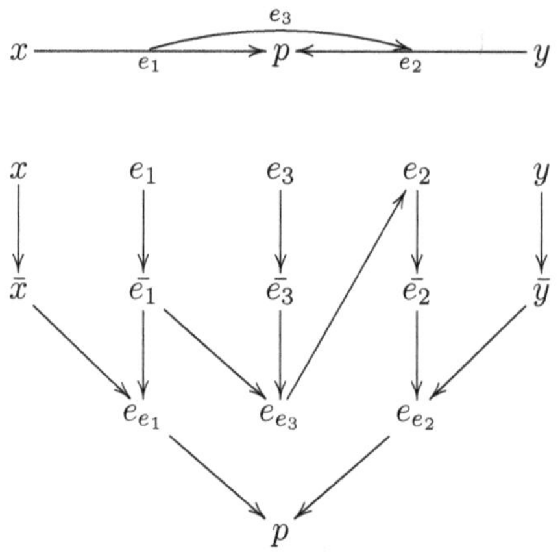

Fig. 4. AF (below) corresponding to AFJA (above)

Example 4.2 To illustrate how the representation works for structures that are further removed from ordinary AFs, consider the AFJA $(\{e\}, \{e \mapsto^E (e, e)\})$. The AF representation will consists of the elements of three sets:

$N = \{e\}$ because e is a named element,

$\overline{N} = \{\bar{e}\}$ because e is *the source* of an edge, and

$N_e = \{e_e, \bar{e}\}$ with both elements included because e is an edge.

This construction, together with the original AFJA, is depicted in Figure 5.

To prove formally that our representation gives the required result, we will show a correspondence theorem linking our semantics for AFJAs with the standard semantics for AFs. For convenience, we will work with a labelling-based formulation of the standard AF semantics. Specifically, given an AF (A, R) and a labelling $f : A \to \{0, {}^1\!/_2, 1\}$ we define $\bar{f}$ as follows, for all $x \in A$:

$$\bar{f}(a) = 1 - max\{f(b) \mid b \in R^-(a)\} \tag{9}$$

We can now formulate the complete semantics for AFs by saying that $f : A \to \{0, {}^1\!/_2, 1\}$ is (AF-)complete whenever we have $f(a) = \bar{f}(a)$ for all

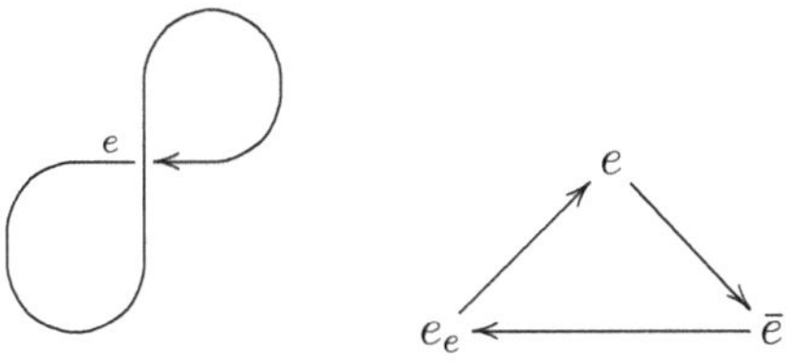

Fig. 5. AFJA (left) and corresponding AF (right).

$a \in A$. It is not hard to see that this definition corresponds to the standard definition of complete semantics for AFs, c.f., the labelling-based formulation given in [5]. Moreover, we can easily express the other classical semantics for AFs in a similar manner, corresponding to the AFJA-semantics in Figure 6. However, for space reasons, we only discuss the complete semantics here (the representation theorem is easily adapted to the other semantics as well).

The next step towards a proof is to establish a link between labellings of arguments in AFJAs and labellings of corresponding AFs. To this end, we construct, for all $\mathcal{A} = (N, E)$ and all $f : N \rightarrow \{0, {}^1\!/_2, 1\}$, a corresponding labelling $f_\rho : A \rightarrow \{0, {}^1\!/_2, 1\}$ for $\tau(\mathcal{A})$. Specifically, for all $x \in A$, we define the value of f_ρ as follows:

$$f_\rho(x) = \begin{cases} f(x) & x \in N \\ 1 - f(n) & x = \bar{n}, n \in N \\ min\{f(n), f(src(n))\} & x = e_n \in N_e \end{cases} \tag{10}$$

This gives us the following representation result.

Theorem 4.3 *For all AFJAs $\mathcal{A} = (N, E)$, we have that $f : N \rightarrow \{0, {}^1\!/_2, 1\}$ is complete for $\mathcal{A}$ if, and only if, f_ρ is complete for $\tau(\mathcal{A}) = (A, R)$.*

Proof. Let $\mathcal{A} = (N, E)$ be an arbitrary AFJA, and let $f = N \rightarrow \{0, {}^1\!/_2, 1\}$ be an arbitrary labelling. Further, let $\tau(\mathcal{A}) = (A, R)$ be the corresponding AF as defined in Definition 4.1 of τ, and f_ρ be the labelling defined in Equation (10) w.r.t. f.

We show that f is complete w.r.t. $\mathcal{A}$ if, and only if f_ρ is complete for $\tau(\mathcal{A})$, that is $f = \bar{f} \Leftrightarrow f_\rho = \bar{f}_\rho$. First we make some observations.

(A) By inspection of the definition of τ, it is easy to verify that the only incoming edge in $\tau(\mathcal{A})$ into arguments of the form $\bar{n}$, for $n \in N$, comes from n. That is, $R^-(\bar{n}) = \{n\}$. By inspecting (10), we can also see that $f_\rho(\bar{n}) = 1 - f_\rho(n) = 1 - f(n)$. Importantly, notice that by (9), we have

$$\bar{f}_\rho(\bar{n}) = 1 - \max_{b \in R^-(\bar{n})} \{f_\rho(b)\} = 1 - f_\rho(n) = f_\rho(\bar{n})$$

(B) By inspecting the definition of τ, we see that $R^-(e_m) = \{\bar{m}, \bar{s}\}$ where

$src(m) = s$. This and (9) yields

$$\bar{f}_\rho(e_m) = 1 - \max_{b \in R^-(e_m)} f_\rho(b) = 1 - \max\{f_\rho(\bar{m}), f_\rho(\bar{s})\}$$

This in turn, by two appeals to (10) and (A), gives

$$\begin{aligned}\bar{f}_\rho(e_m) &= 1 - \max\{1 - f(m), 1 - f(s)\} \\ &= \min\{f(m), f(s)\} = f_\rho(e_m)\end{aligned}$$

$\Rightarrow$) Assume $f = \bar{f}$. We need to show that for every $a \in A$, $f_\rho(a) = \bar{f}_\rho(a)$. By inspecting the definition of τ, we see that either $a \in N$, or otherwise either $a = \bar{n}$ for some $n \in N$, or $a = e_m$ where $m \in N$ and $E(m) \in N \times N$.

- Suppose $x \notin N$. If $x = \bar{n}$ for some $n \in N$, by observation (A) we are done. Otherwise $x = e_m$ for some $m \in N$ and $E(m) \in N \times N$, but then, by observation (B), we are done.
- Otherwise, $x = n \in N$. From (10) and the assumption that f is complete, $f_\rho(n) = f(n) = \bar{f}(n)$.

$$\begin{aligned}\bar{f}(n) &= 1 - \max_{m \in in(x)} \min\{f(m), f(src(m))\} && (2)\\ &= 1 - \max_{m \in in(x)} f_\rho(e_m) && (10)\\ &= 1 - \max_{b \in R^-(n)} f_\rho(b) && (B)\\ &= \bar{f}_\rho(n) && (9)\end{aligned}$$

From the assumption that f is complete for $\mathcal{A}$, we have shown that, for every $a \in A$, $f_\rho(a) = \bar{f}_\rho(a)$, i.e., that f_ρ is complete for $\tau(\mathcal{A})$.

$\Leftarrow$) Follows trivially from $\bar{f}_\rho(n) = \bar{f}(n)$, which we showed above without any appeals to the assumption of completeness of f_ρ.

□

5 Conclusion

This paper has proposed a new formalism for modelling argumentation with justified attacks, extending Dung's theory of argumentation frameworks. The motivation for providing a new formal framework was presented in Section 2 where we briefly stated and defended three design principles that existing formalisms fail to satisfy.

A key point made was that disputes about where attack relations come form should not necessarily be dealt with by attempting to formulate strict rules, maintained externally to the formalism itself, that pin down *the* appropriate way to derive attacks from the internal structure of arguments in a given domain. Instead, we believe the theory of argumentation can benefit greatly from a formalism that supports formal reasoning about attack justifications as meta-arguments, especially if techniques and results developed for AFs can be applied also at the meta-level.

We believe that our AFJA models provide exactly such a formalism, one that extends, clarifies, and simplifies very sensible ideas that can also be found in previous work by Gabbay. The main result of the paper was a representation theorem that shows how results obtained for AFs can indeed be imported to the new setting, with only a linear growth in the number of arguments needed to represent meta-argumentation about attacks as combinatorial dependencies in directed graphs. For this reason especially, we believe the AFJA formalism will make a valuable addition to the toolbox of representation languages employed in the field of formal argumentation.

References

[1] Baroni, P., F. Cerutti, M. Giacomin and G. Guida, *AFRA: argumentation framework with recursive attacks*, Int. J. Approx. Reasoning **52** (2011), pp. 19–37.

[2] Besnard, P., A. J. García, A. Hunter, S. Modgil, H. Prakken, G. R. Simari and F. Toni, *Introduction to structured argumentation*, Argument & Computation **5** (2014), pp. 1–4. URL `http://dx.doi.org/10.1080/19462166.2013.869764`

[3] Boella, G., D. M. Gabbay, L. van der Torre and S. Villata, *Meta-argumentation modelling I: Methodology and techniques*, Studia Logica **93** (2009), pp. 297–355.

[4] Brewka, G. and T. Eiter, *Prioritizing default logic*, in: S. Hölldobler, editor, *Intellectics and Computational Logic (to Wolfgang Bibel on the occasion of his 60th birthday)*, Applied Logic Series **19** (2000), pp. 27–45.

[5] Caminada, M. W. and D. M. Gabbay, *A logical account of formal argumentation*, Studia Logica **93** (2009), pp. 109–145.

[6] Charwat, G., W. Dvok, S. A. Gaggl, J. P. Wallner and S. Woltran, *Methods for solving reasoning problems in abstract argumentation a survey*, Artificial Intelligence **220** (2015), pp. 28 – 63.
URL `http://www.sciencedirect.com/science/article/pii/S0004370214001404`

[7] Dung, P. M., *On the acceptability of arguments and its fundamental role in nonmonotonic reasoning, logic programming and n-person games*, Artificial Intelligence **77** (1995), pp. 321 – 357.

[8] Gabbay, D. M., *Semantics for higher level attacks in extended argumentation frames. part 1: Overview*, Studia Logica: An International Journal for Symbolic Logic **93** (2009), pp. 357–381.
URL `http://www.jstor.org/stable/40587172`

[9] Gabbay, D. M., *Introducing equational semantics for argumentation networks*, in: W. Liu, editor, *Symbolic and Quantitative Approaches to Reasoning with Uncertainty: 11th European Conference, ECSQARU 2011, Belfast, UK, June 29–July 1, 2011. Proceedings*, Springer, 2011 pp. 19–35.

[10] Modgil, S., *Reasoning about preferences in argumentation frameworks*, Artificial Intelligence **173** (2009), pp. 901 – 934.
URL `http://www.sciencedirect.com/science/article/pii/S0004370209000162`

[11] Modgil, S. and T. J. M. Bench-Capon, *Metalevel argumentation*, J. Log. Comput. **21** (2011), pp. 959–1003.
URL `http://dx.doi.org/10.1093/logcom/exq054`

[12] Modgil, S. and H. Prakken, *The ASPIC$^+$ framework for structured argumentation: a tutorial*, Argument & Computation **5** (2014), pp. 31–62.
URL `http://dx.doi.org/10.1080/19462166.2013.869766`

[13] Prakken, H., *An abstract framework for argumentation with structured arguments*, Argument & Computation **1** (2010), pp. 93–124.

[14] Rahwan, I. and G. Simari, editors, "Argumentation in artificial intelligence," Springer, 2009.

Arguments, Responsibilities and Moral Dilemmas in Abductive Default Logic

Sjur K. Dyrkolbotn [1]

Utrecht University, Department of Philosophy and Religious Studies

Truls Pedersen

University of Bergen, Department of Information Science and Media Studies

Jan Broersen

Utrecht University, Department of Philosophy and Religious Studies

Abstract

We assume that an agent is not responsible for rule-induced extensions of its theory about the world; responsibility requires the presence of a choice. This supports the attractive conclusion that responsibility for rule-based agents can only arise when the agent faces a "dilemma" regarding how to apply the rules. Default logic offers precise formulations of this intuition. However, it turns out that existing definitions force us to recognise too many dilemmas when reasoning about rules. Specifically, not all moral conflicts are moral dilemmas; the crucial element of choice is sometimes missing. To address this, we first present a refined definition for normal default theories, before going on to present a generalisation that applies to abstract argumentation frameworks.

1 Introduction

As the gun lobby keeps telling us: guns don't kill, people do. We could not agree more, but would like to add that gun lobbyists probably kill more than most, since they work to uphold deadly rules. Indeed, deadly rules are just like guns; they don't kill people, rule-makers do. This statement might be provocative, but it is structurally similar to the claim made by the gun lobby. Both slogans highlight the importance of morally salient choices; the gun-killer's choice of gun use and the rule-maker's choice of gun regulation. In this article, we address the question of how moral choices like these should be *defined*. For the sake of precision, we provide an example definition in default logic, extending a line of formal work on moral reasoning that was initiated by Horty [5,6].

[1] sjur.dyrkolbotn@uu.nl.

The overarching aim of the article is to contribute to formalising theories about moral responsibility in formal logic. This is becoming an increasingly important research topic in light of the increasing moral salience of intelligent systems in our societies. A key intuition underlying our work in this article is that an agent is not responsible for the rules it has been given, but might be responsible for how it solves conflicts that arise from them. Moreover, if the agent resolves conflicts by weighing the rules in a way that the designers of the rules would not condone, this would appear not to be the designer's fault, but rather the fault of the agent. To unpack these intuitions and make them precise, the notion of a moral choice comes to function as an important anchor, both conceptually and formally.

The definition we provide is novel, departing from previous work on moral reasoning in this context. It is based on the following conceptual premise: moral choices should not be conflated with their indirect rule-induced consequences. If you choose to x and then you y because x triggers a rule saying you should y, you did not (necessarily) choose y. The rule-maker certainly did, but you may not have. We argue that this precept is particularly important in the context of reasoning about responsibility for intelligent systems. Unlike humans, robots do not typically "choose" to follow the rules; they typically have no choice, rules are a physical constraint on their behaviour. This motivates a simple formal definition of moral responsibility relative to default logic, based on our insistence that choices need to be recorded separately from rule-induced consequences.

The structure of the paper is as follows. We begin in section 2 by giving an informal argument to support our ideas about moral choice and responsibility. We do not discuss the vast literature on moral responsibility or the growing literature on machine ethics, but compactly present our starting point and the ideas that motivate our formal work. Then in section 3 we present a formal definition of responsibility which applies to so-called normal default theories. Here we rely on ideas and techniques from argumentation theory to define what it means to make a moral choice, giving rise to a new distinction between moral conflicts and moral dilemmas in the context of default reasoning. We proceed to sketch a theory of responsibility for abstract argumentation in section 4, where responsibility is defined as a modality over argumentation frameworks. This definition lifts the concept of responsibility in such a way that it can be used in any formalism of non-monotonic reasoning that admits an argumentation-based semantics. In section 5, we offer a brief conclusion.

2 Responsibility and choice: a conceptual starting point

We regard a moral rule as an action-directing element of an agent system, possibly one with special significance, but not necessarily an expression of a universally desirable principle. This does not make us relativists and it is not an attack on moral philosophy. Moreover, while we offer the moral realists the courtesy of highlighting our descriptive starting point, we do so with some degree of reservation. This is because we wish to remain sensitive to the possibility that there might not be a clear-cut distinction between descriptive and normative theories about morality. However, since nothing in the current article seems to hinge on this, we feel confident to leave this aspect of meta-ethics behind us for now.

It seems true, in any case, that descriptive moral rules are just like guns; they are not good or bad *in themselves*. It all depends on how you choose to use them. This should not be a controversial claim; if there really is such a thing as a universal moral theory, its object of study is surely good rational agency, not the plethora of descriptive moral rules that might or might not get us there. [2] However, descriptive moral rules do guide our judgements about moral responsibility, relative to a given agent system.

In the following, we take moral responsibility to be a meta-level notion, a notion that we can *apply* when assessing agency against moral rules, not a notion that is inherently dependent on any specific collection of such rules. This sense of responsibility is backward-looking, requiring a form of abductive reasoning about a chain of events leading to some outcome. This is also where the notion of moral choice becomes important, because it helps distinguish between those points in the event chain that can be attributed to agents and those that have to be attributed to rule-makers. As we will see in examples later, this distinction can make a significant difference when reasoning about responsibility.

We remark that we approach moral responsibility from a *normative* perspective, inquiring into what the conditions for moral responsibility *should be*, not what they are taken to be in a given moral community. Furthermore, our work focuses on unpacking what we call *the choice principle*, a constraint on responsibility attribution that we formulate as follows: an agent can be morally responsible for choosing X only if the agent could have chosen differently when X was chosen. We remark that this is weak choice principle, since it only speaks about choices as mental states, without making reference to their possible physical manifestations. There is no requirement, for instance, that there are any consequences of X that could have been prevented by making a different choice. We abstract away from physical aspects of choice on purpose, because they do not seem to play a role in our conceptual argument. A further benefit of doing this is that our choice principle seems to be compatible with physical determinism; it does not appear to have any problem with standard Frankfurt cases (which invokes alternative *possibilities*, apparently intended as physical manifestations of differences between choices). [3]

Since we are interested specifically in the aspects of responsibility that hinge on the choice condition, a single-agent formalism like standard default logic is suitable for our purposes. The formal context is one where we want to know whether an agent is responsible for a formula it derive using default rules. We assume that the agent is not omniscient about the consequences of its reasoning choices. Moreover, we do not model epistemic aspects of agency, with the implicit premise being that the epistemic state of the agent is arbitrary, not that it is logically perfect. This assumption also covers knowledge about the system itself; specifically, we do not assume that the agent knows (or does not know) the (implicit) consequences of applying certain rules. The only assumption we make is that the agent knows whatever it needs to know to

[2] Some might want general principles to express what is good in the universal sense, but those principles would then have little to do with descriptive moral rules. To say that some rules are bad by comparison with a universal principle might be possible, but would not seem particularly informative in itself.

[3] For a detailed philosophical argument that free will and determinism is compatible because free will is not a physical phenomenon, see [7].

apply rules and choose between them in case of conflict.

For the purposes of facilitating a simple formal definition, we assume that agents make all their morally salient reasoning decisions on the basis of explicit moral rules in a propositional, non-schematic, format. With respect to human agency, this is an extreme idealisation, useful for getting at the essence of the phenomenon we wish to highlight. With respect to artificial agency, the assumption might be justified also at a deeper level. Specifically, it is hard to imagine how an intelligent machine could make a moral choice unless there was some kind of explicit moral rule involved (possibly a poorly understood one, formulated in terms of learning principles). Machines still appear to be rule-based systems, our increasing lack of understanding of them notwithstanding. This observation might seem to support a further argument to the effect that machines cannot make morally salient decisions, since meaningful alternatives are lacking. But this is not necessarily the case. Specifically, if a machine resolves a moral conflict, it is *prima facie* plausible to say that it makes a moral choice. This is not the same as saying that the machine is not rule-based; clearly, a conflict between rules is based on rules. Moreover, in a machine, conflicts are also resolved by rules (although these can be very different rules, not directly addressing the rules generating the conflict). Moreover, we are not suggesting that a machine is a moral agent; to have made a moral choice is a *necessary* condition for moral responsibility, not a sufficient one.

To illustrate the importance of choice, consider a self-driving car that kills its passenger by stopping at a red light, allowing a large truck to crash into it from behind. Most people would probably agree that we need to know more before we can conclude that a morally salient choice was made to kill the passenger. This in spite of the fact that the decision would be motivated by a moral rule, resulting in an outcome that the car might in theory also be able to anticipate. Specifically, the car (and its makers) might not have a choice *in the relevant situation*; once the light turns red, the rule-following behaviour might be inescapable. Moreover, this behaviour would hardly be implemented to kill people, but rather to uphold traffic regulations. Hence, additional evidence of wrongdoing would be required in order to hold anyone other than the truck driver responsible.

But now, consider the same scenario again, with the only difference being that the car has an additional algorithm that flags up a dilemma in this situation: "should I obey the traffic rule and kill the passenger or should I break the traffic rule and save the passenger"? If such an algorithm is present, most people would probably approach the case very differently; we would no longer think we were dealing with neutral rule-following, we would think that morally salient choices were being made *by (means of) the machine*. Moreover, when the car decides to stop at the red light, the following conclusion suggests itself: the car has committed murder.

The question of how exactly we should draw distinctions like the one illustrated here seems important and difficult. Simple thought experiments gloss over the difficulty, but highlight the importance. We believe that is fair. Simple examples at least show that those who think there is *not* an important distinction to make, have some explaining to do. This is also were the present article aims to make a contribution; its primary purpose is to help formulate a better foundation for further debate, not to promote a certain view on whether or not machines can be held morally responsible for their behaviour. We believe the only non-trivial assumption we make in this regard is that something like the choice principle is true. If it is, what follows makes a relevant point about moral reasoning, a point that has not to our knowledge been made before.

3 Responsibility and moral dilemmas in default logic

In this section, we develop our ideas more precisely using default logic, a much used formalism in artificial intelligence, capable of representing (and, in its programming variants, implementing) reasoning with rules that can have exceptions. Here is the basic definition:

Definition 3.1 Given a set of propositions Π a default rule is a triple $d : p(d); j(d)/c(d)$ where $p(d), c(d)$ are propositional formulas over Π and $j(d)$ is a set of such formulas. We say that $p(d)$ is the prerequisite of d, $c(d)$ is its conclusion and $j(d)$ is its justification set. A default theory is a pair (T, B) where T is a propositional theory and B is a set of default rules. We define the semantics of default logic as follows:

- Given a (deductively closed) theory T and a rule d, we say that d is active in T, written $a(T, d)$, if $p(d) \in T, \{\neg\phi \mid \phi \in j(d)\} \cap T = \emptyset$ and $c(d) \notin T$. We define $a(T, B) = \{d \in B \mid a(T, d)\}$, the set of defaults active at T.
- Given a default theory (T, B), an argument is a pair $(E, \mathbf{d})$ such that $\mathbf{d} = d_1, \ldots, d_n$ is a sequence of defaults (the reasons supporting the argument) and $E = E_n$ (the content of the argument) is a set of formulas defined by the following recursion: [4]
  · $E_0 = cl(T)$ (where cl is deductive closure in propositional logic),
  · $E_i = \begin{cases} cl(E_{i-1} \cup \{c(d_i)\}) \text{ if } d_i \in a(E_{i-1}, B) \\ \text{undefined otherwise} \end{cases}$

  If $a(E_n, B) = \emptyset$ for some argument $(E_n, \mathbf{d})$ then we say that the argument is an *extension* for B at T. Let $j(E, \mathbf{d}) = \bigcup_{1 \leq i \leq n}\{j(d_i)\}$, the *justification* for the argument $(E, \mathbf{d})$. An extension is an *R-extension* if $\{\neg\phi \mid \phi \in j(E, \mathbf{d})\} \cap E_n = \emptyset$. [5] We use $\mathsf{Ext}(T, B)$ to denote the set of R-extensions of (T, B).
- An *agent* over B is a function $\mathsf{Ag} : 2^{\mathcal{L}} \to 2^{\mathcal{L}}$ such that $\mathsf{Ag}(T) \in \mathsf{Ext}(T, B)$ for all $T \in 2^{\mathcal{L}}$. In short, agents choose, but they don't break the rules.

Notice how we define extensions in terms of arguments. This is different from the original definition of extensions, due to Reiter [11]. However, it is not hard to see that our definition is equivalent to the original definition. It is simply a restatement of it where we keep track of the order in which default rules are applied and formulate Reiter's consistency requirement as a separate closure condition. The use we make of arguments to define extensions is different from the original representation of default logic in terms of argumentation, due to Dung [4]. Our approach is close in spirit, and gives rise to the same semantics, but builds on a different set of arguments. We

[4] It follows that if some E_i from the sequence is undefined, then $E = E_n$ is not defined either, so $(E, \mathbf{d})$ is not an argument.

[5] The R stands for Reiter, who introduced this system of default logic [11].

leave further exploration of the connection between the two approaches for future work. Here we focus instead on the issue of responsibility: what are the necessary and sufficient conditions that should be met before an agent is held responsible for something it derives using default rules?

Let $B, \mathsf{Ag}(T) \models \mathsf{R}\phi$ denote that the agent Ag is responsible for deriving ϕ from (T, B) using the argument $\mathsf{Ag}(T) = (E, \boldsymbol{d}) \in \mathsf{Ext}(T, B)$. Then the avoidance principle for responsibility for rule-based conclusions can be formalised as follows:

$$B, (E, \boldsymbol{d}) \models \mathsf{R}\phi \Rightarrow \exists (E', \boldsymbol{d}') \in \mathsf{Ext}(T, B) : \phi \in E \setminus E' \quad (1)$$

That is, responsibility requires that the formula results from a genuine *choice* made by the agent. Is the requirement also sufficient? We argue that it is not. In section 2, we gave an intuitive argument to this effect, by pointing out how some choices have rule-based consequences, meaning that the responsibility for these consequences rests with the rule-maker, not the agent.

To make this intuition formal, we now define the relation of attack between arguments. We build on the approach in [4], but differ in that we generate attacks from the full content of arguments, not only their final conclusions. Let $\mathcal{A}(T, B)$ denote the set of all arguments at (T, B). Then we define the relation $R = \mathcal{R}(T, B) \subseteq \mathcal{A}(T, B) \times \mathcal{A}(T, B)$:

$$(E, \boldsymbol{d}) R (E', \boldsymbol{d}') \Leftrightarrow E \cap \{\neg\phi \mid \phi \in j(E', \boldsymbol{d}')\} \neq \emptyset \quad (2)$$

In words, we say that x attacks y whenever the content of x contains the negation of a formula that is in the justification set of y.

This gives us, for every (T, B), an argumentation framework $\mathsf{F}(T, B)$ with the arguments $\mathcal{A}(T, B)$ and the attacks $\mathcal{R}(T, B)$ as defined above. If $e = (E, \boldsymbol{d})$ is an argument with $\boldsymbol{d} = d_1, \ldots, d_n$ and $e' = (E', \boldsymbol{d}')$ with $\boldsymbol{d}' = d'_1, \ldots, d_m$ we say that e is a sub-argument of e', written $e \subseteq e'$ if $m \geq n$ and $d_i = d'_i$ for all $1 \leq i \leq n$. For all arguments $e = (E, \boldsymbol{d})$ with $\boldsymbol{d} = d_1, \ldots, d_n$ we let $s(e, d_i)$ denote the sub-argument of e *up to* d_i. That is, $s(e, d_i) = (E', \boldsymbol{d}')$ such that $\boldsymbol{d}' = d_1, \ldots, d_i$ (clearly, this argument is well-defined, c.f., Definition 3.1). The advantage of our way of defining arguments and attacks, compared to how it is done by Dung [4] and in $ASPIC^+$ [8], is that we get additional structure allowing us to be more specific about the relationship that exists between an argument and its set of sub-arguments (this relationship is lost in [4] since attacks are generated by looking only at the final conclusion of the attacking argument). [6] Specifically, we get the following proposition, which will be of great use to us later when when we define responsibility.

Proposition 3.2 *Let* (T, B) *be a default theory and let* $R = \mathcal{R}(T, B)$. *We note the following facts:*

(i) *For all arguments* $e, f, g \in \mathcal{A}(T, B)$ *such that* $e \subseteq f$, *we have (1.a) if* eRg *then* fRg *and (1.b) if* gRe *then* gRf.

[6] It is also worth noting that our framework satisfies the *rationality postulates* of [2]. Moreover, as long as one of the postulates are fulfilled (the sub-argument closure), it is intuitively clear that our way of generating attacks will not produce different results with respect to extension-based semantics for argumentation (although we get additional, helpful, structure for our purposes in this paper).

(ii) *If* (T, B) *is normal (so that* $j(d) = c(d)$ *for all* $d \in B$*), then* R *is irreflexive.*

(iii) *If* $(E, \mathbf{d})$ *is an extension, but not an R-extension, then* $(E, \mathbf{d})R(E, \mathbf{d})$.

Proof. (1) Let $e = (E, \mathbf{d})$, $f = (E', \mathbf{d}')$ such that $e \subseteq f$, and $g = (E'', \mathbf{d}'')$. Since $e \subseteq f$, we have $E \subseteq E'$ and $j(E, \mathbf{d}) \subseteq j(E', \mathbf{d}')$. To prove (1.a), assume eRg, i.e., there is a $\phi \in E$ such that $\neg\phi \in j(E'', \mathbf{d}'')$, then clearly $\phi \in E'$ such that $j(E'', \mathbf{d}'')$ by Equation (2) gives us fRg. To prove (1.b), assume that gRe, then there is a $\phi \in E''$ such that $\neg\phi \in j(E, \mathbf{d})$, but then also $\neg\phi \in j(E', \mathbf{d}')$ so by Equation (2) we have gRf.
(2) Let (B, T) be a normal default theory (recall that T is consistent). Assume towards a contradiction that there is an argument $(E, \mathbf{d}) \in \mathcal{A}(T, B)$ such that $(E, \mathbf{d})R(E, \mathbf{d})$ for $R = \mathcal{R}(T, B)$, i.e., there is a ϕ in E such that $\neg\phi \in j(E, \mathbf{d})$. Since B is normal, $j(E, \mathbf{d}) \subseteq E$, but then E contains a contradiction which was introduced by applying some $d_c \in \mathbf{d}$. Let i be the smallest index such that E_i is inconsistent, c.f., Definition 3.1. That is, E_{i-1} is consistent, and $E_i = c(E_{i-1} \cup \{c(d_i)\})$ is not. This means that d_i was active and that $p(d_i) \in E_{i-1}$ and $\neg j(d_i) \notin E_{i-1}$, but since (T, B) is normal, $j(d_i) = c(d_i)$, so E_{i-1} consistent, contradicting the assumption as desired.
(3) Assume $(E, \mathbf{d})$ is an extension that is not an R-extension. Then by definition R-extension there is some $\phi \in j(E, \mathbf{d})$ such that $\neg\phi \in E$. The claim then follows immediately from Equation 2. □

The fact that the attack relation is irreflexive for normal default theories lets us record an important fact about such theories: all arguments can be developed into R-extensions, there are no arguments that *must be* rejected. We state this formally.

Proposition 3.3 *Given a normal default theory* (T, B)*, let* $e \in \mathcal{A}(T, B)$ *be arbitrary. There is some* $e' \in \mathsf{Ext}(T, B)$ *such that* $e \subseteq e'$.

We now prove a theorem that corresponds to Dung's original instantiation result [4]. Specifically, we show that the notion of a stable set in argumentation corresponds to the notion of an R-extension for default. Recall that S is a *stable set* in an argumentation framework $(\mathcal{A}, \mathcal{R})$ if, and only if, $S \subseteq \mathcal{A}$ with

$$\forall x \in S, y \in \mathcal{A} : y\mathcal{R}x \Rightarrow y \in \mathcal{A} \setminus S \text{ and } \exists z \in S : z\mathcal{R}y \tag{3}$$

Then the instantiation theorem for our representation of default reasoning as argumentation can be stated as follows.

Theorem 3.4 *For all* B, T*, we have*

(i) *If* $(E, \mathbf{d})$ *is an R-extension for* (T, B)*, then there is a stable set* S *of* $\mathsf{F}(T, B)$ *such that* $(E, \mathbf{d}) \in S$.

(ii) *If* S *is a stable extension of* $\mathsf{F}(T, B)$ *then* S *contains at least one R-extension, and all R-extensions in* S *have the same content.*

Proof. To prove (1), we let $e = (E, \mathbf{d})$ be an R-extension. We need to show that there is stable set that contains e. To this end, we define the set S that contains e; every argument with the same content as e that is not attacked by e; and all of their sub-arguments. We first show that S is conflict-free. By Proposition 3.2 we only need to consider maximal arguments. By definition of S these are the arguments that have the same content as e but are not attacked by e. Assume towards contradiction that two such arguments, f and g, attack each other. By definition of S, we know that e does not attack g. This contradicts the assumption that e and f have the same content. To show that S attacks everything outside of S, consider some argument $e' = (E', \boldsymbol{d}') \notin S$. By definition of S we can assume that e' does not have the same content as e (since otherwise e' would be attacked by e). Moreover, we know that e' is not a sub-argument of any member of S (since S is closed under sub-arguments). Hence, there is some rule that has been applied in e' that is not applied in e. We choose the minimal i such that $d'_i \in \boldsymbol{d}'$ is such a rule, i.e., such that $d'_i \notin \boldsymbol{d}$ and $d'_j \in \boldsymbol{d}$ for all $1 \leq j < i$. By minimality of i, we have $p(d'_i) \in E$. By the fact that e is an extension, we have $d'_i \notin a(E, B)$. It follows that there is some $\phi \in j(d'_i)$ such that $\neg\phi \in E$. Hence, e attacks e' as desired.

To prove (2), assume towards contradiction that S is a stable extension that does not contain an R-extension. Since extensions that are not R-extensions attack themselves (Proposition 3.2 point (3)) and S is a stable set it follows that S does not contain any extensions. Let $e = (E, \boldsymbol{d}) \in S$ be a maximal argument, i.e., such that there is no $e' \in S$ with $e \subset e'$. Since e is not an extension, we have $a(E, B) \neq \emptyset$. We let $e^+ = (E^+, \boldsymbol{d}^+)$ be an argument extending e by some $d_{n+1} \in a(E, B)$. Since $e^+ \notin S$ it is attacked by some argument $f' \in S$. Moreover, since $d_{n+1} \in a(E, B)$ we know that the content of f' is not the same as the content of e (since otherwise e would also attack e^+, contradicting that d_{n+1} is active in e). Let $f = (E', \boldsymbol{d}')$ be maximal such that $f' \subseteq f \in S$. Hence, we have two maximal arguments $e, f \in S$ such that $E \neq E'$. This means that there is at least one rule in the sequence $\boldsymbol{d}$ that is not in the sequence $\boldsymbol{d}'$. We choose the *first* rule of this kind encountered along $\boldsymbol{d}$, namely $d_i \in \boldsymbol{d}$ such that $d_i \notin \boldsymbol{d}'$ and $d_j \in \boldsymbol{d}'$ for all $1 \leq j < i$, i.e., elements preceding d_i in $\boldsymbol{d}$ are present *somewhere* in $\boldsymbol{d}'$. Hence, we have $p(d_i) \in E'$ (since this pre-condition was derived in E using only rules that are also present in $\boldsymbol{d}'$). Moreover, since $(f, e) \notin R$ (by $f, e \in S$) it follows that there is no $\phi \in j(d_i)$ such that $\neg\phi \in E'$. Hence, we get $d_i \in a(E', B)$. Let $f^+ = (E'', \boldsymbol{d}'')$ be the argument obtained from f by adding d_i and taking the closure. Since f is maximal we get $f^+ \notin S$ so there must be $g \in S$ that attacks f^+. Hence, g contains $\neg\phi$ for some $\phi \in j(E'', \boldsymbol{d}'')$. Since $f \in S$, we cannot have $\phi \in j(f)$, since then g would attack f, contradicting stability of S. Hence, $\phi \in j(d_i)$. But then g attacks e with g and e both being in S, contradicting that S is stable. To conclude the proof we show that if S contains two R-extensions $(E, \boldsymbol{d})$ and $(E', \boldsymbol{d}')$, then $E = E'$. Assume to the contrary and without loss of generality that there is some $\phi \in E \setminus E'$. It follows that there must be some default $d_i \in \boldsymbol{d}$ that is not applied in $\boldsymbol{d}'$. That is, we must have $d_i \neq d'_j$ for all $d'_j \in \boldsymbol{d}'$. Since $(E', \boldsymbol{d}')$ is an extension, we know that $a(E', B) = \emptyset$. This means that d_i is not active in E'. Hence, it follows that there is a $\phi \in j(d_i)$ such that $\neg\phi \in E'$. It follows that $(E, \boldsymbol{d})$ attacks $(E', \boldsymbol{d}')$, contradicting the assumption that they are in the same stable set. □

Effectively, this theorem gives us an equivalent characterisation of agents and their choices; choosing between extensions is the same as choosing between stable sets of an argumentation framework. Now, according to our conceptual understanding of where responsibility comes from, we would like to demonstrate formally that choosing a stable set amounts to resolving moral dilemmas. But how do we define the moral dilemmas of default reasoning with moral rules? It is tempting to define them in terms of attacks; if defaults are given a deontic interpretation, an attack on an argument can be intuitively recognised as a moral reason not to argue in a certain way on the basis of the rules. Assuming that the agent *could* argue in the way prescribed by some argument, it would seem that an attack on that argument presents the agent with a moral dilemma.

In general, however, an argument might be impossible to accept, on pain of arriving at an inconsistency. In these cases, attacks have a different status; now they encode the derived consequences of moral rules, not any choice for the agent. Specifically, such attacks will bind the choice of any rule-based reasoner that always reasons to an extension. Clearly, the agents defined in Definition 3.1 are reasoners of this type. Hence, some attacks are morally vacuous, as illustrated by the following example.

Example 3.5 Let B be a theory given by

$$1 : \top; a/a \;\; 2 : \top; \neg a/b$$

Here we have the arguments $x = (cl(\{a\}), 1)$ and $y = (cl(\{b\}), 2)$. Moreover, we get $(x, y) \in \mathcal{R}(\emptyset, B)$. This is not a moral conflict, it is simply an encoding of the fact that accepting 2 will not lead to an R-extension. Hence, the attack in this case does not signify that the agent has a choice. Rather, it encodes a rule-bound reasoning step that no agent can resist (modulo our definition of agents as rule-followers).

The example shows that in order to arrive at a definition of moral dilemmas in terms of attacks, we have to prune the attack relation. Looking back at the avoidance principle in Equation 1, it is tempting to try to do so by saying that two arguments, e and e', represent a moral dilemma if both can be expanded to R-extensions and there is some ϕ in one that is negated in the other. In this case, at the very least, they represent a *moral conflict* centred on ϕ. In previous literature, most notably Horty's work on imperatives and defaults [5], moral conflicts of this kind have been conflated with moral dilemmas.

Is such a conflation appropriate? We argue that it is not. The reason is that a moral conflict involving ϕ does *not* necessarily imply a choice for the agent with respect to ϕ. The implication fails just in case ϕ is a rule-induced consequence of a *previous* choice made by the agent. In this case, the agent is certainly responsible for something, but not necessarily for ϕ; while ϕ could have been prevented by the agent, it arose as a consequence from the agent's choice only because of the rules in the system. Intuitively, if the agent did not know this when making the choice, the blame for ϕ rests solely with the rule-maker, not with the agent. In this paper, we do not assume that agents are omniscient and we do not model their epistemic capabilities. Hence, in our setting, an agent is *never* responsible for a formula that arises only as a rule-induced consequence of its choices.[7]

[7] If an agent *does* have the capacity to know and reflect on the fact that some ϕ will arise as a rule-based

In the human realm, the mechanism we identify here is significant to our responsibility attributions. For instance, imagine a young woman wondering whether to enlist in the army. She might have many beliefs about what this might entail, including that she might end up taking lives. Still, if she chooses to enlist and then decides to kill someone – intentionally – on the orders of a superior officer, we would hesitate to say she is morally responsible. At least we would be likely to think she is in a significantly different position than she would be if she had made the relevant choice to kill the person as a civilian. However, if we do not keep track of the relevant choice moments and the differences between them (the choice to enlist and the choice to kill), we could be led to believe that even the choice principle has been fulfilled in this case. This is exactly the kind of conflation we want to avoid in the formal system, to get the right characterisation of moral dilemmas also in situations where locating the relevant choices might be considerably harder than in an intuitive example like this one.

In the world of intelligent systems, where things happen very quickly and many decision steps remain highly opaque to us, the distinction between different choice moments is all the more important. In this setting, it is not a good idea to think about the lack of a relevant choice as a sort of "excuse" that we can address as a separate issue independently from the core definition of responsibility. If we follow this strategy, we are likely to vastly overestimate the moral salience of apparently autonomous choices in chains of harmful events involving machines.

The technical challenge becomes how to pick out exactly those moral conflicts that are also moral dilemmas, because they correspond to moral choices. This cannot be done by simply looking at extensions, we also need to look at reasons. As far as we are aware, this observation has not been made in earlier work. To illustrate the phenomenon from a technical perspective, consider the following example.

Example 3.6 Assume a theory B consisting of the four defaults

$$1 : \top; a/a \;\; 2 : \top; \neg a/\neg a \;\; 3 : a; b/b \;\; 4 : \neg a; \neg b/\neg b$$

Then we have the argument $x = (cl(\{a, b\}), (1, 3))$ resulting from applying rule 1 and then rule 3. Similarly, we get the argument $y = (cl(\{\neg a, \neg b\}), (2, 4))$. It is easy to see that these arguments are R-extensions and that they attack each other. However, on our understanding, these attacks do not correspond to a moral dilemma. Specifically, there is no moral dilemma centred at b. The reason is that once a rule-based agent accepts the argument $x' = (c(\{a\}), 1)$ it has no choice but to apply 3 and accept x, giving b *by a default rule*. Similarly, if the agent accepts $y' = (cl(\{\neg a\}), 2)$, it has no choice but to apply 4 and accept y, again by a default rule. Hence, the only moral dilemma in this scenario is between x' and y', centred on a.

In this example, there is no doubt a moral conflict between b and $\neg b$. Moreover, this moral conflict is not muted because one of the options are ruled out by rule-based reasoning; both conclusions are possible. However, the only dilemma is the choice

consequence of its choice we do not deny that responsibility for ϕ might result. However, this kind of responsibility looks conceptually and technically distinct from the responsibility that arises from making moral choices.

between a and $\neg a$; whatever the agent chooses to do with respect to a will deprive it of choice with respect to b. Moreover, this deprivation of choice does not result from logical necessity; the choice disappeared because of a *default*, a rule imposed by the rule-maker. In a setting like ours, where meta-level reasoning about defaults is not something agents engage in, we believe the only possible conclusion for the example above is that the agent is *not* responsible for deriving b (or $\neg b$). Responsibility begins and ends with a.

How can we generalise this observation? We believe the solution is to use our argumentation representation to pinpoint moral choices precisely, as defined below.

Definition 3.7 Let (T, B) be a default theory and let $e = (E, \boldsymbol{d}) \in \mathcal{A}(T, B)$ be an argument with $\boldsymbol{d} = d_1, \ldots, d_n$. We say that e is a *moral choice*, written $\mathsf{ch}(e)$, if there is some $e' \in \mathcal{R}(T, B)^-(e)$ such that there is no $i < n$ with either $s(d_i, e) \in \mathcal{R}(T, B)^-(e')$ or $e' \in R^- s(d_i, e)$. We use $\mathsf{C}(e)$ to denote the set of moral choices encountered along $\boldsymbol{d}$. That is, $\mathsf{C}(e) = \bigcup_{1 \leq i \leq n} \{s(d_i, e) \mid \mathsf{ch}(s(d_i, e))\}$. In particular, we have $\mathsf{ch}(e)$ if, and only if, $\mathsf{C}(e) \neq \emptyset$.

By Proposition 3.2, if (T, B) is normal, then for every argument e there is an R-extension e' such that $e \subseteq e'$. Hence, for normal theories we know that every moral choice occurs in some R-extension (stable set). This means that in normal default theories, every moral choice is a real moral dilemma; the choice is not forced on the agent by any sub-argument accepted up to that point, and the choice *will* result in an R-extension, provided the agent continues to reason correctly. The latter property is not true for non-normal theories; here some moral choices are blocked because they lead to inconsistency. This further complicates the issue of responsibility, motivating the more abstract definitions provided in section 4.

Already, we can make interesting observations about the case of normal default theories. First, notice that many R-extensions are not themselves moral choices. This is the case, for instance, for both the extensions encountered in the (normal) theory of Example 3.6 above. Hence, we now have a definition that provides a formal reflection of our intuition that defining responsibility with respect to extensions is not sufficiently fine-grained. The adequacy of our definition is further supported by the fact that is is truly a refinement of Principle (1): the existence of multiple R-extensions *implies* that the agent will make at least one moral choice. Specifically, whenever there exist two R-extensions with different content, then every R-extension has at least one moral choice among its sub-arguments. We state this formally.

Proposition 3.8 *Given a default theory (T, B), if there are $(E, \boldsymbol{d}), (E', \boldsymbol{d}') \in \mathsf{Ext}(T, B)$ such that $E \neq E'$, then for every $e \in \mathsf{Ext}(T, B)$ we have $\mathsf{C}(e) \neq \emptyset$.*

Proof. Let $e = (E, \boldsymbol{d})$ and $e' = (E', \boldsymbol{d}')$. By Theorem 3.4, we know that there are stable sets $S_e, S_{e'}$ such that $e \in S_e, e' \in S_{e'}$ and $e' \notin S_e, e \notin S_{e'}$. Let $R = \mathcal{R}(T, B)$. Hence, by stability of S_e and $S_{e'}$, we get $f \in S_e$ with fRe' and $f' \in S_{e'}$ with $f'Re$. We choose f, f' in such a way that we minimize i and j with $fRs(d_i, e')$ and $f'Rs(d_j, e)$. Then the sub-arguments $s(d_i, e'), s(d_j, e)$ are moral choices as desired (f, f' cannot be attacked by smaller sub-arguments of e or e' since then stability of S would see to it that the minimality of f, f', i, j would be contradicted). □

Hence, multiple extensions do indeed arise only when an agent has a choice. For normal default theories, this becomes an equivalence: moral choice implies multiple extensions and vice versa. Specifically, we get the following result.

Theorem 3.9 *Given a normal default theory (T, B), there are $e, f \in \mathsf{Ext}(T, B)$ with different content if, and only if, there is at least one moral choice in $\mathcal{A}(T, B)$.*

Proof. $\Rightarrow$) This follows from Proposition 3.8. $\Leftarrow$) Assume $e = (E, \boldsymbol{d})$ is a moral choice and let $f = (H, \boldsymbol{e}) \in R^{-}(a)$ witness to this fact. By Proposition 3.3 there are R-extension e', f' with $e \subset e', f \subset f'$. We need to show that $e' = (E', \boldsymbol{d}')$ and $f' = (H', \boldsymbol{e}')$ have different content, i.e., $E' \neq H'$. Let $R = \mathcal{R}(T, B)$. Since fRe it follows by Proposition 3.2 that $f'Re'$. Hence, there is some $\phi \in j(e')$ such that $\neg\phi \in H'$. From the fact that e' is an R-extension it follows that $\neg\phi \notin E'$ as desired.□

Although multiple R-extensions only arise when there is at least one moral choice, such choices can appear anywhere in the sequence of reasons supporting an extension, meaning that the extension itself might be as much a consequence of other rules as it is a consequence of the agent's choices. In light of our discussion above, we believe this should influence how we define responsibility. For normal default theories, where we know that every moral choice resolves a real moral dilemma, we believe the following definition is appropriate.

Definition 3.10 Let a normal default theory (T, B) be given. For every argument $e \in \mathcal{A}(B, T)$ we define the *responsibility set* of $e = (E, \boldsymbol{d})$, written $\mathsf{R}(e)$, by induction on $|\mathsf{C}(e)|$ (the number of moral choices encountered along e):

$$\mathsf{R}(e) = \begin{cases} \emptyset \text{ if } |\mathsf{C}(e)| = 0 \\ (d_i \cup \mathsf{R}(s(d_{i-1}, e))) \text{ where } i = max_{d_j \in \mathsf{C}(e)}\{j\} \text{ if } |\mathsf{C}(e)| \geq 1 \end{cases}$$

We write $(T, B), e \models \mathsf{R}\phi$ to indicate $\phi \in \{c(d) \mid d \in \mathsf{R}(e)\}$, where $c(d)$ is the conclusion of the default rule d and $\mathsf{R}(e)$ is defined as above.

It is not hard to see that this definition of responsibility satisfies the prevention constraint specified in Equation (1). Specifically, we get the following simple result (proof omitted).

Theorem 3.11 *If $(T, B), e \models \mathsf{R}\phi$ for some $e \in \mathcal{R}(T, B)$ then there are $(E, \boldsymbol{d}), (E', \boldsymbol{d}') \in \mathsf{Ext}(T, B)$ such that $\phi \in E \setminus E'$.*

Finally, we show that for normal theories, the set of moral choices on an extension suffice to uniquely pick out its content, adding further weight to our claim that moral choices are the roots of responsibility.

Theorem 3.12 *For all normal theories (T, B), if $e, e' \in \mathsf{Ext}(B, T)$ and $\mathsf{C}(e) = \mathsf{C}(e')$ then the content of e is the same as the content of e'.*

Proof. Assume towards contradiction that $e = (E, \boldsymbol{d}), e' = (E', \boldsymbol{d}') \in \mathsf{Ext}(B, T)$ such that $\mathsf{C}(e) = \mathsf{C}(e')$ and $E \neq E'$. This means there is at least one rule in $\boldsymbol{d}$ that is not in $\boldsymbol{d}'$. We choose such a rule with minimal index i, such that $d_i \in \boldsymbol{d}, d_i \notin \boldsymbol{d}'$ and $d_j \in \boldsymbol{d}'$ for all $1 \leq j < i$. Then $p(d_i) \in E'$. Hence, we must have $\neg j(d_i) = \neg c(d_i) \in E'$. Moreover, by the minimality of i we know that e' does not attack any sub-argument of $s(d_i, e)$ and that no sub-argument of $s(d_i, e)$ can attack e' (in either case, e' would attack itself, so it would not be an R-extension). □

In terms of argumentation theory, what this result tells us is that the moral choices in a stable set uniquely determine that set. This observation is the key to the generalisation presented in the next section, where we lift the notion of responsibility defined herein to the level of abstract argumentation. This definition can then be applied to *any* system of default reasoning that instantiates some argumentation framework. [8]

4 Responsibility in abstract argumentation

We assume given an argumentation framework $\mathsf{F} = (\mathcal{A}, \mathcal{R})$ with $\mathcal{R} \subseteq \mathcal{A} \times \mathcal{A}$ (see, e.g., [10] for further details). We interpret this as a deontic structure, such that if a attacks b then accepting a is a moral reason not to accept b. In this context, we define an agent as a function $\mathsf{Ag} : \mathcal{A} \to \{1, 0, \frac{1}{2}\}$. Intuitively, it means that the agent accepts and argument just in case it assigns 1 to it, it rejects it if it assigns 0, and it withholds judgement if it assigns $\frac{1}{2}$ (for further details on why this third semantic status should be included, see, e.g., [3]). We want to know when an agent is responsible for assigning a given value to an argument.

To talk about what follows from an agent's choice of extension, we use the following modal language, $\mathcal{L}$:

$$\phi := p \mid \phi \to \phi \mid \neg\phi \mid \mathsf{R}\phi$$

where $p \in \mathcal{A}$ is an argument. The formula $\mathsf{R}\phi$ should be understood as expressing the fact that the agent is responsible for concluding the formula ϕ. We want to generalise the ideas developed in section 3, so that responsibility arises from a moral choice made by the agent. Hence. a natural starting point is to ask: what are the moral dilemmas in F? If we try to adapt the definition provided for default logic, we need to first define a notion corresponding to the sub-argument relation that we defined for arguments based on their internal content as derivations in default logic (c.f., Definition 3.1). Since arguments in abstract argumentation have no internal structure whatsoever, a straightforward generalisation is therefore blocked.

Instead, we will capture the intuition behind the sub-argument relation in terms of the combinatorial properties of $\mathcal{A}$. To achieve this, we first think more closely about why we introduced the sub-argument relation in the first place. In short, the reason was the following: for all arguments, we found among its sub-arguments a set of moral choices that uniquely determined the semantic status of that argument. Specifically, the purpose of introducing the notion of a sub-argument was to capture the phenomenon that arises when a sub-argument suffices to *force* the acceptance of a unique super-argument. Specifically, if an argument is accepted, the responsibility of the agent appears co-equal with the responsibility it has for the minimal sub-argument that enforces that argument. To arrive at the total responsibility of an agent for an argument, we then iterated this to account also for the responsibility the agent had with respect to the minimally enforcing sub-argument of this sub-argument and so on all the way back to the starting point (c.f., Definition 3.10).

[8] For the issue of instantiation generally, see [1,9].

Adopting now a more abstract perspective on this iterative process, we see that the crucial aspect of what we did was to identify the reasoning steps that the agent *had to take*, given its previous choices. Some steps were *induced* and therefore did not give rise to responsibility. The notion of induced choices is in fact well-known in argumentation, as a technical notion used to extend certain kinds of partial extensions of argumentation frameworks. In the following, we use the labelling-based formulation due to Caminada, see, e.g., [3].

Specifically, let $\pi : \mathcal{A} \to \{1, 0, \frac{1}{2}\}$ be some partial assignment of boolean values to the arguments of F (such that all arguments not receiving a boolean value are assigned $\frac{1}{2}$). Then we define the rule-based closure of π as $\Gamma(\pi)$ defined as follows:

$$\Gamma(\pi)(x) = 1 - \max_{y \in R^-(x)} \{\pi_{i-1}(y)\} \tag{4}$$

If $\Gamma(\pi)$ admits a least fixed point $\bar{\pi}$ (with $\Gamma(\bar{\pi}) = \bar{\pi}$), then $\bar{\pi}$ is the rule-based closure of π. The intuition for us is that the rule-based closure of π extends π according to rules that do not require the agent to resolve any moral dilemmas; the closure depends only on rules, not on choices. If π is incoherent (not *admissible* in argumentation jargon), then applying rules to extend it will result in an inconsistency, in which case the operator Γ has no fixed point and $\bar{\pi}$ is not defined. In all other cases, we claim that $\bar{\pi}$ is morally equivalent to π, in the sense that all moral choices that have a bearing on $\bar{\pi}$ have already been made at π. We remark that a coherent $\pi : \mathcal{A} \to \{1, 0, \frac{1}{2}\}$ corresponds to a stable set (obtained by collecting all arguments assigned the value 1) if, and only if, $\bar{\pi}$ is boolean-valued. There are also many other semantics for argumentation, but in this paper, the distinctions between them are not important: we define our notion of responsibility so that it works for all coherent π, meaning that it works for all the "classic" argumentation semantics, including those defined in Dung's original paper [4].

We now use the definition of rule-based closure to define a class of assignments around every assignment π, collecting all those assignments that are morally equivalent to it. From the point of view of an agent $\mathsf{Ag} : \mathcal{A} \to \{1, 0, \frac{1}{2}\}$, this is the collection of all the morally equivalent choices that the agent could have made at $\mathcal{A}$.

$$D(\pi) = \{\pi' \mid \bar{\pi}' = \bar{\pi}\} \tag{5}$$

π is incoherent if $D(\pi) = \emptyset$. Intuitively, $D(\pi)$ gives us all assignments that correspond to a given way of making moral choices, a given way of resolving moral dilemmas.

The characterisation of moral choices allows us to define a semantics for the language $\mathcal{L}$. Specifically, we define the following inductive extension of $\bar{\pi}$ to all of $\mathcal{L}$.

$$\bar{\pi}(x) = \begin{cases} \bar{\pi}(x) & \text{if } x \in \mathcal{A} \\ 1 - \bar{\pi}(\phi) \text{ if } x = \neg\phi \\ min\{1, 1 - (\pi(\phi) - \pi(\psi))\} & \text{if } x = \phi \to \psi \\ min(1, min_{\pi' \in D(\pi)}\{\pi'(\phi)\}) & \text{if } x = \mathsf{R}\phi \end{cases} \tag{6}$$

We write $\mathsf{F}, \pi \models \phi$ if, and only if, $\bar{\pi}(\phi) = 1$. This gives us a three-valued modal logic. We leave an in-depth exploration of our logic for future work, but note some of its validities. Specifically, we record that the R-modality is a peculiar kind of modality that is either degenerate (when $D(\pi) = \emptyset$) or else $S5$. More interestingly, it satisfies some special properties arising from the underlying structure of F, including the following (we omit the proof).

Proposition 4.1 *For all* F *and all agents such that* $\mathsf{Ag}(\mathsf{F}) = \pi$ *we have:*

- $\mathsf{F}, \pi \models \mathsf{R}(\phi \wedge \neg\phi)$ *if, and only if,* $\bar{\pi}$ *is not defined (meaning* π *is incoherent) ("You are responsible for a contradiction if, and only if, you reason incorrectly").*
- *There is a* p *such that* $\mathsf{F}, \pi \models \neg \mathsf{R}p$ *if, and only if,* $\bar{\pi}$ *is defined ("You reason correctly if, and only if, there is some proposition you are not responsible for").*
- $\mathsf{F}, \pi \models \mathsf{R}(p \wedge \neg p) \to \mathsf{R}\phi$ *("The person who is responsible for a contradiction is responsible for everything").*
- $\mathsf{F}, \emptyset \models \neg\mathsf{R}(\phi \to \neg\phi)$ *if, and only if,* ϕ *is made true by all coherent* π *(*ϕ *is sceptically accepted in* F*). Notice especially the indirect way of expressing this in* $\mathcal{L}$*.*
- $\mathsf{F}, \pi \models \mathsf{R}(\phi \to \phi)$ *("You are responsible for reasoning correctly").*
- $\mathsf{F}, \pi \models (\phi \wedge \neg\mathsf{R}\psi) \to \mathsf{R}(\neg\phi \to \phi)$ *("If you reason correctly and accept a formula, you are responsible for not rejecting it").*

In future work, we would like to explore this modal logic further, but for now we are satisfied with concluding that it seems like an intuitively reasonable formalisation of responsibility for single-agent argumentation.

5 Conclusion

We have studied responsibility for rule-based reasoning, starting from the premise that an agent can only be held responsible for its conclusion if that conclusion reflects a choice made by the agent. To formally pin-point the morally salient choices made by a rule-based reasoner, we used an argumentation-based representation of the semantics of default logic. This enabled us to specify the sub-set of arguments that correspond to moral choices for the agent. We observed that for normal default theories, every moral choice corresponds to a real moral dilemma without a pre-determined answer. Hence, for such theories, we could define responsibility in a way that matched our intuitions. To generalise this to any kind of default theory, we abstracted away from the internal structure of arguments and proceeded to sketch a theory of responsibility that applies to argumentation frameworks directly. In future work, we would like to study the resulting logic of responsibility in more depth and examine instantiations of it, including both non-normal default theories and other frameworks for non-monotonic reasoning which includes an element of choice. In addition, we would like to extend the treatment of responsibility given here to take into account information about priorities and preferences over rules and arguments.

References

[1] Besnard, P., A. J. García, A. Hunter, S. Modgil, H. Prakken, G. R. Simari and F. Toni, *Introduction to structured argumentation*, Argument & Computation **5** (2014), pp. 1–4.
URL http://dx.doi.org/10.1080/19462166.2013.869764

[2] Caminada, M. and L. Amgoud, *On the evaluation of argumentation formalisms*, Artificial Intelligence **171** (2007), pp. 286 – 310.

[3] Caminada, M. W. and D. M. Gabbay, *A logical account of formal argumentation*, Studia Logica **93** (2009), pp. 109–145.

[4] Dung, P. M., *On the acceptability of arguments and its fundamental role in nonmonotonic reasoning, logic programming and n-person games*, Artificial Intelligence **77** (1995), pp. 321 – 357.

[5] Horty, J. F., *Moral dilemmas and nonmonotonic logic*, Journal of Philosophical Logic **23** (1994), pp. 35–65.

[6] Horty, J. F., "Reasons as Defaults," Oxford University Press, 2012.

[7] List, C., *Free will, determinism, and the possibility of doing otherwise*, Noûs **48** (2014), pp. 156–178.

[8] Modgil, S. and H. Prakken, *The* ASPIC^{+} *framework for structured argumentation: a tutorial*, Argument & Computation **5** (2014), pp. 31–62.
URL http://dx.doi.org/10.1080/19462166.2013.869766

[9] Prakken, H., *An abstract framework for argumentation with structured arguments*, Argument & Computation **1** (2010), pp. 93–124.

[10] Rahwan, I. and G. Simari, editors, "Argumentation in artificial intelligence," Springer, 2009.

[11] Reiter, R., *A logic for default reasoning*, Artificial intelligence **13** (1980), pp. 81–132.

Basic Action Deontic Logic

Alessandro Giordani [1]

Catholic University of Milan
Largo A. Gemelli, 1
20123 Milan, Italy

Ilaria Canavotto

Munich Center for Mathematical Philosophy, LMU Munich
Geschwister-Scholl-Platz 1
D-80539 München

Abstract

The aim of this paper is to introduce a system of dynamic deontic logic in which the main problems related to the definition of deontic concepts, especially those emerging from a standard analysis of permission in terms of possibility of doing an action without incurring in a violation of the law, are solved. The basic idea is to introduce two crucial distinctions allowing us to differentiate (i) what is ideal with respect to a given code, which fixes the types of action that are abstractly prescribed, and what is ideal with respect to the specific situation in which the agent acts, and (ii) the transitions associated with actions and the results of actions, which can obtain even without the action being performed.

Keywords: dynamic deontic logic; deontic paradoxes; ought-to-be logic; ought-to-do logic.

1 Introduction

Systems of deontic logic aim at modeling our intuitions concerning prescriptive concepts, such as prohibition, permission, and obligation, so as to provide appropriate formal frameworks for analyzing deontic problems, conceiving deontically constrained procedures, and assessing existing deontic systems. It is well-known that different kinds of deontic systems can be introduced in the light of the position one assumes with respect to the following non-exclusive options:

(i) developing a deontic logic of states [1,7,14] (ought-to-be logic, *sein-sollen* logic) or carrying the analysis to a deontic logic of actions [5,9,12] (ought-to-do logic, *tun-sollen* logic);

[1] alessandro.giordani@unicatt.it - ilaria.canavotto@gmail.com.

(ii) developing a static logic of actions [4,9,10] (where what is crucial is to characterize the structure of a system of actions and their basic properties) or carrying the analysis to a dynamic logic of actions [6,8,11] (where it is also crucial to characterize the sequential composition of actions and the properties of such sequences).

It is also well-known that, while the descriptive power of systems of dynamic logic of actions allows us both to solve some traditional paradoxes and to highlight important distinctions which would be otherwise neglected, these systems are still subject to difficulties [2,11], thus appearing inadequate to account for our basic deontic judgements.

The aim of this paper is to introduce a system of dynamic deontic logic in which the main problems related to the definition of deontic concepts, especially those emerging from a standard analysis of permission in terms of possibility of doing an action without incurring in a violation of the law, are solved. Our proposal is based on the idea that, in order to account for the intuitions which generate the paradoxes, more distinctions than those which can be drawn within a standard dynamic deontic system are to be made. In particular, we think that it is crucial to consider (i) a distinction between what is ideal with respect to a given code, i.e., the abstract ideal allowing us to determine the types of action which are permitted or prohibited, and what is ideal with respect to a specific situation, i.e., the concrete ideal determined by the context of the agent [3,7]; and (ii) a distinction between the transitions associated with an action and the result of the action, which possibly obtains without the action being performed. Accordingly, we propose a system constituted of

- an *ontic part*, which includes both a logic of states and a logic of actions, where states are represented, as usual, by sets of possible worlds, and actions, more precisely action types, are represented by relations between worlds;
- a *deontic part*, which includes both a logic of an abstract deontic ideal, represented by a set of worlds satisfying the prescriptions of a code, and an actual deontic ideal, represented by an ordering of the worlds accessible by performing some action.

In this way, we hope to provide a deeper perspective on what is prescribed in a certain context, by constructing a very general modal system for handling traditional problems. The plan of the paper is then as follows. In the next section, we briefly discuss the basic intuitions that our system aims at capturing as they emerge from a discussion of the main deontic paradoxes derivable in a dynamic logic of action. In section 3 we introduce our system of deontic logic of states and actions. Finally, in the last section, we define four groups of deontic concepts and provide solutions to the problems discussed in section 2.

2 Difficulties in defining deontic concepts

In a dynamic deontic logic, where action terms can be combined by using suitable operators, like negation ($\bar{\cdot}$), alternative execution ($\sqcup$), simultaneous execution ($\sqcap$), and sequential execution (;), the deontic operators of prohibition, permission, and obligation can be defined in terms of a propositional constant I, representing an ideal state of law satisfaction, and of the dynamic operator $[\cdot]$, which takes an action term α and a formula φ and returns a new formula $[\alpha]\varphi$, stating that all ways of doing α lead to a φ-state. In fact, an action is

(i) prohibited iff it necessarily results in a violation of the law ($F(\alpha) := [\alpha]\neg I$)

(ii) permitted iff it is not prohibited ($P(\alpha) := \neg[\alpha]\neg I$)

(iii) obligatory iff not doing it is prohibited ($O(\alpha) := [\bar{\alpha}]\neg I$)

Although these definitions seem to be unproblematic, together with some intuitive principles on the action operators, they imply several counter-intuitive conclusions. We especially focus on three groups.

Group 1: standard paradoxes of obligation and permission.

- Ross's paradox: $O(\alpha) \to O(\alpha \sqcup \beta)$ (if it is obligatory to mail a letter, then it is obligatory to mail-the-letter-or-burn-it).
- Permission paradox: $P(\alpha) \to P(\alpha \sqcup \beta)$ (if it is permitted to mail a letter, then it is permitted to mail-the-letter-or-burn-it).

Group 2: paradoxes of permission and prohibition of sequential actions.

- van der Mayden's paradox: $\neg[\alpha]\neg P(\beta) \to P(\alpha;\beta)$ (if there is a way of shooting the president after which it is permitted to remain silent, then it is permitted to shoot-the-president-and-then-remain-silent)
- Anglberger's paradox: $F(\alpha) \to [\alpha]F(\beta)$ (if it is forbidden to shoot the president, then shooting the president necessarily leads to a state in which remaining silent is forbidden).

Group 3: contrary to duties obligations [3].

Paradoxes of group 1 can be avoided by introducing strong notions of obligation and permission, according to which, for an action to be obliged or permitted, it is necessary both that no way of performing it leads to a state of violation and that there is at least a way to perform it which does not lead to a state of violation. Paradoxes of group 2 are more difficult to solve. If we think of an action as characterized by a starting state, a final state, and a transition leading from the first to the second state, then these paradoxes can be seen as the result of disregarding the deontic relevance of the starting state and the process of an action. To be sure, van der Mayden's paradox follows from neglecting the difference between the fact that the final state is safe and the fact that the transition which leads to this state is safe, in the sense that no step in the transition infringes the law, or fails to be the best the agent can do from a deontic perspective, given the initial conditions. Similarly, Anglberger's paradox follows from neglecting the difference between the absolutely ideal states, in which no norm is violated, and the relatively ideal states, in which the best conditions realizable by the agent in the actual conditions are in fact realized. Interestingly, once these distinctions are taken into account, also paradoxes of group 3 turn out to find a solution (but more on this below).

3 Action deontic logic

The language $\mathcal{L}$ of the system **ADL** of action deontic logic contains a set $Tm(\mathcal{L})$ of terms and a set $Fm(\mathcal{L})$ of formulas. Assuming a standard distinction between action types and individual actions, let A be a countable set of action types variables. Then $Tm(\mathcal{L})$ is defined according to the following grammar:

$\alpha ::= a_i \mid 1 \mid \overline{\alpha} \mid \alpha \sqcup \beta \mid \alpha \sqcap \beta \mid \alpha ; \beta$ where $a_i \in A$

Intuitively, 1 is the action type instantiated by any action whatsoever; $\bar{\alpha}$ is the action type instantiated by any action which does not instantiate the type α; $\alpha \sqcup \beta$ is the action type instantiated by any action which instantiates either the type α or the type β or both; $\alpha \sqcap \beta$ is the action type instantiated by any action which instantiates the types α and β in parallel; $\alpha ; \beta$ is the action type instantiated by any action which instantiates the types α and β in sequence. We assume that an individual action can instantiate different action types. Accordingly, when we say that an action is a token of a_i we do not exclude the possibility that it is also a token of a different type a_j.

Turning to the set of formulas of $\mathcal{L}$, let P be a countable set of propositional variables. Then $Fm(\mathcal{L})$ is defined according to the following grammar:

$\varphi ::= p \mid \neg\varphi \mid \varphi \wedge \varphi \mid \Box\varphi \mid [\alpha]\varphi \mid \mathbf{R}(\alpha) \mid [\uparrow]\varphi \mid I$ where $p \in P$, and $\alpha \in Tm(\mathcal{L})$.

The other connectives and the dual modal operators, $\Diamond\varphi$, $\langle\alpha\rangle\,\varphi$, $\langle\uparrow\rangle\,\varphi$, are defined as usual. The intended interpretation of the modal formulas is as follows: "$\Box\varphi$" says that φ holds in any possible world; "$[\alpha]\varphi$" says that φ holds in any world that can be accessed by performing action α, i.e., that φ holds as a consequence of α; "$\mathbf{R}(\alpha)$" says that the state which is the result of action α is realized[2]; "$[\uparrow]\varphi$" says that φ holds in all the best worlds that can be accessed by performing some action; and, finally, "I" says that the ideal of deontic perfection is realized. It is worth noting that, since 1 is the action type instantiated by any action, "$\langle 1\rangle\,\varphi$" says that φ can be realized by doing an action. Hence, the crucial distinction between *what is possible* and *what is realizable* is captured by the distinction between $\Diamond\varphi$ and $\langle 1\rangle\,\varphi$.

3.1 Semantics

The conceptual framework we adopt is based on the following notion of frame.

Definition 3.1 frame for $\mathcal{L}(\mathbf{ADL})$.
A frame for $\mathcal{L}(\mathbf{ADL})$ is a tuple $F = \langle W, R, \{R_w \mid w \in W\}, r, S, Ideal\rangle$

As mentioned above, frames for $\mathcal{L}(\mathbf{ADL})$ can be subdivided into two parts.

Ontic part: $\langle W, R, \{R_w \mid w \in W\}, r\rangle$, where
(i) $R : W \rightarrow \wp(W)$
(ii) $R_w : Tm(\mathcal{L}) \rightarrow \wp(W)$, for all $w \in W$
(iii) $r : Tm(\mathcal{L}) \rightarrow \wp(W)$

[2] Hence, the formulas $\mathbf{R}(\alpha)$ and $[\alpha]\varphi$ allow us to capture von Wright's distinction between the result and the consequences of an action [13].

We assume that an agent is endowed with a set of primitive actions and think of these actions as ways of obtaining specific resulting states, represented as subsets of a set of possible worlds W. Since the same result can be obtained in different ways, every primitive action corresponds to a set of transitions between worlds in W[3]. More specifically, R, R_w and r are characterized by the following conditions.

Conditions on R

(a) $w \in R(w)$
(b) $v \in R(w) \Rightarrow R(v) = R(w)$

Hence, R models a standard $S5$ notion of ontic modality[4]

Conditions on R_w:

(a) $R_w(\alpha \sqcup \beta) = R_w(\alpha) \cup R_w(\beta)$
(b) $R_w(\alpha; \beta) = \bigcup_{v \in R_w(\alpha)} R_v(\beta)$
(c) $R_w(\alpha) \subseteq R(w)$

Here, R_w is a function that, for each action term, returns the outcomes of the transitions associated with the action performed at w, so that $R_w(\alpha)$ is the set of worlds that are accessible by doing α at w. While conditions (a) and (b) characterize the notions of alternative and sequential actions, (c) captures the intuition that every *realizable state* is a *possible state*. Hence, R and R_w allow us to account for the distinction between what is possible and what is realizable by acting at a world. In fact, it might be the case that reaching a world is beyond the power of the agent, even if that world is possible.

Conditions on r:

(a) $r(\bar{\alpha}) = W - r(\alpha)$
(b) $r(\alpha \sqcap \beta) = r(\alpha) \cap r(\beta)$
(c) $r(\alpha \sqcup \beta) = r(\alpha) \cup r(\beta)$
(d) $r(\alpha) \subseteq r(1)$
(e) $r(\alpha; \beta) \subseteq r(\beta)$
(f) $R_w(\alpha) \subseteq r(\alpha)$
(g) $R(w) \cap r(\alpha) \subseteq r(\beta) \Rightarrow R_w(\alpha) \subseteq R_w(\beta)$
(h) $w \in r(\alpha) \Rightarrow R_w(1) \cap r(\beta) \subseteq r(\alpha; \beta)$

Here, r is a function that, for each action term, returns the state corresponding to the *result* of the action, so that $r(\alpha)$ is the result of α. The conditions connect the intuitive algebra of action results to a corresponding algebra on sets and connect actions with their results. Intuitively:

(a) realizing $\bar{\alpha}$ coincides with not realizing α;
(b) realizing $\alpha \sqcap \beta$ coincides with realizing both α and β;
(c) realizing $\alpha \sqcup \beta$ coincides with realizing either α or β;
(d) realizing any action α is a way of realizing action 1;

[3] Notice that we use the terms "world" and "state" for expressing different concepts, while in the literature about transition systems they are interchangeable with each other. In particular, we use "world" for the complete state which can be reached by performing an action (hence, a world w is an element of W), and "state" for the state of affairs that is the result of an action, as in [13] (hence, a state is in general a subset of W, i.e. a set of worlds).

[4] Using a universal modality would simplify the semantics, but the use of an $S5$ modality gives us a more flexible framework, since the stock of necessary states of affairs can change across the worlds.

(e) realizing any sequence $\alpha; \beta$ is a way of realizing the last action β.

Finally, every realized action realizes its result, by (f); every action whose result involves the result of another action counts as a realization of the latter action, by (g); and, if the result of β is realized after the result of α, then the result of $\alpha; \beta$ is realized as well, by (h).

It is important to note that $r(\alpha)$ does not coincide with $\bigcup_{w \in W} R_w(\alpha)$, since we allow for the possibility that a state of affairs, which is the result of an action, obtains even if no action has brought it about. Indeed, it is possible for a door to be open, even if it was not opened by an agent. As a consequence, $r(1)$, which is W, does not coincide with $\bigcup_{w \in W} R_w(1)$, which is the set of worlds the agent can reach by performing some actions. In addition, we do not assume that $R(w)$ coincides with $R_w(1)$, since, as mentioned above, we allow for a difference between what is possible at a world and what is achievable by acting at it. This is crucial to account for cases where the ideal of perfection, although possible, is not realizable by performing any action.

Deontic part: $\langle W, R, S, Ideal \rangle$, where
(i) $S : W \to \wp(W)$
(ii) $Ideal \subseteq W$

We introduce a deontic function S on W, so that $S(w)$ is the set of the best accessible worlds relative to w, which are the worlds where the conditional ideal that can be achieved in w is realized. In contrast, $Ideal$ is the subset of W containing the best possible worlds from a deontic point of view, which are the worlds where the ideal of deontic perfection is realized.

Conditions on S:	**Conditions on $Ideal$:**
(a) $\varnothing \neq S(w)$	(a) $R(w) \cap Ideal \neq \emptyset$
(b) $S(w) \subseteq R_w(1)$	(b) $R_w(1) \cap Ideal \subseteq S(w)$
(c) $v \in S(w) \Rightarrow S(v) \subseteq S(w)$	(c) $R_w(1) \cap Ideal \neq \emptyset \Rightarrow S(w) \subseteq Ideal$

According to the conditions on S, the set of worlds that can be accessed by the agent always contains a non-empty subset of realizable best options, such that the best options that are accessible by acting in a world that can be reached by w are accessible by w itself. According to the conditions on $Ideal$, the set of accessible worlds always contains a non-empty subset of best possible options. In addition, no accessible world is strictly better, according to S, than any world in $Ideal$, which coincides with the set of the best options if some ideal world is accessible. It is worth noting that a conditional ideal is achievable even if the ideal of perfection cannot be possibly achieved, since $R_w(1) \cap S(w) = S(w)$ is non-empty even if $R_w(1) \cap Ideal$ is empty.

Definition 3.2 model for $\mathcal{L}(\mathbf{ADL})$.
A model for $\mathcal{L}(\mathbf{ADL})$ is a pair $M = \langle F, V \rangle$, where (i) F is a frame for $\mathcal{L}(\mathbf{ADL})$ and (ii) V is a function that maps propositional variables in $\wp(W)$.

Definition 3.3 truth in a model for $\mathcal{L}(\mathbf{ADL})$. The definition of truth is as follows:

$M, w \models p_i \Leftrightarrow w \in V(p_i)$
$M, w \models \neg\varphi \Leftrightarrow M, w \not\models \varphi$
$M, w \models \varphi \wedge \psi \Leftrightarrow M, w \models \varphi$ and $M, w \models \psi$
$M, w \models \Box\varphi \Leftrightarrow \forall v \in W(v \in R(w) \Rightarrow M, v \models \varphi)$
$M, w \models [\alpha]\varphi \Leftrightarrow \forall v \in W(v \in R_w(\alpha) \Rightarrow M, v \models \varphi)$
$M, w \models \mathbf{R}(\alpha) \Leftrightarrow w \in r(\alpha)$
$M, w \models [\uparrow]\varphi \Leftrightarrow \forall v \in W(v \in S(w) \Rightarrow M, v \models \varphi)$
$M, w \models I \Leftrightarrow w \in Ideal$

3.2 Axiomatization

The system **ADL** is defined by the following axioms and rules. The first three groups of axioms take into account the pure modal part of the system, while groups 4, 5 and 6 characterize actions and their results. On the way, we define deontic operators in the Andersonian style.

Group 1: axioms for $\Box$
$\Box$K: $\Box(\varphi \to \psi) \to (\Box\varphi \to \Box\psi)$
$\Box$T: $\Box\varphi \to \varphi$
$\Box$5: $\Diamond\varphi \to \Box\Diamond\varphi$
$\Box$R: $\varphi \,/\, \Box\varphi$

Group 2: axioms for $[\uparrow]$
$[\uparrow]$K: $[\uparrow](\varphi \to \psi) \to ([\uparrow]\varphi \to [\uparrow]\psi)$
$[\uparrow]$D: $[\uparrow]\varphi \to \langle\uparrow\rangle\, \varphi$
$[\uparrow]$4: $[\uparrow]\varphi \to [\uparrow][\uparrow]\varphi$
$[\uparrow]$I: $[1]\varphi \to [\uparrow]\varphi$

Group 3: axioms for I
I1: $\Diamond I$
I2: $[\uparrow]\varphi \to [1](I \to \varphi)$
I3: $\langle 1\rangle\, I \to [\uparrow]I$

Definition 3.4 Deontic operators on states based on I.
$[I]\varphi := \Box(I \to \varphi)$ and $\langle I\rangle\, \varphi := \Diamond(I \wedge \varphi)$.

$[I]\varphi$ is a standard concept of obligation for states[5], as proposed in [1]. It is not difficult to see that $[I]$ is a $KD45$ modality, since we can derive:

(i) $[I](\varphi \to \psi) \to ([I]\varphi \to [I]\psi)$
(ii) $[I]\varphi \to \langle I\rangle\, \varphi$
(iii) $[I]\varphi \to [I][I]\varphi$
(iv) $\langle I\rangle\, \varphi \to [I]\,\langle I\rangle\, \varphi$
(v) $\varphi/[I]\varphi$

The fundamental distinction we want to highlight here concerns $\langle I\rangle\, \varphi$ and $\langle\uparrow\rangle\, \varphi$. While $\langle I\rangle\, \varphi$ states that φ holds in some ideal world, $\langle\uparrow\rangle\, \varphi$ states that φ holds in some of the best accessible worlds. As we will see, this distinction gives rise to two different operators of permission.

Let us now introduce the axioms concerning actions and their results.

---

[5] Letting $O\varphi$ be $[I]\varphi$ and $P\varphi$ be $\langle I\rangle\, \varphi$, the choice of an $S5$ modal logic gives us theorems like $O\varphi \to \Box O\varphi$ and $P\varphi \to \Box P\varphi$. In our setting, these principles are justified by the intended interpretation of a formula like $[I]\varphi$. I is an ideal state determined by a specific legal code, and we assume that the distinction between what is prescribed and what is not prescribed is also fixed by that same code. Hence, given that "$O\varphi$" is interpreted as φ is prescribed by the code that fixes I, the previous principles turn out to be intuitive, since it is impossible to change what is prescribed according to the code without changing that code as well.

Group 4: axioms for $[\alpha]$

$[\alpha]$K: $[\alpha](\varphi \to \psi) \to ([\alpha]\varphi \to [\alpha]\psi)$ $\quad$ $[\alpha]2$: $[\alpha;\beta]\varphi \leftrightarrow [\alpha][\beta]\varphi$

$[\alpha]1$: $[\alpha]\varphi \wedge [\beta]\varphi \to [\alpha \sqcup \beta]\varphi$ $\quad$ $[\alpha]3$: $\Box\varphi \to [\alpha]\varphi$

Group 5: axioms for **R**

R1: $\mathbf{R}(\alpha) \leftrightarrow \neg\mathbf{R}(\bar{\alpha})$ $\quad$ R5: $\mathbf{R}(\alpha;\beta) \to \mathbf{R}(\beta)$

R2: $\mathbf{R}(\alpha \sqcap \beta) \leftrightarrow \mathbf{R}(\alpha) \wedge \mathbf{R}(\beta)$ $\quad$ R6: $[\alpha]\mathbf{R}(\alpha)$

R3: $\mathbf{R}(\alpha \sqcup \beta) \leftrightarrow \mathbf{R}(\alpha) \vee \mathbf{R}(\beta)$ $\quad$ R7: $\Box(\mathbf{R}(\alpha) \to \mathbf{R}(\beta)) \to ([\beta]\varphi \to [\alpha]\varphi)$

R4: $\mathbf{R}(\alpha) \to \mathbf{R}(1)$ $\quad$ R8: $\mathbf{R}(\alpha) \to [1](\mathbf{R}(\beta) \to \mathbf{R}(\alpha;\beta))$

These groups of axioms take into account the operations on actions and results and the connections between actions and results, which is further clarified by the following facts.

(1) $[\bar{\bar{\alpha}}]\varphi \leftrightarrow [\alpha]\varphi$ $\quad$ (5) $[\overline{\alpha_1}]\varphi \vee [\overline{\alpha_2}]\varphi \to [\overline{\alpha_1 \sqcup \alpha_2}]\varphi$

(2) $[\alpha_1]\varphi \vee [\alpha_2]\varphi \to [\alpha_1 \sqcap \alpha_2]\varphi$ $\quad$ (6) $\langle\alpha\rangle\top \wedge [\alpha]\varphi \to \langle\alpha\rangle\varphi$

(3) $[\alpha \sqcup \beta]\varphi \to [\alpha]\varphi \wedge [\beta]\varphi$ $\quad$ (7) $\langle\alpha\rangle\top \to \langle\alpha\rangle\mathbf{R}(\alpha)$

(4) $[\overline{\alpha_1 \sqcap \alpha_2}]\varphi \leftrightarrow [\overline{\alpha_1}]\varphi \wedge [\overline{\alpha_2}]\varphi$ $\quad$ (8) $[1]\varphi \to [\alpha]\varphi$

Proof. Let us prove (4).

$\mathbf{R}(\bar{\alpha} \sqcup \bar{\beta}) \leftrightarrow \mathbf{R}(\bar{\alpha}) \vee \mathbf{R}(\bar{\beta})$, by R3

$\mathbf{R}(\bar{\alpha} \sqcup \bar{\beta}) \leftrightarrow \neg\mathbf{R}(\alpha) \vee \neg\mathbf{R}(\beta)$, by R1

$\mathbf{R}(\bar{\alpha} \sqcup \bar{\beta}) \leftrightarrow \neg\mathbf{R}(\alpha \sqcap \beta)$, by R2

$\mathbf{R}(\bar{\alpha} \sqcup \bar{\beta}) \leftrightarrow \mathbf{R}(\overline{\alpha \sqcap \beta})$, by R1

$[\overline{\alpha \sqcap \beta}]\varphi \leftrightarrow [\bar{\alpha} \sqcup \bar{\beta}]\varphi$, by R7

$[\overline{\alpha \sqcap \beta}]\varphi \leftrightarrow [\bar{\alpha}]\varphi \wedge [\bar{\beta}]\varphi$, by (3), and $[\alpha]$**1** $\Box$

Since (1-5) are derivable, our system is powerful enough to interpret the system proposed by Meyer in [8], except for the axiom on the negation of sequential actions. In addition, since (7) is derivable, within **ADL** the performability of an action, expressed by $\langle\alpha\rangle\top$, is to be distinguished from the possibility of the result of the action, i.e., $\Diamond\mathbf{R}(\alpha)$. In fact, while $\langle\alpha\rangle\top \to \langle\alpha\rangle\mathbf{R}(\alpha)$, and, hence, $\langle\alpha\rangle\top \to \Diamond\mathbf{R}(\alpha)$, it is possible that $\Diamond\mathbf{R}(\alpha)$ even if α is not performable. Finally, in this system two intuitive concepts of inclusion between actions or action results are definable.

Definition 3.5 inclusions.

(i) $\beta \sqsubseteq \alpha := [\alpha]\mathbf{R}(\beta)$.

(ii) $\beta \sqsubseteq_R \alpha := \Box(\mathbf{R}(\alpha) \to \mathbf{R}(\beta))$.

As it is easy to check, both $\sqsubseteq$ and $\sqsubseteq_R$ are preorders. As it will become clear below, the introduction of these preorders allows us to represent actions that, while being optimal in their results, are not permitted, due to the fact that they also realize what is prohibited during their course.

3.3 Characterization

The system **ADL** is sound and strongly complete with respect to the class of models introduced above. Soundness is straightforward. Completeness is proved by a canonicity argument. Let us first define $w/\Box := \{\varphi \mid \Box\varphi \in w\}$; $w/[\uparrow] := \{\varphi \mid [\uparrow]\varphi \in w\}$; $w/[\alpha] = \{\varphi \mid [\alpha]\varphi \in w\}$.

Definition 3.6 canonical model for $\mathcal{L}(\mathbf{ADL})$. The canonical model for $\mathcal{L}(\mathbf{ADL})$ is the tuple

$M_C = \langle W, R, S, Ideal, \{R_w \mid w \in W\}, r, V\rangle$, where

(1) W is the set of maximal consistent sets of formulas
(2) $R : W \to \wp(W)$ is such that $v \in R(w) \Leftrightarrow w/\Box \subseteq v$
(3) $S : W \to \wp(W)$ is such that $v \in S(w) \Leftrightarrow w/[\uparrow] \subseteq v$
(4) $Ideal = \{w \mid I \in w\} \subseteq W$
(5) $R_w : Tm(\mathcal{L}) \to \wp(W)$ is such that $v \in R_w(\alpha) \Leftrightarrow w/[\alpha] \subseteq v$
(6) $r : Tm(\mathcal{L}) \to \wp(W)$ is such that $v \in r(\alpha) \Leftrightarrow \mathbf{R}(\alpha) \in v$
(7) $V : P \to \wp(W)$ is such that $v \in V(p) \Leftrightarrow p \in v$

For reason of space, we omit the proofs of the following lemmas.

Lemma 3.7 (*Truth Lemma*): $M_C, w \models \varphi \Leftrightarrow \varphi \in w$.

Lemma 3.8 (*Model Lemma*): M_C *is a model for* $\mathcal{L}(\mathbf{ADL})$.

They essentially follow from the definitions of R, S, $Ideal$, R_w, r and from the correspondence between axioms of **ADL** and conditions on models for $\mathcal{L}(\mathbf{ADL})$.

4 Deontic concepts and paradoxes

At this point, we can introduce the definition of four different kinds of deontic concepts[6].

Definition 4.1 deontic concepts on states and actions.

Group 1: ideal on states.	**Group 2**: ideal on results.
1. $\mathbf{P}(\varphi) := \langle I\rangle\,\varphi$	1. $\mathbf{P}(\mathbf{R}(\alpha)) := \langle I\rangle\,\mathbf{R}(\alpha)$
2. $\mathbf{F}(\varphi) := [I]\neg\varphi$	2. $\mathbf{F}(\mathbf{R}(\alpha)) := [I]\neg\mathbf{R}(\alpha)$
3. $\mathbf{O}(\varphi) := [I]\varphi$	3. $\mathbf{O}(\mathbf{R}(\alpha)) := [I]\mathbf{R}(\alpha)$
4. $\mathbf{P}^S(\varphi) := \Diamond\varphi \wedge \Box(\varphi \to I)$	4. $\mathbf{P}^S(\mathbf{R}(\alpha)) := \Diamond\mathbf{R}(\alpha) \wedge \Box(\mathbf{R}(\alpha) \to I)$
Group 3: ideal on actions.	**Group 4**: conditional on results.
1. $\mathbf{P}!(\alpha) := \langle\alpha\rangle\,I$	1. $\mathbf{P}(\alpha) := \langle\uparrow\rangle\,\mathbf{R}(\alpha)$
2. $\mathbf{F}!(\alpha) := \neg\,\langle\alpha\rangle\,I$	2. $\mathbf{F}(\alpha) := \neg\,\langle\uparrow\rangle\,\mathbf{R}(\alpha)$
3. $\mathbf{O}!(\alpha) := \neg\,\langle\bar{\alpha}\rangle\,I$	3. $\mathbf{O}(\alpha) := \neg\,\langle\uparrow\rangle\,\mathbf{R}(\bar{\alpha})$
4. $\mathbf{P}!^S(\alpha) := \langle\alpha\rangle\,I \wedge [\alpha]I$	4. $\mathbf{P}^S(\alpha) := \langle\uparrow\rangle\,\mathbf{R}(\alpha) \wedge [\uparrow]\mathbf{R}(\alpha)$

The definition of the conditional deontic concepts can be justified by considering the following equivalences.

$$M, w \models \mathbf{F}(\alpha) \;\Leftrightarrow\; M, w \models \neg\,\langle\uparrow\rangle\,\mathbf{R}(\alpha)$$
$$M, w \models \mathbf{F}(\alpha) \;\Leftrightarrow\; \forall v \in W(v \in S(w) \Rightarrow M, v \not\models \mathbf{R}(\alpha))$$

[6] Concepts in Group 2 are specific instances of concepts in Group 1. They characterize deontic concepts on actions in terms of action results and are of interest when compared with concepts in Group 3 and Group 4.

$M, w \models \mathbf{F}(\alpha) \Leftrightarrow \forall v \in W(v \in S(w) \Rightarrow v \notin r(\alpha))$
$M, w \models \mathbf{F}(\alpha) \Leftrightarrow r(\alpha) \cap S(w) = \varnothing$

Hence, an action is conditionally prohibited provided that its result only holds in worlds that are worse than the best accessible worlds. Similarly, an action is conditionally permitted (obliged) when its result holds is some (all) of the best accessible worlds.

Fact 4.2 *Relations between different deontic concepts.*
(1) $\mathbf{P}!(\alpha) \rightarrow \mathbf{P}(\alpha)$
(2) $\mathbf{P}!(\alpha) \wedge [\alpha]\varphi \rightarrow \mathbf{P}(\varphi)$, *and so* $\mathbf{P}!(\alpha) \rightarrow \mathbf{P}(\mathbf{R}(\alpha))$
(3) $\langle 1 \rangle I \wedge \langle \uparrow \rangle \varphi \rightarrow \langle I \rangle \varphi$, *by* I3, *and so* $\langle 1 \rangle I \wedge \mathbf{P}(\alpha) \rightarrow \mathbf{P}(\mathbf{R}(\alpha))$

As expected, (1) all ideally permitted actions are conditionally permitted and (2) both the result and all the consequences of ideally permitted actions are ideally permitted states. In addition, (3) provided that the ideal can be accessed, the result of conditionally permitted actions are ideally permitted. By contrast, it can be proved that not all actions that are conditionally permitted are ideally permitted. Thus, conditional prescription can be effective even in cases where no action is ideally permitted.

Fact 4.3 *Permission and inclusion.*
(1) $\mathbf{P}!(\alpha) \wedge \beta \sqsubseteq \alpha \Rightarrow \mathbf{P}(\beta)$
(2) $\mathbf{P}(\alpha) \wedge \beta \sqsubseteq_R \alpha \Rightarrow \mathbf{P}(\beta)$

Accordingly, actions including conditionally prohibited actions are prohibited.

Now, our claim is that the best way for capturing the intuitions discussed in section 2 is to use conditional deontic concepts. Thus, we assume them to provide a solution to the three groups of paradoxes mentioned above.

4.1 Paradoxes on standard prescriptions

Within **ADL** standard paradoxes concerning the conditional notions of obligation and permission can be solved in two different ways. Firstly, we can opt for using notions of strong permission and obligation as in [6]. Secondly, and more interestingly, we can define two specific notions of choice permission and choice obligation:

- choice permission: $\mathbf{P}(\alpha + \beta) := \langle \uparrow \rangle \mathbf{R}(\alpha) \wedge \langle \uparrow \rangle \mathbf{R}(\beta)$
- choice obligation: $\mathbf{O}(\alpha + \beta) := \mathbf{O}(\alpha \sqcup \beta) \wedge \mathbf{P}(\alpha + \beta)$

It is then not difficult to see that:

$\vdash_{\mathbf{ADL}} \mathbf{P}(\alpha + \beta) \rightarrow \mathbf{P}(\alpha) \wedge \mathbf{P}(\beta)$; $\quad \nvdash_{\mathbf{ADL}} \mathbf{P}(\alpha) \rightarrow \mathbf{P}(\alpha + \beta)$
$\vdash_{\mathbf{ADL}} \mathbf{O}(\alpha + \beta) \rightarrow \mathbf{P}(\alpha + \beta)$; $\quad \nvdash_{\mathbf{ADL}} \mathbf{O}(\alpha) \rightarrow \mathbf{O}(\alpha + \beta)$

The present solution seems to be more intuitive insofar as both strong permission and strong obligation require that there is no way we can violate the law if we act according to what is strongly permitted or obliged, while ordinary choices can be risky: we are ordinarily allowed to choose between alternative actions even if there are ways of performing such actions that lead to a violation of the law.

4.2 Paradoxes on prescriptions on sequential actions

Within **ADL** paradoxes concerning obligation and permission of sequential actions, when these concepts are fixed according to the conditional definition, find an insightful solution.

As to van der Mayden's paradox, note that both $\langle\alpha\rangle\,\mathbf{P}(\beta) \rightarrow \mathbf{P}(\alpha;\beta)$ and the stronger $\mathbf{P}(\alpha) \wedge \langle\alpha\rangle\,\mathbf{P}(\beta) \rightarrow \mathbf{P}(\alpha;\beta)$ can fail. Consider the following model:

1) $W = R(w) = R(v) = R(u) = R(x) = \{w, v, u, x\}$
2) $R_w(\alpha) = \{v\}$; $R_v(\alpha) = R_u(\alpha) = R_x(\alpha) = \varnothing$
3) $R_w(1) = \{v, u, x\}$; $R_v(1) = R_u(1) = \{u\}$; $R_x(1) = \{x\}$
4) $S(w) = S(x) = \{x\} = Ideal$; $S(v) = S(u) = \{u\}$
5) $r(\alpha) = \{v, x\}$; $r(\beta) = r(\alpha;\beta) = \{u\}$

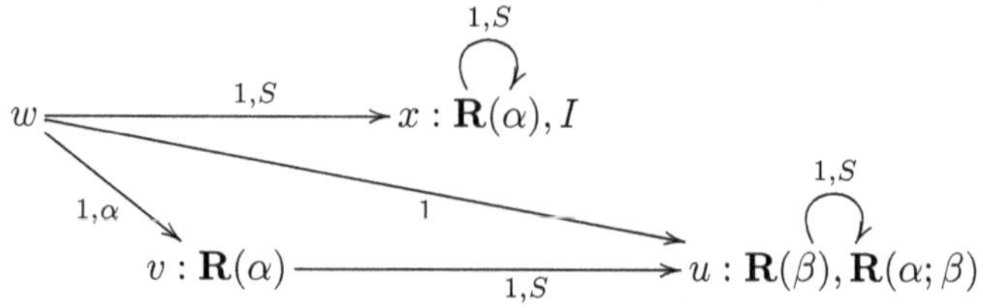

In this model, $w \models \langle\uparrow\rangle\,\mathbf{R}(\alpha)$ and $w \models \langle\alpha\rangle\,\langle\uparrow\rangle\,\mathbf{R}(\beta)$, but $w \not\models \langle\uparrow\rangle\,\mathbf{R}(\alpha;\beta)$, whence the conclusion. The failure of these principles is due to the fact that, even when α is permitted, $\langle\alpha\rangle\,\mathbf{P}(\beta)$ is not sufficient for $\mathbf{P}(\alpha;\beta)$, since the world we land on by performing α at w may not be one of best options of w. In the previous model, β is permitted in v because the $R(\beta)$-world u is among the best options achievable from v. Still, since this is not sufficient to obtain that u is also among the best options achievable from w, $\alpha;\beta$ is not permitted in w. In addition, note that the converse of the first principle also fails, since $u \models \langle\uparrow\rangle\,\mathbf{R}(\alpha;\beta)$, but $u \not\models \langle\uparrow\rangle\,\mathbf{R}(\alpha)$.

As to Anglberger's paradox, note that both $\mathbf{F}(\alpha) \rightarrow \mathbf{F}(\alpha;\beta)$ and $\mathbf{F}(\alpha) \rightarrow [\alpha]\mathbf{F}(\beta)$ can fail. Consider the following model:

1) $W = R(w) = R(v) = R(u) = R(x) = \{w, v, u, x\}$
2) $R_w(\alpha) = \{v\}$; $R_v(\alpha) = R_u(\alpha) = R_x(\alpha) = \varnothing$;
3) $R_w(1) = \{v, u, x\}$; $R_v(1) = R_u(1) = \{u\}$; $R_x(1) = \{x\}$
4) $S(w) = S(v) = S(u) = \{u\} = Ideal$; $S(x) = \{x\}$
5) $r(\alpha) = \{v\}$; $r(\beta) = r(\alpha;\beta) = \{u\}$

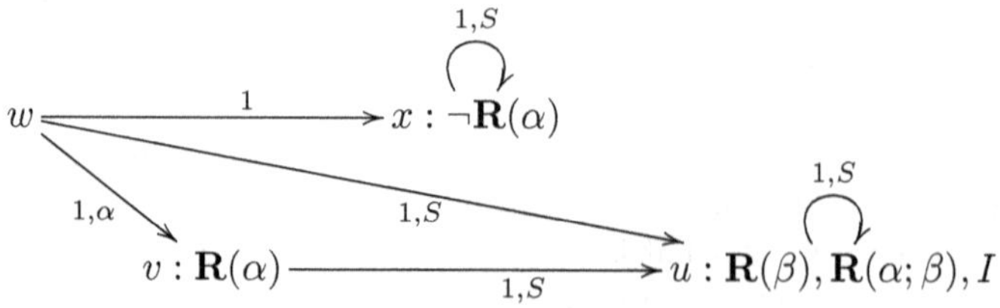

In this model, $w \not\models \langle\uparrow\rangle\,\mathbf{R}(\alpha)$, but $w \models \langle\alpha\rangle\,\langle\uparrow\rangle\,\mathbf{R}(\beta)$ and $w \models \langle\uparrow\rangle\,\mathbf{R}(\alpha;\beta)$, and the conclusion follows. The failure of these principles is due to the fact that, for α to be prohibited, it is sufficient that α makes the deontic condition of the reference world worse than any of the best accessible worlds. Still, this is not sufficient to exclude that doing $\alpha;\beta$ leads to one of these best accessible worlds.

4.3 Contrary to duty obligations

As a final application, let us consider cases of contrary to duty obligations instantiating these classical schemas:

It ought to be that φ, but $\neg\varphi$	It ought to be that φ, but $\neg\varphi$
It ought to be that if φ then $\neg R(\alpha)$	It ought to be that if φ then $\neg R(\alpha)$
If $\neg\varphi$, then it ought to be that $R(\alpha)$	It ought to be that if $\neg\varphi$ then $R(\alpha)$

In our framework, the most intuitive analysis is:

$[I]\varphi \wedge \neg\varphi$	$[I]\varphi \wedge \neg\varphi$
$[I](\varphi \rightarrow \neg R(\alpha))$	$[I](\varphi \rightarrow \neg R(\alpha))$
$\Box(\neg\varphi \rightarrow [\uparrow]R(\alpha))$	$[\uparrow](\neg\varphi \rightarrow R(\alpha))$

In both cases, we obtain that $[I]\neg R(\alpha)$ and $[\uparrow]R(\alpha)$. Still, no contradiction follows, since in any situation in which the result of α is prohibited, according to the law, the obligation to do α is only conditional. Finally, note that the present interpretation of the conditional leading to a contrary to duty obligation validates both

FD: factual detachment and **DD**: deontic detachment

$$\frac{\Box(\varphi \rightarrow [\uparrow]\psi) \quad \varphi}{[\uparrow]\psi} \qquad \frac{\Box(\varphi \rightarrow [\uparrow]\psi) \quad [\uparrow]\varphi}{[\uparrow]\psi}$$

which is one of the desiderata proposed in [3].

5 Conclusion

In this paper, we have presented a general system of deontic logic of actions in which the main problems related to the definition of deontic concepts in a dynamic framework can be overcome. The solutions we have proposed are based on the introduction of a group of conditional deontic concepts, according to which what is permitted, prohibited and obligatory depends on the best states that the agent can realize, given the conditions in which she is acting. The conceptual apparatus encoded in our system, which allows us to capture these new concepts, includes a twofold distinction on the ontic level. First, a distinction between what is possible and what is realizable by performing an action; and, second, a distinction between the result associated with an action and the consequences of that action. Being based on this conceptually rich framework, our system gives us the possibility of systematically bringing together and comparing in an innovative way Andersonian deontic concepts on states as well as on results of actions, ideal deontic concepts on actions *à la*

Meyer, and conditional deontic concepts on actions. We have shown that the availability of both ideal deontic concepts on states and conditional deontic concepts on actions provides us with a natural solution to the paradoxes of contrary to duty obligations. What is more, the introduction of conditional deontic concepts allows us to define original notions of choice permission and choice obligation that, while not being subject to standard paradoxes, take into account the riskiness of choices. Finally, besides not incurring in paradoxes concerning the sequential execution of actions, the new deontic concepts provides us with a way of making sure that, even in states in which the ideal of deontic perfection is not realizable, the actions of the agent can be deontically qualified in a non-trivial way.

Acknowledgments. We would like to thank the referees of DEON 2016 for helpful comments and for highlighting an important issue in a previous version of this paper.

References

[1] Anderson, A.R. (1958). A Reduction of Deontic Logic to Alethic Modal Logic. Mind, 67: 100–103.

[2] Anglberger, A. (2008). Dynamic Deontic Logic and Its Paradoxes. Studia Logica, 89: 427–435.

[3] Carmo, J. & Jones, A.J.I. (2002). Deontic Logic and Contrary-to-Duties. In Gabbay, D.M. and Guenthner, F. (Eds). Handbook of philosophical logic. Dordrecht: Springer 2002: 265–343.

[4] Castro, P.F. & Maibaum, T.S.E. (2009). Deontic Action Logic, Atomic Boolean Algebras and Fault-tolerance, Journal of Applied Logic, 7: 441–466.

[5] Castro, P.F. & Kulicki, P. (2014). Deontic Logics Based on Boolean Algebra. In Tryputz, R. (Ed.), Krister Segerberg on Logic of Actions. Dordrecht: Springer 2014: 85–117.

[6] Dignum, F., Meyer, J.J. & Wieringa, R.J. (1996). Free Choice and Contextually Permitted Actions. Studia Logica, 57: 193–220.

[7] Jones A.J.I. & Pörn, I. (1985). Ideality, Sub-ideality and Deontic Logic. Synthese, 65: 275–290.

[8] Meyer, J.-J. Ch. (1988). A Different Approach to Deontic Logic: Deontic Logic Viewed as Variant of Dynamic Logic. Notre Dame Journal of Formal Logic, 29: 109–136.

[9] Segerberg, K. (1982). A Deontic Logic of Action. Studia Logica, 41: 269–282.

[10] Trypuz, R., & Kulicki, P. (2015). On deontic Action Logics Based on Boolean Algebra. Journal of Logic and Computation, 25: 1241–1260.

[11] van Der Meyden, R. (1986). The Dynamic Logic of Permission. Journal of Logic and Computation, 6: 465–479.

[12] von Wright, G.H. (1951). Deontic Logic. Mind, 237: 1–15.

[13] von Wright, G.H. (1963). Norm and Action: A Logical Inquiry. London: Routledge & Kegan Paul.

[14] von Wright, G.H (1971). Deontic Logic and the Theory of Conditions. In Hilpinen, R. (Ed.). Deontic Logic: Introductory and Systematic Readings. Dordrecht: Riedel 1971: 159–177.

Sequence Semantics for Norms and Obligations

Guido Governatori[1], Francesco Olivieri[2]
Erica Calardo[3] and Antonino Rotolo[3]

[1] *Data61, CSIRO, Brisbane, Australia*
[2] *University of Verona, Verona, Italy*
[3] *CIRSFID, University of Bologna, Bologna, Italy*

Abstract

This paper presents a new version of the sequence semantics presented at DEON 2014. This new version allows us for a capturing the distinction between logic of obligations and logic of norms. Several axiom schemata are discussed, while soundness and completeness results are proved.

Keywords: Deontic systems, Neighbourhood semantics, Logic of norms.

1 Introduction

Most of the work in deontic logic has focused on the study of the concepts of obligation, permission, prohibition and related notions, but little attention has been dedicated on how these prescriptions are generated within a normative system. [1] The general idea of norms is that they describe conditions under which some behaviours are deemed as 'legal'. In the simplest case, a behaviour can be described by an obligation (or a prohibition, or a permission), but often norms additionally specify what are the consequences of not complying with them, and what sanctions follow from violations and whether such sanctions compensate for the violations.

To address the above issues, Governatori and Rotolo [12] presented a Gentzen style sequent system to describe a non classical operator ($\otimes$) which models chains of obligations and compensatory obligations. The interpretation of a chain like $a \otimes b \otimes c$ is that a is obligatory, but if it is violated (i.e., $\neg a$ holds), then b is the new obligation (and b compensates for the violation of a); again, if the obligation of b is violated as well, then c is obligatory (and so on).

As we argued in [12, 8], the logic of $\otimes$ offers a proof-theoretic approach to normative reasoning (and in particular, CTD reasoning), which, as done by [18, 17] in the context of Input/Output Logic, follows the principle "no logic of norms without attention to the normative systems in which they occur" [16].

---

[1] A normative system can be understood as a, possibly hierarchically structured, set of norms and mechanisms that systematically interplay for deriving deontic prescriptions in force in a given situation.

This idea draws inspiration from the pioneering works in [20] and [1], and focuses on the fact that normative conclusions derive form of norms as interplaying together in normative systems. Indeed, it is essential in this perspective to distinguish prescriptive and permissive norms from obligations and permissions [3, 10]: the latter ones are merely the effects of the application of norms.

While Input/Output approach mainly works by imposing some constraints on the manipulation of conditional norms, the $\otimes$-logic uses $\otimes$-chains to express the logical structures (norms) that generate actual obligations and permissions. In [4], we proposed a model-theoretic semantics (called *sequence semantics*) for the $\otimes$-logic, that addresses the problem identified in [7] that affects most of the existing approaches for the representation of norms, in particular compensatory obligations, using 'standard' possible world semantics. A compensatory obligation is a sub-class of a contrary-to-duty obligation, where the violation of the primary obligation is compensated by the fulfilment of the secondary obligation. Compensatory obligations can be modelled by $\otimes$-chains. As we have already discussed, an expression like $a \otimes b$ means that a is obligatory, but its violation is compensated by b or, in other terms, it is obligatory to do b to compensate the violation of the obligation of a. Thus, a situation where a does not hold (or $\neg a$ holds) and b holds is still deemed as a 'legal' situation. Accordingly, when we use a 'standard' possible world semantics, there is a deontically accessible world where $\neg a$ holds, but this implies, according to the usual evaluation conditions for permission (something is permitted, if there is a deontically accessible world where it holds), that $\neg a$ is permitted. However, we have the norm modelling the compensatory obligation that states that a is obligatory (and if it were not, then there would be no need for b to compensate for such a violation since, there would be no violation of the obligation of a to begin with). The sequence semantics solves this problem by establishing that to have an obligation, we must have a norm generating the obligation itself (where a norm is represented by an $\otimes$-chain), and not simply that something is obligatory because it holds in all the deontically accessible worlds.

The work of the present paper completes the picture in three points.

- We extend sequence semantics and split the treatment of $\otimes$-chains and obligations; the intuition is that chains are the generators of obligations and permissions, we hence semantically separate structures interpreting norms from those interpreting obligations and permissions.
- We add $\oplus$-sequences to express ordering among explicit permissions [8]; as for $\otimes$, given the chain $a \oplus b$, we can proceed through the $\oplus$-chain to obtain the derivation of $\mathsf{P}b$. However, permissions cannot be violated. Consequently, it does not make sense to obtain $\mathsf{P}b$ from $a \oplus b$ and $\neg a$. Here, the reason to proceed in the chain is rather that the normative system allows us to prove $\mathsf{O}\neg a$;
- We systematically study several options for the axiomatisation of $\otimes$ and $\oplus$.

The layout of the paper is as follows: in Section 2 we introduce the language of our logics. In Section 3 we progressively introduce axioms for the deontic operators to axiomatise more expressive deontic logics with and without interaction between the operators, and we discuss some intuition behind the axiomatisation. In Section 4 we provide the definitions of sequence semantics to cover the case of weak and strong permission. Soundness and completeness of the various deontic logic with the novel semantics are proved in Section 5. Finally, a short discussion of related work and further work (Section 6) concludes the paper.

2 Language

The language consists of a countable set of atomic formulae. Well-formed-formulae are then defined using the typical Boolean connectives, the n-ary connectives $\otimes$ and $\oplus$, and the modal (deontic) operators O for obligation and P for permission. The intended reading of $\otimes$ is that it encodes a sequence of obligations where each obligation is meant to compensate the violation of the previous obligation. The intuition behind $\oplus$ is instead meant to model ordered lists of permissions, i.e., a preference order among different permissions [8].

Let $\mathcal{L}$ be a language consisting of a countable set of propositional letters $Prop = \{p_1, p_2, \ldots\}$, the propositional constant $\bot$, round brackets, the boolean connective $\rightarrow$, the unary operators O and P, the set of n-ary operators $\otimes^n$ for $n \in \mathbb{N}^+$ and the set of n-ary operators $\oplus^n$ for $n \in \mathbb{N}^+$. We shall refer to the language where $\oplus$ does not occur as $\mathcal{L}^{\otimes}$, and the language where $\otimes$ does not occur as $\mathcal{L}^{\oplus}$. There is no technical difficulty in avoiding that $\otimes$ and $\oplus$ be binary operators: the reason why we define them as n-ary ones is mainly conceptual and is meant to exclude the nesting of $\otimes$- and $\oplus$-expressions. Consider $a \otimes \neg(b \otimes c) \otimes d$. The expression $\neg(b \otimes c)$ means either that b is not obligatory or that it is so but c does not compensate the violation of $\mathsf{O}b$. What does it mean this as a compensation of the violation of $\mathsf{O}a$? Also, what is the meaning of $a \otimes (b \oplus c) \otimes d$?

Definition 2.1 [Well Formed Formulae] Well formed formulae (wffs) are defined as follows:

- Any propositional letter $p \in Prop$ and $\bot$ are wffs;
- If a and b are wffs, then $a \rightarrow b$ is a wff;
- If a is a wff and no operator $\otimes^m$, $\oplus^m$, O and P occurs in a, then $\mathsf{O}a$ and $\mathsf{P}a$ are a wff;
- If $a_1, \ldots, a_n$ are wffs and no operator $\otimes^m$, $\oplus^m$, O and P occurs in any of them, then $a_1 \otimes^n \cdots \otimes^n a_n$ and $a_1 \oplus^n \cdots \oplus^n a_n$ are a wff, where $n \in \mathbb{N}^+$;[2]
- Nothing else is a wff.

We use *WFF* to denote the set of well formed formulae.

Other Boolean operators are defined in the standard way, in particular $\neg a =_{def} a \rightarrow \bot$ and $\top =_{def} \bot \rightarrow \bot$.

We use $\odot$ to refer to either $\otimes$ or $\oplus$. Accordingly, we say that any formula $a_1 \odot \cdots \odot a_n$ is an $\odot$-*chain*; also the negation of an $\odot$-chain is an $\odot$-chain. The formation rules allow us to have $\odot$-chains of any (finite) length, and the arity of

[2] We use the prefix forms $\otimes^1 a$ and $\oplus^1 a$ for the case of $n = 1$.

the operator is equal to number of elements in the chain; we thus drop the index m from $\odot^m$. Moreover, we use the prefix notation $\bigodot_{i=1}^n a_i$ for $a_1 \odot \cdots \odot a_n$.

3 Logics for $\otimes$ and $\oplus$

The aim of this section is to discuss the intuitions behind some principles governing the behaviour and the interactions of the various deontic operators. These principles are captured by axioms or inference rules.

3.1 Basic Axiomatisation

In this paper, we assume classical propositional logic, CPC, as the underlying logic on which all the deontic logics we examine are based.

The first principle is that of syntax independence or, in other terms, that the deontic operators are closed under logical equivalence. To this end, all the logics have the following inference rules:

$$\frac{a \equiv b}{\mathsf{O}a \equiv \mathsf{O}b}\,\text{O-RE} \qquad \frac{a \equiv b}{\mathsf{P}a \equiv \mathsf{P}b}\,\text{P-RE}$$

$$\frac{\bigwedge_{i=1}^n (a_i \equiv b_i)}{\bigotimes_{i=1}^n a_i \equiv \bigotimes_{i=1}^n b_i}\,\otimes\text{-RE} \qquad \frac{\bigwedge_{i=1}^n (a_i \equiv b_i)}{\bigoplus_{i=1}^n a_i \equiv \bigoplus_{i=1}^n b_i}$$

Consider the $\odot$ chain $a \odot b \odot a \odot c$. If $\odot$ is $\otimes$, the meaning of the chain above is that a is obligatory, but if a is violated (meaning that $\neg a$ holds) then b is obligatory. If also b is violated, then a becomes obligatory. But we already know that we will incur in the violation of it, since $\neg a$ holds. Accordingly, we have the obligation of c. However, this is the meaning of the $\otimes$-chain: $a \otimes b \otimes c$.

If $\odot$ is $\oplus$, the intuitive reading of $a \odot b \odot a \odot c$ is that a should be permitted unless (for other reasons) a is forbidden; in such a case b is permitted. However, if also b is forbidden, then a is permitted. Nevertheless, we have already established that this is not possible, since a is forbidden, we thus have the permission of c. Again, this is what is encoded by the $\oplus$-chain $a \oplus b \oplus c$.

The above example shows that duplications of formulae in $\odot$-chains do not contribute to the meaning of the chains themselves. This motivates us to adopt the following axioms to remove (resp., introduce) an element from (to) a chain if an equivalent formula occurs on the left of it.

$$\bigotimes_{i=1}^{n} a_i \equiv \bigotimes_{i=1}^{k-1} a_i \otimes \bigotimes_{i=k+1}^{n} a_i \text{ where } a_j \equiv a_k,\ j < k \qquad (\otimes\textbf{-contraction})$$

$$\bigoplus_{i=1}^{n} a_i \equiv \bigoplus_{i=1}^{k-1} a_i \oplus \bigoplus_{i=k+1}^{n} a_i \text{ where } a_j \equiv a_k,\ j < k \qquad (\oplus\textbf{-contraction})$$

The minimal logics resulting from the above axioms and inference rules are $\mathsf{E}^{\otimes}$ when the language is restricted to $\mathcal{L}^{\otimes}$, $\mathsf{E}^{\oplus}$ for $\mathcal{L}^{\oplus}$, and $\mathsf{E}^{\otimes\oplus}$ for $\mathcal{L}$.

3.2 Deontic Axioms

The logics presented in the previous section are minimal, and besides the intended deontic reading of the operators, they do not not provide any 'genuine' deontic principle. In the present section, we introduce axioms to model the relationships between O and P; specifically, the axioms lay down the conditions under which the various operators are consistent.

The first axiom defines the duality of obligation and permission.

$$\mathsf{P}a \equiv \neg\mathsf{O}\neg a \qquad (\mathsf{OP}\text{-}\mathbf{duality})$$

This axiom implies the reading of permission as weak permission, i.e., the lack of the obligation of the contrary.

$$\mathsf{O}a \rightarrow \mathsf{P}a \qquad (\mathsf{O}\text{-}\mathsf{P})$$

Axiom O-P, is the standard **D** axiom of modal/deontic logic. This axiom can have different meanings depending on whether O and P are the dual of each other. If they are, the axiom is trivially equivalent to the following one:

$$\mathsf{O}a \rightarrow \neg\mathsf{O}\neg a \qquad (\mathbf{D}\text{-}\mathsf{O})$$

The axiom states the external consistency of a normative system: a normative system is externally consistent if no formula is obligatory and forbidden at the same time. If O and P are independent modalities, then Axiom O-P establishes the consistency between obligations and permissions, while Axiom **D**-O must be assumed to guarantee the external consistency of obligations.

Internal consistency of obligation is the property that no obligation is self-inconsistent; this is expressed by:

$$\neg\mathsf{O}\bot \qquad (\overline{\mathbf{P}}\text{-}\mathsf{O})$$

Finally, when obligation and permission are not dual, while the consistency between obligation and permission is covered by Axiom O-P, we have yet to cover the consistency between prohibition and permission. To this end, we can use one direction of the duality, namely:

$$\mathsf{O}a \rightarrow \neg\mathsf{P}\neg a \qquad (\mathsf{O}\neg\mathsf{P})$$

The axioms we consider hitherto focus on consistency principles for O and P. The next axioms provide consistency principles for $\odot$-chains.

Given that we use classical propositional logic as the underlying logic, it is not possible that an $\odot$-chain and its negation hold at the same time. What about when $\odot$-chains like $a \odot b \odot c$ and $\neg(a \odot b)$ hold. In case $\odot$ is $\otimes$, the first chain states that a is obligatory and its violation is compensated by b, which in turn is itself obligatory and it is compensated by c. The second expression states that 'either it is not the case that a is obligatory, but if it is so, then its violation is not compensated by b'. Accordingly, the combination of the two expressions should result in a contradiction (a similar argument can be made for $\oplus$-chains). To ensure this, we must assume the following axioms that allow us to derive, given a chain, all its sub-chains with the same initial element(s).

$$a_1 \otimes \cdots \otimes a_n \rightarrow a_1 \otimes \cdots \otimes a_{n-1},\ n \geq 2 \qquad (\otimes\text{-}\mathbf{shortening})$$
$$a_1 \oplus \cdots \oplus a_n \rightarrow a_1 \oplus \cdots \oplus a_{n-1},\ n \geq 2 \qquad (\oplus\text{-}\mathbf{shortening})$$

While any combination of the axioms presented in this section can be added to any of the minimal logics of the previous section, we focus on two options that we believe are meaningful for the representation of norms. For the first option, we call the resulting logic $\mathsf{D}^{\otimes}$, we consider O and P as dual, and it extends $\mathsf{E}^{\otimes}$ with OP-**duality**, $\overline{\mathbf{P}}$-O and $\otimes$-**shortening**. For the second option, we reject the duality of O and P, essentially taking the strong permission stance, and we assume $\mathsf{E}^{\otimes\oplus}$ plus all axioms presented in this section with the exclusion of OP-**duality**. We use $\mathsf{D}^{\otimes\oplus}$ for the resulting logic.

3.3 Axioms for $\otimes$ and O

In this section, we address the relationships between $\otimes$ and O; we thus focus on axioms for extending $\mathsf{D}^{\otimes}$ (though the axioms are suitable for extensions of $\mathsf{D}^{\otimes\oplus}$). As we have repeatedly argued, $\otimes$-chains are meant to generate obligations. In particular, we have seen that the first element of an $\otimes$-chain is obligatory. This is formalised by the following axiom:

$$a_1 \otimes \cdots \otimes a_n \to \mathsf{O}a_1. \qquad (\otimes\text{-}\mathsf{O})$$

Furthermore, we say that if the negation of the first element does not hold, we can infer the obligation of the second element. Formally

$$a_1 \otimes \cdots \otimes a_n \wedge \neg a_1 \to \mathsf{O}a_2. \qquad (1)$$

Moreover, we argued that we can repeat the same procedure. This leads us to generalise (1) for the axiom that expresses the detachment principle for $\otimes$-chains and factual statements about the opposites of the first k elements of an $\otimes$-chain.

$$a_1 \otimes \cdots \otimes a_n \wedge \bigwedge_{i=1}^{k<n} \neg a_i \to \mathsf{O}a_{k+1} \qquad (\mathsf{O}\text{-}\textbf{detachment})$$

A possible intuition behind this schema is that it can be used to determine which are the obligations that can be complied with. For example, since $\neg a_1$ holds, then we know that it is no longer possible to comply with the obligation of a_1. In a similar way, we could ask what are the parts of norms which are effective in a particular situation. In this case, instead of detaching an obligation we could detach an $\otimes$-chain. Accordingly, we formulate the following axiom:

$$a_1 \otimes \cdots \otimes a_n \wedge \neg a_1 \to a_2 \otimes \cdots \otimes a_n \qquad (\otimes\text{-}\textbf{detachment})$$

where $a_2 \otimes \cdots \otimes a_n$ does non contain a_1 or formulae equivalent to it.

Notice that, contrary to what we did for (1), there is no need to generalise $\otimes$-**detachment** to a version where we consider the negation of the first k elements of the $\otimes$-chain since

$$a_1 \otimes \cdots \otimes a_n \wedge \bigwedge_{i=1}^{k<n} \neg a_i \to a_{k+1} \otimes \cdots \otimes a_n \qquad (2)$$

is derivable from k applications of $\otimes$-**detachment**; hence, there is no need to take (2) as an axiom. Furthermore, in case Axiom $\otimes$-**detachment** holds, it is possible to use (1) to detach O from an $\otimes$-chain instead of O-**detachment** which would then be derivable from $\otimes$-**detachment** and (1).

The attentive reader will not fail to observe that the above detachment axioms do not explicitly mention that the negations of the first k elements of an $\otimes$-chain are violations. The next few axioms address this aspect:

$$a_1 \otimes \cdots \otimes a_n \wedge \bigwedge_{i=1}^{k<n} (\mathsf{O}a_i \wedge \neg a_i) \rightarrow \mathsf{O}a_{k+1} \qquad (\textbf{O-violation-detachment})$$

$$a_1 \otimes \cdots \otimes a_n \wedge \mathsf{O}a_1 \wedge \neg a_1 \rightarrow a_2 \otimes \cdots \otimes a_n \qquad (\boldsymbol{\otimes}\textbf{-violation-detachment})$$

Axioms **O-violation-detachment** and **⊗-violation-detachment** are the immediate counterpart of Axioms **O-detachment** and **⊗-detachment** just including the violation condition in the their antecedent (and we can repeat the argument about the possible axiom combination for their counterparts).

The question is now what are the differences between the cases with or without the explicit violations. Suppose, we have the $\otimes$-chains

$$a \otimes b \qquad\qquad \neg a \otimes c$$

Applying $\otimes$-O and **D**-O results in a contradiction. Suppose that a normative system is equipped with some mechanisms (as it is the case of real life normative systems) to resolve conflicts like this (maybe, using some form of preferences over norms).[3] Also, for the sake of the example, the resolution prefers the first $\otimes$-chain to the second one, and that the first norm has been complied with, that is a holds. Then, we can ask what the obligations in force are.

On the one hand, one can argue that the norm prescribing the second $\otimes$-chain is still effective and thus it is able to generate obligations, but since the first option ($\neg$) would produce a violation, then we can settled for the second option, and we can hence derive $\mathsf{O}c$ from it. If one subscribes to this interpretation, then Axioms **O-detachment** and, eventually, **⊗-detachment** are to be assumed. On the other hand, it is possible to argue that when a norm overrides other norms, then the norms that are overridden are no longer effective. Accordingly, in the case under analysis, a is not a violation of the second $\otimes$-chain, and then there is no ground to proceed with the derivation of $\mathsf{O}c$. But, if $\neg a$ holds instead of a, then we have a violation of the first $\otimes$-chain: we can apply $\otimes$-O to conclude $\mathsf{O}a$, and then **O-violation-detachment** to obtain $\mathsf{O}b$. Hence, the axioms suitable for modelling this intuition are **O-violation-detachment** and, eventually, **⊗-violation-detachment** in case one wants to derive which sub-chains are effective after violations.

Notice that the logic of $\otimes$ was devised to grasp the ideas of violation and compensation: for this reason, we do *not* commit to any reading in which, given $a \otimes b$, the fact $\neg a$ prevents the derivation of $\mathsf{O}a$. If this were the case, we would not have any violation at all. On the contrary, $\mathsf{O}b$ is precisely meant to compensate for the effects of the non legal situation described by $\mathsf{O}a \wedge \neg a$. To further illustrate the idea behind compensatory obligations, consider a situation where $\neg a$ and b hold. Suppose, that you have the norm $a \otimes b$. Here, we can derive the obligations $\mathsf{O}a$ and $\mathsf{O}b$, the first of which is violated, and such a

[3] It is beyond the scope of the present paper to discuss mechanisms to resolve conflicts, the focus of the paper is to propose which combinations of formulae result in conflicts, the reader interested in some solutions using the $\otimes\oplus$-logic can consult [8].

violation triggers the second obligation, i.e., $\mathsf{O}b$, whose fulfilment compensates the violation. Accordingly, the situation, while not ideal, can be still considered compliant with the norm. Suppose that instead of $a \otimes b$ we have two norms $\otimes^1 a$, and $\otimes^1 b$. Similarly, we derive the obligations $\mathsf{O}a$ and $\mathsf{O}b$. However, $\mathsf{O}b$ does not depend on having the violation of $\mathsf{O}a$, nor does it compensate for that violation. Thus, in the last case, $\mathsf{O}a$ is an obligation that cannot be compensated for, and $\mathsf{O}b$ is in force even when we comply with $\mathsf{O}a$.

3.4 Axioms for $\otimes$, $\oplus$, O, P

We now turn our attention to the study of the relationships between $\oplus$-chains and permissions. The basic Axiom $\oplus$-P states that the first element of a permissive chain is a permission.

$$a_1 \oplus \cdots \oplus a_n \to \mathsf{P}a_1 \tag{$\oplus$-P}$$

As we have seen, the intuitive reading of $a \oplus b \oplus c$ is that a should be permitted, but if it is not, then b should be permitted and, if even b is not permitted, then, finally, c should be permitted. Consequently, we formulate the following axioms for detaching a permission from a permissive chain, and for detaching a permissive sub-chain.

$$a_1 \oplus \cdots \oplus a_n \wedge \bigwedge_{i=1}^{k<n} \neg\mathsf{P}a_i \to \mathsf{P}a_{k+1} \tag{P-detachment}$$

$$a_1 \oplus \cdots \oplus a_n \wedge \neg\mathsf{P}a_1 \to a_2 \oplus \cdots \oplus a_n \tag{$\oplus$-detachment}$$

The considerations we made about the choice of axioms for $\otimes$ and O apply for the axioms relating P and $\oplus$ as well.

If we assume the obligation-permission and prohibition-permission consistency principles, i.e., Axioms O-P and $\mathsf{O}\neg\mathsf{P}$, then the axioms in the previous section and the axioms above suffice to describe the relationships among the various deontic operators. In absence of such axioms, several variations of the axioms are possible to maintain consistency between obligations and permissions.

$$a_1 \otimes \cdots \otimes a_n \wedge \neg\mathsf{P}\neg a_1 \to \mathsf{O}a_1 \tag{3}$$

$$a_1 \oplus \cdots \oplus a_n \wedge \neg\mathsf{O}\neg a_1 \to \mathsf{P}a_1. \tag{4}$$

In the situation where a norm holds while the permission of contrary of the first element (of the chain) does not, (3) allows us to determine that the first element is mandatory. Symmetrically, (4) derives the first element of a permissive chain as a permission whereas its contrary is not mandatory. Similar combinations can be used for the detachment axioms we have proposed. For instance, we can integrate the obligation-permission consistency in Axiom **O-violation-detachment** to obtain

$$a_1 \otimes \cdots \otimes a_n \wedge \bigwedge_{i=1}^{k<n} \neg a_i \wedge \neg\mathsf{P}\neg a_{k+1} \to \mathsf{O}a_{k+1} \tag{5}$$

or we integrate the prohibition-permission in (4) resulting in

$$a_1 \oplus \cdots \oplus a_n \wedge \mathsf{O}\neg a_1 \to a_2 \oplus \cdots \oplus a_n. \tag{6}$$

Notice that (3)–(6) (and similar extensions of the various detachment axioms) are derived when Axioms O-P and O¬P as well as the corresponding detachment axioms hold.

3.5 Logics

In this paper, we shall prove completeness results for three groups of systems, as outlined in the table below.

Basic Systems	
$\mathsf{E}^{\otimes}$	CPC + O-RE + ⊗-RE + **⊗-contraction**
$\mathsf{E}^{\oplus}$	CPC + P-RE + ⊕-RE + **⊕-contraction**
$\mathsf{E}^{\otimes\oplus}$	$\mathsf{E}^{\otimes}$ + $\mathsf{E}^{\oplus}$
Basic Deontic Systems	
$\mathsf{D}^{\otimes}$	$\mathsf{E}^{\otimes}$ + OP-**duality** + O-P + **$\overline{\mathbf{P}}$**-O + **⊗-shortening**
$\mathsf{D}^{\otimes\oplus}$	$\mathsf{E}^{\otimes\oplus}$ + O-P + **$\overline{\mathbf{P}}$**-O + **D**-O + O¬P + **⊗-shortening** + **⊕-shortening**
$\mathsf{D}^{\mathsf{O}\otimes}$	$\mathsf{D}^{\otimes}$ + ⊗-O
Basic Full Deontic System	
$\mathsf{D}^{\mathsf{OP}\otimes\oplus}$	$\mathsf{D}^{\otimes\oplus}$ + ⊗-O

Besides these systems, in Section 5 we shall also analyse systems extending $\mathsf{D}^{\mathsf{OP}\otimes\oplus}$ with combinations of detachments axioms (including ⊕-P).

4 Sequence Semantics

Sequence semantics is an extension of neighbourhood semantics. The extension is twofold: (1) we introduce a second neighbourhood-like function, and (2) the new function generates a set of sequences of sets of possible worlds instead of set of sets of possible worlds. This extension allows us to provide a clean semantic representation of ⊙-chains.

Before introducing the semantics, we provide some technical definitions for the operation of *s-zipping*, i.e., the removal of repetitions or redundancies occurring in sequences of sets of worlds. This operation is required to capture the intuition described for the ⊙-shortening axioms.

Definition 4.1 Let $X = \langle X_1, \ldots, X_n \rangle$ be such that $X_i \in 2^W$ $(1 \leq i \leq n)$. Y is *s-zipped from* X iff Y is obtained from X by applying the following operation: for $1 \leq k \leq n$, if $X_j = X_k$ and $j < k$, delete X_k from the sequence.

Definition 4.2 A set S of sequences of sets of possible worlds is closed under s-zipping iff if $X \in S$, then (i) for all Y such that X is s-zipped from Y, $Y \in S$; and (ii) for all Z such that Z is s-zipped from X, $Z \in S$.

Closure under s-zipping essentially determines classes of equivalences for ⊙-chain based on Axioms ⊗-**shortening** and ⊕-**shortening**.

The next three definitions provide the basic scaffolding for sequence semantics: frame, valuation, and model.

Definition 4.3 A *sequence frame* is a structure $\mathcal{F} = \langle W, \mathcal{C}, \mathcal{N} \rangle$, where

- W is a non empty set of possible worlds,
- $\mathcal{C}$ is a function with signature $W \to 2^{(2^W)^n}$ such that for every world w, every $X \in \mathcal{C}_w$ is closed under s-zipping.
- $\mathcal{N}$ is a function with signature $W \to 2^{2^W}$.

Definition 4.4 A *sequence model* is a structure $\mathcal{M} = \langle \mathcal{F}, V \rangle$, where
- $\mathcal{F}$ is a sequence frame, and
- V is a valuation function, $V\colon \mathit{Prop} \to 2^W$.

Definition 4.5 The valuation function for a sequence model is a follows:
- usual for atoms and boolean conditions,
- $w \models \odot_{i=1}^{n} a_i$ iff $\langle \|a_1\|_V, \dots, \|a_n\|_V \rangle \in \mathcal{C}_w$,
- $w \models \Box a$ iff $\|a\|_V \in \mathcal{N}_w$.

Sequence models are meant to be used for the combination of a deontic operator (in this paper $\Box$ ranges over O and P) and the corresponding $\odot$-chain operator ($\otimes$ and $\oplus$, respectively). We are going to use sequence models for the logics where we consider only $\otimes$ and O, and P is defined as the dual of O.

The next three definitions extend sequences semantics to the case of two sets of independent combinations of $\odot$ and the corresponding unary deontic operator.

Definition 4.6 A *bi-sequence frame* is a structure $\mathcal{F} = \langle W, \mathcal{C}^{\mathsf{O}}, \mathcal{C}^{\mathsf{P}}, \mathcal{N}^{\mathsf{O}}, \mathcal{N}^{\mathsf{P}} \rangle$, where
- W is a non empty set of possible worlds;
- $\mathcal{C}^{\mathsf{O}}$ and $\mathcal{C}^{\mathsf{P}}$ are two functions with signature $W \to 2^{(2^W)^n}$, such that for every world $w \in W$, for every $X \in \mathcal{C}^{\mathsf{O}}_w$ and $Y \in \mathcal{C}^{\mathsf{P}}_w$, X and Y are closed under s-zipping;
- $\mathcal{N}^{\mathsf{O}}$ and $\mathcal{N}^{\mathsf{P}}$ are two functions with signature $W \to 2^{2^W}$.

Definition 4.7 A *bi-sequence model* is a structure $\mathcal{M} = \langle \mathcal{F}, V \rangle$, where
- $\mathcal{F}$ is a bi-sequence frame, and
- V is a valuation function, $V\colon \mathit{Prop} \to 2^W$.

Definition 4.8 The valuation function for a bi-sequence model is as follows:
- usual for atoms and boolean conditions,
- $w \models a_1 \otimes \cdots \otimes a_n$ iff $\langle \|a_1\|_V, \dots, \|a_n\|_V \rangle \in \mathcal{C}^{\mathsf{O}}_w$,
- $w \models a_1 \oplus \cdots \oplus a_n$ iff $\langle \|a_1\|_V, \dots, \|a_n\|_V \rangle \in \mathcal{C}^{\mathsf{P}}_w$,
- $w \models \mathsf{O}a$ iff $\|a\|_V \in \mathcal{N}^{\mathsf{O}}_w$,
- $w \models \mathsf{P}a$ iff $\|a\|_V \in \mathcal{N}^{\mathsf{P}}_w$.

5 Soundness and Completeness

In this section we study the soundness and completeness of the logics defined in Section 3.5. Completeness is based on adaptation of the standard Lindenbaum's construction for modal (deontic) neighbourhood semantics.

Definition 5.1 [$\mathcal{L}$-maximality] A set w is $\mathcal{L}$-maximal iff for any formula a of $\mathcal{L}$, either $a \in w$, or $\neg a \in w$.

Lemma 5.2 (Lindenbaum's Lemma) *Any consistent set w of formulae in the language $\mathcal{L}$ can be extended to a consistent $\mathcal{L}$-maximal set w^+.*

Proof. Let $a_1, a_2, \ldots$ be an enumeration of all the possible formulae in $\mathcal{L}$.

- $w_0 := w$;
- $w_{n+1} := w_n \cup \{a_n\}$ if its closure under the axioms and rules of S is consistent, $w \cup \{\neg a_n\}$ otherwise;
- $w^+ := \bigcup_{n \geq 0} w_n$. □

5.1 Basic classical systems: $\mathsf{E}^{\otimes}$, $\mathsf{E}^{\oplus}$

The construction of a sequence canonical model is as follows.

Definition 5.3 [$\mathsf{E}^{\otimes}$-Canonical Models] A sequence canonical model $\mathcal{M} = \langle W, \mathcal{C}, \mathcal{N}, V\rangle$ for a system S in the language $\mathcal{L}^{\otimes}$ (where $\mathsf{S} \supseteq \mathsf{E}^{\otimes}$) is defined as follows:

1. W is the set of all the $\mathcal{L}^{\otimes}$-maximal consistent sets.
2. For any propositional letter $p \in \mathit{Prop}$, $\|p\|_V := |p|_{\mathsf{S}}$, where $|p|_{\mathsf{S}} := \{w \in W \mid p \in w\}$.
3. Let $\mathcal{C} := \bigcup_{w \in W} \mathcal{C}_w$, where, for each $w \in W$, $\mathcal{C}_w := \{\langle \|a_1\|_V, \ldots, \|a_n\|_V\rangle \mid \bigotimes_{i=1}^{n} a_i \in w\}$, where each a_i is a meta-variable for a Boolean formula.
4. Let $\mathcal{N} := \bigcup_{w \in W} \mathcal{N}_w$ where for each world w, $\mathcal{N}_w := \{\|a_i\|_V \mid \mathsf{O}a_i \in w\}$.

Any canonical model for a logic extending $\mathsf{E}^{\oplus}$, on the other hand, would be exactly the same, but for condition (3), to be changed as to read: Let $\mathcal{C} := \bigcup_{w \in W} \mathcal{C}_w$, where, for each $w \in W$, $\mathcal{C}_w := \{\langle \|a_1\|_V, \ldots, \|a_n\|_V\rangle \mid \bigoplus_{i=1}^{n} a_i \in w\}$, where each a_i is a meta-variable for a Boolean formula.

Lemma 5.4 (Truth Lemma for Canonical Sequence Models) *If $\mathcal{M} = \langle W, \mathcal{C}, \mathcal{N}, V\rangle$ is canonical for S, where $\mathsf{S} \supseteq \mathsf{E}^{\otimes}$ or $\mathsf{S} \supseteq \mathsf{E}^{\oplus}$, then for any $w \in W$ and for any formula A, $A \in w$ iff $w \models A$.*

Proof. Given the construction of the canonical model, this proof is easy and can be given by induction on the length of an expression A. We consider only some relevant cases.

Assume A has the form $a_1 \otimes \cdots \otimes a_n$. If $A \in w$, by definition of the canonical model, then there is a sequence $\langle \|a_1\|_V, \ldots, \|a_n\|_V\rangle \in \mathcal{C}_w$. Following from the semantic clauses given to evaluate $\otimes$-formulae, it holds that $w \models a_1 \otimes \cdots \otimes a_n$. For the opposite direction, assume that $w \models a_1 \otimes \cdots \otimes a_n$. By definition, there is $\mathcal{C}_w$ which contains an ordered n-tuple $\langle \|a_1\|_V, \ldots, \|a_n\|_V\rangle$ and by construction $a_1 \otimes \cdots \otimes a_n \in w$. Clearly the same argument holds in the case of operator $\oplus$.

If, on the other hand, A has the form $\mathsf{O}b$ and $\mathsf{O}b \in w$, then $\|b\|_V \in \mathcal{N}_w$ by construction, and by definition $w \models \mathsf{O}b$. Conversely, if $w \models \mathsf{O}b$, then $\|b\|_V \in \mathcal{N}_w$ and, by construction of $\mathcal{N}$, $\mathsf{O}b \in w$. □

It is easy to verify that the canonical model exists, it is not empty, and it is a sequence semantics model. Consider any formula $A \notin \mathsf{S}$ such that $\mathsf{S} \supseteq \mathsf{E}^{\otimes}, \mathsf{S} \supseteq \mathsf{E}^{\oplus}$; $\{\neg A\}$ is consistent and it can be extended to a maximal set w such that for some canonical model, $w \in W$. By Lemma 5.4, $w \not\models A$. That $\mathcal{C}_w$ is closed under zipping follows immediately from the Lindembaum construction.

Corollary 5.5 (Completeness of $\mathsf{E}^{\otimes}$ and $\mathsf{E}^{\oplus}$) *The systems $\mathsf{E}^{\otimes}$ and $\mathsf{E}^{\oplus}$ are sound and complete with respect to the class of sequence frames.*

Definition 5.6 [Bi-sequence Canonical Models] A bi-sequence canonical model $\mathcal{M} = \langle W, \mathcal{C}^{\mathsf{O}}, \mathcal{C}^{\mathsf{P}}, \mathcal{N}^{\mathsf{O}}, \mathcal{N}^{\mathsf{P}}, V\rangle$ for a system S in $\mathcal{L}^{\otimes\oplus}$ (where $\mathsf{S} \supseteq \mathsf{E}^{\otimes\oplus}$) is defined as follows:

1. W is the set of all the $\mathcal{L}^{\otimes\oplus}$-maximal consistent sets.
2. For any propositional letter $p \in Prop$, $\|p\|_V := |p|_{\mathsf{S}}$, where $|p|_{\mathsf{S}} := \{w \in W \mid p \in w\}$.
3. Let $\mathcal{C}^{\mathsf{O}} := \bigcup_{w\in W} \mathcal{C}^{\mathsf{O}}_w$, where for each $w \in W$, $\mathcal{C}^{\mathsf{O}}_w := \{\langle \|a_1\|_V, \ldots, \|a_n\|_V\rangle \mid \bigotimes_{i=1}^{n} a_i \in w\}$, where each a_i is a meta-variable for a Boolean formula.
4. Let $\mathcal{C}^{\mathsf{P}} := \bigcup_{w\in W} \mathcal{C}^{\mathsf{P}}_w$, where for each $w \in W$, $\mathcal{C}^{\mathsf{P}}_w := \{\langle \|a_1\|_V, \ldots, \|a_n\|_V\rangle \mid \bigoplus_{i=1}^{n} a_i\}$, where each a_i is a meta-variable for a Boolean formula.
5. Let $\mathcal{N}^{\mathsf{O}} := \bigcup_{w\in W} \mathcal{N}^{\mathsf{O}}_w$ where for each world w, $\mathcal{N}^{\mathsf{O}}_w := \{\|a_i\|_V \mid \mathsf{O}a_i \in w\}$.
6. Let $\mathcal{N}^{\mathsf{P}} := \bigcup_{w\in W} \mathcal{N}^{\mathsf{P}}_w$ where for each world w, $\mathcal{N}^{\mathsf{P}}_w := \{\|a_i\|_V \mid \mathsf{P}a_i \in w\}$.

Lemma 5.7 (Truth Lemma for Canonical Bi-sequence Models) *If $\mathcal{M} = \langle W, \mathcal{C}^{\mathsf{O}}, \mathcal{C}^{\mathsf{P}}, \mathcal{N}^{\mathsf{O}}, \mathcal{N}^{\mathsf{P}}, V\rangle$ is canonical for S, where $\mathsf{S} \supseteq \mathsf{E}^{\otimes\oplus}$, then for any $w \in W$ and for any formula A, $A \in w$ iff $w \models A$.*

Since the modal operators do not interact with each other, we can state:

Corollary 5.8 (Completeness of $\mathsf{E}^{\otimes\oplus}$) *The system $\mathsf{E}^{\otimes\oplus}$ is sound and complete with respect to the class of bi-sequence frames.*

5.2 Deontic Systems

Theorem 5.9 (Completeness of $\mathsf{D}^{\otimes}$) *The frame of a canonical model for $\mathsf{D}^{\otimes}$, as defined in Definition 5.3, has the following properties. For any $w \in W$,*

1. *$X \in \mathcal{N}_w$ if and only if $-X \notin \mathcal{N}_w$. (see* OP**-duality**, O-P *and* **D**-O)
2. *$\emptyset \notin \mathcal{N}_w$ (see* $\overline{\mathbf{P}}$-O)
3. *$\langle X_1, \ldots, X_n\rangle \in \mathcal{C}_w$ for $n \geq 2$ then $\langle X_1, \ldots, X_{n-1}\rangle \in \mathcal{C}_w$ (see* $\otimes$**-shortening**)

Proof.

1. $X \in \mathcal{N}_w$ iff $X = \|a\|_V$ for some $\mathsf{O}a \in w$, i.e., iff $\neg\mathsf{O}\neg a \in w$, $\mathsf{O}\neg a \notin w$, $-\|a\|_V \notin \mathcal{N}_w$.
2. Assume by reductio that $\emptyset \in \mathcal{N}_w$. Then $w \models \mathsf{O}\bot$, $\mathsf{O}\bot \in w$, reaching a contradiction.
3. Assume $\langle \|a\|_1, \ldots, \|a_n\|\rangle \in \mathcal{C}_w$. By construction it means that $\bigotimes_{i=1}^{n} a_i \in w$ and by $\otimes$**-shortening**, $\bigotimes_{i=1}^{n-1} a_i \in w$, thus $\langle \|a\|_1, \ldots, \|a_{n-1}\|\rangle \in \mathcal{C}_w$. □

Theorem 5.10 (Completeness of $\mathsf{D}^{\mathsf{O}\otimes}$) *The frame of a canonical model for $\mathsf{D}^{\mathsf{O}\otimes}$ (Definition 5.3) has the properties expressed in Theorem 5.9 and the following: For any world w, if $\langle X_1, \ldots, X_n\rangle \in \mathcal{C}_w$ then $X_1 \in \mathcal{N}_w$ (see* $\otimes$-O)

Proof. If $\langle X_1, \ldots, X_n\rangle \in \mathcal{C}_w$, then there are n formulae such that for $1 \leq i \leq n$, $X_i = \|a_i\|_V$ and $a_1 \otimes \cdots \otimes a_n \in w$. By Axiom $\otimes$-O, $\mathsf{O}a_1 \in w$ and hence $\|a_1\|_V \in \mathcal{N}_w$. □

Theorem 5.11 (Completeness of $\mathsf{D}^{\otimes\oplus}$) *The frame of a canonical model for $\mathsf{D}^{\otimes\oplus}$, as defined in Definition 5.6, has the following properties. For any $w \in W$,*

1. $\mathcal{N}_w^{\mathsf{P}} \supseteq \mathcal{N}_w^{\mathsf{O}}$ *(see* O-P*)*
2. $X \in \mathcal{N}_w^{\mathsf{O}}$ *implies* $-X \notin \mathcal{N}_w^{\mathsf{O}}$ *(see* **D**-O*)*
3. $\emptyset \notin \mathcal{N}_w^{\mathsf{O}}$ *(see* $\overline{\mathbf{P}}$-O*)*
4. $X \in \mathcal{N}_w^{\mathsf{O}}$ *implies* $-X \notin \mathcal{N}_w^{\mathsf{P}}$ *(see* O¬P*)*
5. $\langle X_1, \ldots, X_n\rangle \in \mathcal{C}_w^{\mathsf{O}}$ *for* $n \geq 2$ *then* $\langle X_1, \ldots, X_{n-1}\rangle \in \mathcal{C}_w^{\mathsf{O}}$ *(see* $\otimes$-**shortening***)*
6. $\langle X_1, \ldots, X_n\rangle \in \mathcal{C}_w^{\mathsf{P}}$ *for* $n \geq 2$ *then* $\langle X_1, \ldots, X_{n-1}\rangle \in \mathcal{C}_w^{\mathsf{P}}$ *(see* $\oplus$-**shortening***)*

Proof. Recall that $\mathsf{D}^{\otimes\oplus} = \mathsf{E}^{\otimes\oplus} +$ O-P $+ \overline{\mathbf{P}}$-O $+$ **D**-O $+$ O¬P $+ \otimes$-**shortening** $+$ $\oplus$-**shortening**; remember that the operator P is not defined as a dual of O.

1. Assume $\|a\|_V \in \mathcal{N}_w^{\mathsf{O}}$, then $\mathsf{O}a \in w$ and, by O-P, $\mathsf{P}a \in w$. Hence $\|a\|_V \in \mathcal{N}_w^{\mathsf{P}}$.
2. Assume $X \in \mathcal{N}_w^{\mathsf{O}}$ for some $w \in W$, then, by construction, there is some formula $\mathsf{O}a \in w$ and $X = \|a\|_V$. By **D**-O and **MP**, $\neg\mathsf{O}\neg a \in w$, i.e., $\mathsf{O}\neg a \notin w$, $\not\models_w^V \mathsf{O}\neg a$, $\|\neg a\|_V \notin \mathcal{N}_w^{\mathsf{O}}$, hence $-\|a\|_V \notin \mathcal{N}_w^{\mathsf{O}}$.
3. See the proof of Theorem 5.9.
4. Assume $X \in \mathcal{N}_w^{\mathsf{O}}$; by Definition 5.6 $X = \|a\|_V$ for some a such that $\mathsf{O}a \in w$. Then, by O¬P, $\neg\mathsf{P}\neg a \in w$, $\mathsf{P}\neg a \notin w$, hence $\|a\|_V \notin \mathcal{N}_w^{\mathsf{P}}$.
5. See the proof of Theorem 5.9.
6. See the proof of Theorem 5.9. □

5.3 Extended Deontic Systems

In what follows we shall prove completeness results for various systems by adding 6 detachment schemata that combine the modal operators introduced.

Theorem 5.12 (Completeness of $\mathsf{D}^{\mathsf{OP}\otimes\oplus}$) *The canonical frame (see Definition 5.6) for the logic $\mathsf{D}^{\mathsf{OP}\otimes\oplus}$ has the properties stated in Theorem 5.11 plus: For any world w if $\langle X_1, \ldots, X_n\rangle \in \mathcal{C}_w^{\mathsf{O}}$ then $X_1 \in \mathcal{N}_w^{\mathsf{O}}$ (see $\otimes$-O).*

Proof. See the proof of Theorem 5.10. □

Theorem 5.13 *Let* S *be a system such that* $\mathsf{S} \supseteq \mathsf{D}^{\mathsf{OP}\otimes\oplus}$. *If* S *contains any of the axioms listed below, the canonical frame enjoys the corresponding property: For any world w*

1. O-**detachment***:*
   If $\langle X_1, \ldots, X_n\rangle \in \mathcal{C}_w^{\mathsf{O}}$ *and* $w \notin X_i$ *for* $1 \leq i \leq k$ *and* $k < n$, *then* $X_{k+1} \in \mathcal{N}_w^{\mathsf{O}}$.
2. $\otimes$-**detachment***:*
   If $\langle X_1, \ldots, X_n\rangle \in \mathcal{C}_w^{\mathsf{O}}$ *and* $w \notin X_1$, *then* $\langle X_2, \ldots, X_n\rangle \in \mathcal{C}_w^{\mathsf{O}}$.
3. O-**violation-detachment***:*
   If $\langle X_1, \ldots, X_n\rangle \in \mathcal{C}_w^{\mathsf{O}}$ *and, for* $1 \leq i \leq k$ *and* $k < n$, $w \notin X_i$ *and* $X_i \in \mathcal{N}_w^{\mathsf{O}}$, *then* $X_{k+1} \in \mathcal{N}_w^{\mathsf{O}}$.
4. $\otimes$-**violation-detachment***:*
   If $\langle X_1, \ldots, X_n\rangle \in \mathcal{C}_w^{\mathsf{O}}$ *and* $X_1 \in \mathcal{N}_w^{\mathsf{O}}$ *and* $w \notin X_1$, *then* $\langle X_2, \ldots, X_n\rangle \in \mathcal{C}_w^{\mathsf{O}}$.
5. $\oplus$-P*:*
   If $\langle X_1, \ldots, X_n\rangle \in \mathcal{C}_w^{\mathsf{P}}$ *then* $X_1 \in \mathcal{N}_w^{\mathsf{P}}$.
6. P-**detachment***:*

If $\langle X_1, \ldots, X_n \rangle \in \mathcal{C}_w^{\mathsf{P}}$ *and* $X_i \notin \mathcal{N}_w^{\mathsf{P}}$ *for* $1 \leq i \leq k < n$, *then* $X_{k+1} \in \mathcal{N}_w^{\mathsf{P}}$.

7. $\oplus$**-detachment**: *If* $\langle X_1, \ldots, X_n \rangle \in \mathcal{C}_w^{\mathsf{P}}$ *and* $X_i \notin \mathcal{N}_w^{\mathsf{P}}$ *for* $1 \leq i \leq k$ *and* $k < n$, *then* $\langle X_{k+1}, \ldots, X_n \rangle \in \mathcal{C}_w^{\mathsf{P}}$.

Proof. Again, the proof is very straightforward and it follows closely the syntactical structure of the schemata. Notice that the fact $\langle X_1, \ldots, X_n \rangle \in \mathcal{C}_w^{\mathsf{O}}$ always implies that for $1 \leq i \leq n$ formulae $X_i = \|a_i\|_V$.

1. If $\langle X_1, \ldots, X_n \rangle \in \mathcal{C}_w^{\mathsf{O}}$ and $w \notin X_i$ for $1 \leq i \leq k$ with $k < n$, then for $1 \leq i \leq n$ formulae it holds that $X_i = \|a_i\|_V$, $a_1 \otimes \cdots \otimes a_n \in w$, $a_i \notin w$ for $1 \leq i \leq k$, hence $\bigwedge_{i=1}^{k} \neg a_i \in w$. Thus, by **O-detachment**, $\mathsf{O}a_{k+1} \in w$ and $\|a_{k+1}\|_V \in \mathcal{N}_w^{\mathsf{O}}$.
2. If $\langle \|a_1\|_V, \ldots, \|a_n\|_V \rangle \in \mathcal{C}_w^{\mathsf{O}}$ and $w \notin \|a_1\|_V$, then $a_1 \otimes \cdots \otimes a_n \in w$ and $\neg a_1 \in w$, thus, by $\otimes$**-detachment**, $a_2 \otimes \cdots \otimes a_n \in w$ and $\langle \|a_2\|_V, \ldots, \|a_n\|_V \rangle \in \mathcal{C}_w^{\mathsf{O}}$.
3. Assume $\langle \|a_1\|_V, \ldots, \|a_n\|_V \rangle \in \mathcal{C}_w^{\mathsf{O}}$ and, for $1 \leq i \leq k$ with $k < n$, $w \notin \|a_i\|_V$ and $\|a_i\|_V \in \mathcal{N}_w^{\mathsf{O}}$. Then $a_1 \otimes \cdots \otimes a_n \in w$, $\bigwedge_{i=1}^{k} \neg a_i \in w$, and $\bigwedge_{i=1}^{k} \mathsf{O}a_i \in w$. By classical propositional logic $\bigwedge_{i=1}^{k} (\mathsf{O}a_i \wedge \neg a_i) \in w$ and, by **O-violation-detachment**, $\mathsf{O}a_{k+1} \in w$ and $\|a_{k+1}\|_V \in \mathcal{N}_w^{\mathsf{O}}$.
4. Assume $\langle \|a_1\|_V, \ldots, \|a_n\|_V \rangle \in \mathcal{C}_w^{\mathsf{O}}$, $w \notin \|a_1\|_V$, and $\|a_1\|_V \in \mathcal{N}_w^{\mathsf{O}}$. Then $a_1 \otimes \cdots \otimes a_n \in w$, $\neg a_1 \in w$ and $\mathsf{O}a_1 \in w$ and, by $\otimes$**-violation-detachment**, $a_2 \otimes \cdots \otimes a_n \in w$ and $\langle \|a_2\|_V, \ldots, \|a_n\|_V \rangle \in \mathcal{C}_w^{\mathsf{O}}$.
5. See Theorem 5.10.
6. Assume $\langle \|a_1\|_V, \ldots, \|a_n\|_V \rangle \in \mathcal{C}_w^{\mathsf{P}}$ and, for $1 \leq i \leq k$ with $k < n$, $\|a_i\|_V \notin \mathcal{N}_w^{\mathsf{P}}$. Then $a_1 \oplus \cdots \oplus a_n \in w$ and $\bigwedge_{i=1}^{k} \neg \mathsf{P}a_i \in w$ and, by **P-detachment**, $\mathsf{P}a_{k+1} \in w$, implying that $\|a_{k+1}\|_V \in \mathcal{N}_w^{\mathsf{P}}$.
7. Assume $\langle \|a_1\|_V, \ldots, \|a_n\|_V \rangle \in \mathcal{C}_w^{\mathsf{P}}$ and, for $1 \leq i \leq k$ with $k < n$, $\|a_i\|_V \notin \mathcal{N}_w^{\mathsf{P}}$. Then $a_1 \oplus \cdots \oplus a_n \in w$ and $\bigwedge_{i=1}^{k} \neg \mathsf{P}a_i \in w$ and, by $\oplus$**-detachment**, $a_{k+1} \oplus \cdots \oplus a_n \in w$ and hence $\langle \|a_{k+1}\|_V, \ldots, \|a_n\|_V \rangle \in \mathcal{C}_w^{\mathsf{P}}$. □

6 Conclusions and Related Work

The deontic logic literature on CTD reasoning is vast. However, two fundamental mainstreams have emerged as particularly interesting.

A first line of inquiry is mainly semantic-based. Moving from well-known studies on dyadic obligations, CTD reasoning is interpreted in settings with ideality or preference orderings on possible worlds or states [15]. The value of this approach is that the semantic structures involved are rather flexible: different deontic logics can thus be obtained. This semantic approach has been fruitfully renewed in the '90s, for instance by [19, 21], and most recently by works such as [14, 2], which have confirmed the vitality of this line of inquiry. However, most of these approaches are based on 'standard' possible world semantics with the risk of being affected by the paradox advanced in [7].

While the original systems for $\otimes$ were mainly motivated by modelling CTD reasoning [12, 4], in this paper we have broadened our analysis by extending chains to permissions and by generically dealing with compensations and violations. Indeed, we accept different types of O-detachment, either allowing for the derivation of all obligations from any $\otimes$-chain, or only the subsequent ones in the chains with respect to the ones that have been violated. Our aim was to provide the semantics analysis for several axioms (principles) for the novel operators $\otimes$ and $\oplus$ and how they can be used to generated obligations and permissions. In this paper, we did not study what combinations of axioms are suitable to model different interpretations for different intuitions for the various deontic notions. This study is left to future investigations.

The second mainstream is mostly proof-theoretic. Examples, among others, are various systems springing from Input/Output Logic [18, 17] and the $\otimes$-logic originally proposed in [12]. The logic for $\otimes$ proved to be flexible for several applied domains, such as in business process modelling [13], normative multi-agent systems [6, 9], temporal deontic reasoning [11], and reasoning about different types of defeasible permission [8].

This paper completes the effort in [4] and offers a systematic semantic study of the $\otimes$ and $\oplus$ operators originally introduced in [12] and [8]. We showed that suitable axiomatisations can be characterised in a class of structures extending neighbourhood frames with sequences of sets of worlds. In this perspective, our contribution may offer useful insights for establishing connections between the proof-theoretic and model theoretic approaches to CTD reasoning. Also, we have shown that the semantic structures can easily keep separate structures interpreting norms from those interpreting obligations and permissions, thus mirroring the difference between $\otimes$ and $\oplus$ operators from O and P.

A number of open research issues are left for future work. Among others, we plan to explore decidability questions using, for example, the filtration methods. The fact that neighbourhoods contain sequences of sets of worlds instead of sets is not expected to make the task significantly harder than the one in standard neighbourhood semantics for modal logics.

Second, we intend to study richer deontic logic. For example, we could extend rule RM for O (i.e., $a \to b / \mathsf{O}a \to \mathsf{O}b$), this would allow us to determine that the combination of $a \otimes c$, $b \otimes d$, where $a \to \neg b$, results in a contradiction. In this case, the semantic condition to add is that $\mathcal{N}$ is supplemented. Similarly, we may study what are the $\odot$ counterpart of axioms like **M**, **C** an so on. [5] shows how to provide a generalisation of rule RM to the case of $\otimes$.

Third, [9] investigates how to characterise different degrees and types of goal-like mental attitudes of agents (including obligation) with chain operators. We plan to explore the use of sequence semantics to provide axioms (and corresponding semantic conditions) that correspond to the mechanisms governing the goal-like attitudes and their interactions.

Finally, we expect to enrich the language and to further explore the meaning of the nesting of $\otimes$- and $\oplus$-expressions, thus having formulae like $a \otimes \neg(b \otimes c) \otimes d$. As we have said, the meaning of those formulae is not clear. However, a semantic analysis of them in the sequence semantics can clarify the issue. Indeed, in the current language we can evaluate in any world w formulae like $\neg(a \otimes b)$, which semantically means that there is no sequence $\langle \|a\|_V, \|b\|_V \rangle \in \mathcal{C}^{\mathsf{O}}_w$. Conceptually, this means that there is no norm stating that a is obligatory and that the violation of this primary obligation generates an obligation b. Accordingly, the truth at w of $a \otimes \neg(b \otimes c) \otimes d$ means that there exists a norm stating that a is obligatory, but either b does not compensate a or, otherwise, c does not compensate b, and d compensates what compensates a, whatever it is.

References

[1] Alchourrón, C.E., *Philosophical foundations of deontic logic and the logic of defeasible conditionals*, in: J.-J. Meyer and R. Wieringa, editors, *Deontic Logic in Computer Science*, Wiley, 1993 pp. 43–84.

[2] van Benthem, J., D. Grossi and F. Liu, *Priority structures in deontic logic*, Theoria **80** (2014), pp. 116–152.

[3] Boella, G., G. Pigozzi and L. van der Torre, *A normative framework for norm change*, in: C. Sierra, C. Castelfranchi, K.S. Decker and J.S. Sichman, editors, *AAMAS 2009*, (2009), pp. 169–176.

[4] Calardo, E., G. Governatori and A. Rotolo, *A preference-based semantics for CTD reasoning*, in: F. Cariani, D. Grossi, J. Meheus and X. Parent, editors, *DEON 2014*, LNCS 8554, (2014), pp. 49–64.

[5] Calardo, E., G. Governatori and A. Rotolo, *Sequence semantics for modelling reason-based preferences*, Fundamenta Informaticae (2016).

[6] Dastani, M., G. Governatori, A. Rotolo and L. van der Torre, *Programming cognitive agents in defeasible logic*, in: G. Sutcliffe and A. Voronkov, editors, *LPAR 2005*, LNAI 3835, (2005), pp. 621–636.

[7] Governatori, G., *Thou shalt is not you will*, in: K. Atkinson, editor, *Proceedings of the Fifteenth International Conference on Artificial Intelligence and Law*, (2015), pp. 63–68.

[8] Governatori, G., F. Olivieri, A. Rotolo and S. Scannapieco, *Computing strong and weak permissions in defeasible logic*, Journal of Philosophical Logic **42** (2013), pp. 799–829.

[9] Governatori, G., F. Olivieri, S. Scannapieco, A. Rotolo and M. Cristani, *The rational behind the concept of goal*, Theory and Practice of Logic Programming **16** (2016), pp. 296–324.

[10] Governatori, G. and A. Rotolo, *Changing legal systems: legal abrogations and annulments in defeasible logic*, Logic Journal of IGPL **18** (2010), pp. 157–194.

[11] Governatori, G. and A. Rotolo, *Justice delayed is justice denied: logics for a temporal account of reparations and legal compliance*, in: J. Leite, P. Torroni, T. Ågotnes, G. Boella and L. van der Torre, editors, *CLIMA XII*, LNCS 6814, (2011), pp. 364–382.

[12] Governatori, G. and A. Rotolo, *Logic of violations: a Gentzen system for reasoning with contrary-to-duty obligations*, Australasian Journal of Logic **4** (2006), pp. 193–215.

[13] Governatori, G. and S. Sadiq, *The journey to business process compliance*, in: J. Cardoso and W. van der Aalst, editors, *Handbook of Research on Business Process Modeling*, IGI Global, 2009 pp. 426–454.

[14] Hansen, J., *Conflicting imperatives and dyadic deontic logic*, Journal of Applied Logic **3** (2005), pp. 484–511.

[15] Hansson, B., *An analysis of some deontic logics*, Noûs **3** (1969), pp. 373–398.

[16] Makinson, D., *On a fundamental problem of deontic logic*, in: P. McNamara and H. Prakken, editors, *Norms, Logics, and Information Systems*, IOS Press, 1999.

[17] Makinson, D. and L. van der Torre, *Constraints for input/output logics*, Journal of Philosophical Logic **30** (2001), pp. 155–185.

[18] Makinson, D. and L. van der Torre, *Input/output logics*, Journal of Philosophical Logic **29** (2000), pp. 383–408.

[19] Prakken, H. and M.J. Sergot, *Contrary-to-duty obligations*, Studia Logica **57** (1996), pp. 91–115.

[20] Stenius, E., *Principles of a logic of normative systems*, Acta Philosophica Fennica **16** (1963), pp. 247–260.

[21] van der Torre, L., *Reasoning about obligations: defeasibility in preference-based deontic logic*, PhD thesis, Erasmus University Rotterdam, 1997.

To Do Something Else

Fengkui Ju

School of Philosophy, Beijing Normal University, Beijing, China
fengkui.ju@bnu.edu.cn

Jan van Eijck

Centrum Wiskunde & Informatica, Amsterdam, The Netherlands
Institute for Logic, Language & Computation, University of Amsterdam
jve@cwi.nl

Abstract

This paper presents two deontic logics following an old idea: normative notions can be defined in terms of the consequences of performing actions. The two deontic logics are based on two special propositional dynamic logics; they interpret actions as sets of state sequences and have a process modality. The difference between the two deontic logics is that they contain different formalizations of refraining to do an action. Both of the two deontic logics have a propositional constant for marking the bad states. The normative notions are expressed by use of the process modality and this propositional constant.

Keywords: deontic logic, dynamic logic, process modality, negative action

1 Background

There is an old idea in the field of deontic logic: an action is *prohibited* if doing it would cause a bad thing; it is *permitted* if performing the action is possible without causing a bad thing; it is *obligated* if refraining to do it would cause a bad thing. This idea is intuitive in some sense; the point of it is that the three fundamental normative notions, *prohibition*, *permission* and *obligation*, can be defined in terms of the consequences of doing actions. According to [4], this idea can be traced back to Leibniz.

[1] and [9] independently develop this idea along similar lines. The resulting deontic logic has a modal operator $\Box$, the classical alethic modality whose dual is $\Diamond$. It also has a propositional constant $\mathcal{V}$ which intuitively means that what morality prescribes has been violated. The three normative notions are defined as follows: $\Box(\phi \to \mathcal{V})$ says that the proposition ϕ is prohibited, $\Diamond(\phi \wedge \neg\mathcal{V})$ says that ϕ is permitted and $\Box(\neg\phi \to \mathcal{V})$ says that ϕ is obligated. This logic applies deontic operators to propositions and does not really analyze actions. As mentioned in the literature, e.g., [10], this approach leads to quite a few problems.

Starting from the same idea, [11] proposes a different approach with emphasis on the analysis of actions in terms of their postconditions. In his dynamic logic $[\alpha]\phi$ expresses that no matter how the action α is performed, ϕ will be the case afterwards. The dual of $[\alpha]\phi$ is $\langle\alpha\rangle\phi$, which expresses that there is a way to perform α s.t. ϕ will be the case after α is done. The logic presented by [11] has a propositional constant $\mathcal{V}$ saying, again, that this is a *undesirable* state. By use of $[\alpha]\phi$ and $\mathcal{V}$, the three normative notions can be expressed: $[\alpha]\mathcal{V}$, meaning that α is prohibited, $\langle\alpha\rangle\neg\mathcal{V}$ indicating that α is permitted and $[\overline{\alpha}]\mathcal{V}$ denoting that α is obligated. By $\overline{\alpha}$, [11] intends to express this: to perform $\overline{\alpha}$ is to refrain from doing α. This work applies deontic operators to actions and many problems with previous deontic logics are avoided this way. [11] is a seminal paper that has given rise to a class of *dynamic deontic logics* following this approach.

There are two problems with [11]. The first one concerns the three normative notions. Whether an action α is prohibited/permitted/obligated or not is completely determined by whether the output of performing α is undesirable or not, and has nothing to do with what happens during the performance of α. As pointed out by [15], this is problematic, because it entails that while killing the president is prohibited, killing him and then surrendering to the police may not be, that while smoking in this room is not permitted, smoking in this room and then leaving it may be permitted, that while rescuing the injured and then calling an ambulance is obligated, rescuing the injured may not be. None of this sounds reasonable.

The second problem with [11] lies in how it technically deals with $\overline{\alpha}$. It presents a complicated semantics for actions. In short, it firstly assigns each action a so called s-trace-set; then it links each s-trace-set to a binary relation. In this way each action is interpreted as a binary relation. Essentially, this is like the standard semantics for actions from propositional dynamic logic (PDL). Under the semantics defined by [11], although $\overline{\alpha}$ is not the complement of α, still the behaviour of $\overline{\alpha}$ is not quite in line with the intuition of refraining from α. Firstly, the intersection of the interpretations of $\overline{\alpha}$ and α is not always empty, which would mean that in some states there may be ways to refrain from α while at the same time doing α. Secondly, the intersection of the interpretations of $\overline{\alpha}$ and $\alpha;\beta$ is not always empty, which would mean that in some cases, performing $\alpha;\beta$ is a way to refrain from doing α. This runs counter to our intuition about refraining from doing an action.

Indeed, [11] shows clear awareness of the requirement that $\overline{\alpha}$ and α should be disjoint and that $\overline{\alpha}$ and $\alpha;\beta$ should be disjoint as well. The correspondence between actions and s-trace-sets was designed to achieve this, but the assignment of binary relations to s-trace-sets results in some crucial information loss.

Dynamic logics in the style of PDL interpret actions as binary relations and can not deal with the progressive behaviour of actions. To solve this problem, so-called *process logics* take the intermediate states of doing actions into consideration and view actions as sets of sequences of states. Based on a process logic from [12], [15] proposes a deontic logic which aims to handle *free choice permission* and *lack-of-prohibition permission* in one setting. The sentence "you can use my pen or pencil" involves the former permission and "you can use his pen or pencil" involves the latter permission. The first sentence gives the addressee the permission to use *the* pen, but the second one does not. To see that the latter is the case, imagine a situation where the speaker of the second sentence is just reporting something by this sentence, and he/she knows that the owner of the pen and pencil allows the addressee to use the pen or pencil but does not know exactly which. Unlike [11], [15] does not introduce undesirable states, but uses undesirable transitions instead. The resulting logic allows description of the states during execution of actions and it avoids the first problem with [11]. However, the focus is on permission only, and there is no attempt to deal with refraining to do an action or with obligation. [13] extends the logic in [15] by introducing two dynamic operators: one adds and another removes desirable transitions. The two operators are used to model the dynamics of the so called *policies*, which are on what is and what is not permitted.

Realizing that the formalization of refraining to do an action in [11] is problematic, [2] and [14] present alternative proposals, both based on a relational semantics for actions. The motivation of [2] is that the formalization in [11] can not be easily generalized to encompass iteration and converse of actions. [2] views $\overline{\alpha}$ as a constrained complement of α: $\overline{\alpha}$ is not the complement of α w.r.t. the universal relation, but the complement of α w.r.t. the set consisting of all the transitions resulting from performing actions constructed without use of $\overline{}$. Under this treatment, the intersection of the interpretations of $\overline{\alpha}$ and α is always empty; however, the problem with the intersection of the interpretations of $\overline{\alpha}$ and $\alpha;\beta$ remains: this intersection might not be empty. [14] thinks that the sentence "you are permitted either to eat the dessert or not" has different meaning from "you are permitted either to kiss me or not", as the latter implies that the addressee may kiss the speaker but the former does not. The two sentences turn out equivalent. To remedy this, [14] interprets $\overline{\alpha}$ in a so called *stratified* way. Firstly, for any atomic action a with the interpretation R_a, it defines $R_{\overline{a}}$, the interpretation of $\overline{a}$, in the following way: a transition (w,u) is in $R_{\overline{a}}$ if and only if (w,u) is not in R_a but (w,x) is in R_a for some x; then by four inductive rules taken from [17], it defines the interpretation of $\overline{\alpha}$ for any compound action α. However, this approach suffers from the same problem as [11]: neither the intersection of $\overline{\alpha}$ and α nor the intersection of $\overline{\alpha}$ and $\alpha;\beta$ is always empty.

It is our aim in this paper to propose two deontic logics that follow the general approach of [11] but resolve the problems mentioned above.

2 Two Challenges

Two challenges are crucial in dynamic deontic logics: how to formalize refraining to do an action and how to handle the normative notions. We here state our ideas for these two issues, as a prelude to the two deontic logics to be presented below.

To refrain to do an action is *to do something else*. We think that to do something else meets the principle of symmetry: if doing α is doing something else than β, then doing β is also doing something else than α. We also think it is reasonable to impose the principle of perfect tense: deeds that are done remain done forever. In other words, for any action, if the agent has done it, then he/she will always have done it. Under the two principles, we do not have many choices in analyzing to do something else.

Let's look at an example. Let a and b be two different actions. Fix a start point. When would we say that the agent has done something else than $a; b$? Clearly, if the agent has done a, he/she has done something else than b. By the principle of the perfect tense, if he/she has done $a; b$, he/she has done something else than b. By the symmetry principle of to do something else, if he/she has done b, he/she has done something else than $a; b$. We can not say that if the agent has done a, he/she has done something else than $a; b$. Why? Because if an agent has done $a; b$ she has done a, by the principle of perfect tense. So if she has done a then it cannot be the case that she has done something else than $a; b$. We must therefore conclude that doing b is doing something else than doing $a; b$, but doing a is not doing something else than doing $a; b$.

About the issue of normative notions, we propose a sharpened version of the old idea mentioned in the previous section. There are a class of states, a group of people and an agent who might not belong to this group. The agent doing an action at a state might change this state to a different one. Some states are *bad* and others are *fine* for this group. An action of the agent is *prohibited* at a state *relative to* this group if the state will be bad at some point during any performance of this action. An action is *permitted* at a state if the state will always be fine during some performance of this action. An action is *obligated* at a state if the state will be bad at some point during any performance of *anything else*.

Next, how to formalize these ideas? In process logics such as those of [12] and [3], atomic actions are interpreted as sets of state sequences which might not be binary relations. [7] presents a simple process logic where atomic actions are viewed as binary relations and the action constructors of composition, union and iteration are treated in the usual way. We will follow this to formalize the notion of to do something else. Actually we will work this out in two different ways. As a follow-up to [7], [6] proposes two *process modalities* to describe what happens during execution of actions. One of them is called the $\forall\exists$ process modality. Below, we will use this modality plus a propositional constant to express the three normative notions.

3 A Deontic Logic Based on Process Theory

Let Π_0 be a *finite* set of atomic actions and Φ_0 a countable set of atomic propositions. Let a range over Π_0 and p over Φ_0. The sets Π_{PDL} of actions and Φ_{PDDL} of propositions are defined as follows:

$$\alpha ::= a \mid 0 \mid (\alpha;\alpha) \mid (\alpha \cup \alpha) \mid \alpha^*$$
$$\phi ::= p \mid \top \mid \mathfrak{b} \mid \neg\phi \mid (\phi \wedge \phi) \mid \|\alpha\|\phi$$

Here in "Φ_{PDDL}", "P" is for "process" and "DDL" for "dynamic deontic logic". 0 is the *impossible* action. $\mathfrak{b}$ means that this is a *bad* state. $\mathfrak{f}$, this is a *fine* state, is defined as $\neg\mathfrak{b}$. $\|\alpha\|\phi$ indicates that for any way to perform α, ϕ will be the case at some point in the process. The dual $\langle\!\langle\alpha\rangle\!\rangle\phi$ of $\|\alpha\|\phi$ is defined as $\neg\|\alpha\|\neg\phi$, which says that there is a way to perform α s.t. ϕ will be the case at all the points in the process. $F\alpha$, α is prohibited, is defined as $\|\alpha\|\mathfrak{b}$; it means that no matter how to perform α, the state will be bad at some point in the process. $P\alpha$, α is permitted, is defined as $\langle\!\langle\alpha\rangle\!\rangle\mathfrak{f}$; it means that there is a way to perform α s.t. the state will always be fine in the process. Other standard syntactic abbreviations apply here.

In next section, for any action α in Π_{PDL}, we will specify a β in Π_{PDL} and claim that to do something else but α is to do β. The special action 0 will be needed there. After that we will specify the formula saying that it is obligated to perform α.

$\mathfrak{M} = (W, \{R_a \mid a \in \Pi_0\}, B, V)$ is a model if

1. W is a nonempty set of states
2. for any $a \in \Pi_0$, $R_a \subseteq W \times W$, and for any $a, b \in \Pi_0$, $R_a \cap R_b = \varnothing$
3. $B \subseteq W$
4. V is a function from Φ_0 to 2^W

Atomic relations are pairwise disjoint. This constraint guarantees that syntactically different atomic actions are genuinely different. B is a set of bad states. $\overline{B}$, the complement of B, is the set of fine states. Note that there is no constraint on B; it could be the whole universe and could also be the empty set. A model is just a so called *interpreted labeled transition system* with the constraint that the relations are pairwise disjoint, plus a set of bad states.

Fix a model $\mathfrak{M} = (W, \{R_a \mid a \in \Pi_0\}, B, V)$. Define $\mathcal{R} = \bigcup\{R_a \mid a \in \Pi_0\}$. A sequence $w_0 \dots w_n$ of states is called a *trace* if $w_0\mathcal{R}\dots\mathcal{R}w_n$. Specially, for any $w \in W$, w is a trace. A trace represents a transition sequence made by doing a series of basic actions. A special trace w means doing nothing. Let $\mathcal{T}$ be the set of traces. Define a partial binary function ext on $\mathcal{T}$ as follows: $ext(u_0 \dots u_n, v_0 \dots v_m)$ equals $u_0 \dots u_n v_1 \dots v_m$ if $u_n = v_0$, otherwise it is undefined. Let S and T be two sets of traces. Define a function $\otimes$, called *fusion*, like this: $S \otimes T = \{ext(\kappa, \lambda) \mid \kappa \in S \ \&\ \lambda \in T$, and $ext(\kappa, \lambda)$ is defined $\}$. Each action α is interpreted as a set S_α of traces in the following way:

1. $S_a = R_a$
2. $S_{\beta;\gamma} = S_\beta \otimes S_\gamma$
3. $S_{\beta\cup\gamma} = S_\beta \cup S_\gamma$
4. $S_{\alpha^*} = W \cup S_\alpha \cup S_{\alpha;\alpha} \cup \dots$

This semantics for actions is called *trace semantics.* This semantics has the following feature: for any basic actions $a_1, \dots, a_n$, all the traces in $S_{a_1;\dots;a_n}$ contain $n+1$ states, provided it is given that $S_{a_1;\dots;a_n}$ is not empty.

$\mathfrak{M}, w \Vdash \phi$, ϕ being true at w in $\mathfrak{M}$, is defined as follows:

1. $\mathfrak{M}, w \Vdash p \Leftrightarrow w \in V(p)$
2. $\mathfrak{M}, w \Vdash \top$ always holds
3. $\mathfrak{M}, w \Vdash \mathfrak{b} \Leftrightarrow w \in B$
4. $\mathfrak{M}, w \Vdash \neg\phi \Leftrightarrow$ not $\mathfrak{M}, w \Vdash \phi$
5. $\mathfrak{M}, w \Vdash (\phi \wedge \psi) \Leftrightarrow \mathfrak{M}, w \Vdash \phi$ and $\mathfrak{M}, w \Vdash \psi$
6. $\mathfrak{M}, w \Vdash \|\alpha\|\phi \Leftrightarrow$ for any trace $w_0 \dots w_n$, if $w_0 = w$ and $w_0 \dots w_n \in S_\alpha$, then $\mathfrak{M}, w_i \Vdash \phi$ for some i s.t. $1 \le i \le n$

Recall the definitions of $F\alpha$ and $P\alpha$ above. It can be verified that

7. $\mathfrak{M}, w \Vdash \mathfrak{f} \Leftrightarrow w \in \overline{B}$
8. $\mathfrak{M}, w \Vdash \langle\!\langle\alpha\rangle\!\rangle\phi \Leftrightarrow$ there is a trace $w_0 \dots w_n$ s.t. $w_0 = w$, $w_0 \dots w_n \in S_\alpha$ and $\mathfrak{M}, w_i \Vdash \phi$ for any i s.t. $1 \le i \le n$
9. $\mathfrak{M}, w \Vdash F\alpha \Leftrightarrow$ for any trace $w_0 \dots w_n$, if $w_0 = w$ and $w_0 \dots w_n \in S_\alpha$, then $\mathfrak{M}, w_i \Vdash \mathfrak{b}$ for some i s.t. $1 \le i \le n$
10. $\mathfrak{M}, w \Vdash P\alpha \Leftrightarrow$ there is a trace $w_0 \dots w_n$ s.t. $w_0 = w$, $w_0 \dots w_n \in S_\alpha$ and $\mathfrak{M}, w_i \Vdash \mathfrak{f}$ for any i s.t. $1 \le i \le n$

Note that the semantics views the ending point of doing α as a point during the process of doing α but does not view the starting point as a point of the process.

The notions of *validity* and *satisfiability* are defined as usual. This logic is called PDDL. Illustrations of this logic will be given in **section 5** after we make it clear which formula expresses the obligation to do α.

4 To Do Something Else

In this section, we provide a formalization for the notion of *to do something else* following the idea stated in **section 2**.

A finite sequence of atomic actions is called a *computation sequence*, abbreviated as *seq*. The empty seq is denoted by ϵ and the set of seqs denoted by $\mathcal{CS}$. Each seq corresponds to a composition of atomic actions and seqs are understood by their corresponding actions. For any sets Δ and Θ of seqs, let $\Delta;\Theta = \{\gamma\delta \mid \gamma \in \Delta \ \&\ \delta \in \Theta\}$. $CS(\alpha)$, the set of the seqs of α, is defined as follows:

1. $CS(a) = \{a\}$
2. $CS(0) = \varnothing$
3. $CS(\alpha;\beta) = CS(\alpha); CS(\beta)$
4. $CS(\alpha \cup \beta) = CS(\alpha) \cup CS(\beta)$
5. $CS(\alpha^*) = \{\epsilon\} \cup CS(\alpha) \cup CS(\alpha;\alpha) \cup \dots$

Each seq of α represents a way to perform α. α is an *empty action* if $CS(\alpha) = \varnothing$. In the sequel, for any seq σ and set Δ of seqs, we use $\sigma\Delta$ to denote the set $\{\sigma\tau \mid \tau \in \Delta\}$. For any model, define S_ϵ, the interpretation of ϵ in this model, as the whole universe. It can be shown that $S_\alpha = \bigcup\{S_\sigma \mid \sigma \in CS(\alpha)\}$.

In the semantics defined in last section, atomic actions are interpreted as pairwise disjoint binary relations and compound actions are interpreted as sets of traces. As a result, the following proposition holds (assume again that we have fixed a model $\mathfrak{M}$, with traces computed in that model):

Proposition 4.1 *For any α and β, if $CS(\alpha) \cap CS(\beta) = \varnothing$, then $S_\alpha \cap S_\beta = \varnothing$.*

Proof. Assume $S_\alpha \cap S_\beta \neq \varnothing$. Let $w_0 \dots w_n$ be a trace in $S_\alpha \cap S_\beta$. Then there is a seq $a_1 \dots a_n$ in $CS(\alpha)$ and a seq $b_1 \dots b_n$ in $CS(\beta)$ s.t. $w_0 \dots w_n$ is in $S_{a_1;\dots;a_n}$ and $S_{b_1;\dots;b_n}$. Then for any i s.t. $1 \leq i \leq n$, $w_{i-1}w_i$ is in S_{a_i} and S_{b_i}. As atomic actions are pairwise disjoint, $a_i = b_i$ for any i s.t. $1 \leq i \leq n$. Then $a_1 \dots a_n = b_1 \dots b_n$. This means $CS(\alpha) \cap CS(\beta) \neq \varnothing$. □

This is a crucial fact for this work.

Let $\sqsubseteq$ denote the relation of *initial segment* for sequences and $\sqsupseteq$ the converse of $\sqsubseteq$, called *extension.*

Definition 4.2 [Mutual extension, x-difference] Let σ and τ be two seqs. Then $\sigma \approx \tau$ if if $\sigma \sqsubseteq \tau$ or $\tau \sqsubseteq \sigma$. Call this the relation of mutual extension. Say that σ is x-different from τ if $\sigma \not\approx \tau$.

For example, ac is x-different from ab, but a is not x-different from ab, as $a \sqsubseteq ab$. cab is also x-different from ab, as $ab \not\sqsubseteq cab$ and $cab \not\sqsubseteq ab$, although ab is a segment of cab. Here are some basic facts about the relation of x-difference. As ϵ is an initial segment of any seq, no seq is x-different from ϵ. x-difference is closed under extension: if $\sigma \not\approx \tau$ and $\tau \sqsubseteq \tau'$, then $\sigma \not\approx \tau'$. The relation of mutual extension is closed under initial segment: if $\sigma \approx \tau$ and $\tau' \sqsubseteq \tau$, then $\sigma \approx \tau'$. If σ is x-different from τ, then there is no way to extend σ s.t. the extension of σ is identical to τ, and there is also no way to extend τ s.t. the extension of τ is identical to σ. The notion of x-difference is intuitively understood as follows. Assume that σ is *x-different* from τ. Then there is no moment during the performance of σ at which the agent has done τ, and there is also no moment after the performance of σ at which the agent has done τ, no matter what he/she does afterwards.

For any actions α and β, α is *x-different* from β, $\alpha \not\approx \beta$, if for any seqs $\sigma \in CS(\alpha)$ and $\tau \in CS(\beta)$, $\sigma \not\approx \tau$. The relation of x-difference for actions formalizes the word "else" in the imperatives such as "don't watch cartoons anymore and do something else". β is *something else* but α if β is x-different from α. Note that given an action α, there might be many actions each of which is something else. For example, both b and c are something else for a. This means that the relation of x-different itself is not enough to handle the notion of to do something else, as the latter also involves a quantifier over actions. Luckily, for any α, among the actions which are something else, there is a *greatest* one in the sense that it is the union of all of them. This lets us deal with the notion of to do something else without introducing any quantifier over actions.

Definition 4.3 [The function of opposite] Let Δ be a set of seqs. $\widetilde{\Delta}$, the opposite of Δ, is defined as the set $\{\tau \mid \tau \not\approx \sigma \textit{ for any } \sigma \in \Delta\}$.

$\widetilde{\Delta}$ is always closed under extension; this is an important feature of the function of opposite. Opposite is different from complement: $\widetilde{\Delta}$ is always a subset of $\overline{\Delta}$, but not vice versa. Here is a counter-example: let $\Delta = \{ab\}$; then $a \in \overline{\Delta}$ but $a \notin \widetilde{\Delta}$. Opposite has certain connection with complement. Define Δ^T as the set of the seqs which are x-equal to some seq in Δ. Δ^T is called the *tree* generated from Δ. It can be seen that $\widetilde{\Delta} = \overline{\Delta^T}$. About Δ^T, there is a different way to look at it. Let Δ' be the smallest set which contains Δ and is closed under extension, and Δ'' the smallest set containing Δ' which is closed under initial segments. It can be verified that $\Delta'' = \Delta^T$. This result will be used later. Note that Δ^T might not be closed under extension.

The following proposition specifies some important properties of the function of opposite:

Proposition 4.4

1. $\Delta \cap \widetilde{\Delta} = \varnothing$
2. $\widetilde{\Delta} \cap (\Delta;\Theta) = \varnothing$
3. $\widetilde{\Delta \cup \Theta} = \widetilde{\Delta} \cap \widetilde{\Theta}$
4. $\Delta \subseteq \widetilde{\widetilde{\Delta}}$
5. $\widetilde{\Delta;\Theta} \sqsubseteq \widetilde{\Delta} \cup (\Delta;\widetilde{\Theta})$ *if* $\Theta \neq \varnothing$
6. $\widetilde{\Delta} \subseteq \widetilde{\Delta;\Theta}$

Proof.

1. This is easy to show.

2. By the sixth item of this proposition, $\widetilde{\Delta} \subseteq \widetilde{\Delta;\Theta}$. As $\widetilde{\Delta;\Theta} \subseteq \overline{\Delta;\Theta}$, $\widetilde{\Delta} \subseteq \overline{\Delta;\Theta}$. Then $\widetilde{\Delta} \cap (\Delta;\Theta) = \varnothing$.

3. $\sigma \in \widetilde{\Delta \cup \Theta} \Leftrightarrow \sigma \not\approx \tau$ for any $\tau \in \Delta \cup \Theta \Leftrightarrow \sigma \not\approx \tau$ for any $\tau \in \Delta$ and $\sigma \not\approx \tau$ for any $\tau \in \Theta \Leftrightarrow \sigma \in \widetilde{\Delta}$ and $\sigma \in \widetilde{\Theta}$.

4. Let $\sigma \in \Delta$. Assume $\sigma \notin \widetilde{\widetilde{\Delta}}$. Then there is a $\tau \in \widetilde{\Delta}$ s.t. $\sigma \approx \tau$. This is impossible.

5. Let $\sigma \in \widetilde{\Delta;\Theta}$. Then $\sigma \not\approx \tau$ for any $\tau \in \Delta;\Theta$. Assume $\sigma \notin \widetilde{\Delta}$. We want to show $\sigma \in (\Delta;\widetilde{\Theta})$. Then there is a $\kappa \in \Delta$ s.t. $\sigma \sqsubseteq \kappa$ or $\kappa \sqsubseteq \sigma$. Assume $\sigma \sqsubseteq \kappa$. Let $x \in \Theta$, as $\Theta \neq \varnothing$. Then $\kappa x \in \Delta;\Theta$. As $\sigma \sqsubseteq \kappa$, $\sigma \sqsubseteq \kappa x$. Then $\sigma \approx \kappa x$. This is impossible, as $\sigma \in \widetilde{\Delta;\Theta}$. Then $\kappa \sqsubseteq \sigma$. Let $\sigma = \kappa\lambda$. We want to show $\lambda \in \widetilde{\Theta}$. Assume not. Then there is a $\tau \in \Theta$ s.t. $\lambda \approx \tau$. Then $\kappa\lambda \approx \kappa\tau$. Then $\kappa\tau \in \Delta;\Theta$. Then $\kappa\lambda \notin \widetilde{\Delta;\Theta}$. This is impossible. Then $\lambda \in \widetilde{\Theta}$. Then $\kappa\lambda \in (\Delta;\widetilde{\Theta})$, that is, $\sigma \in (\Delta;\widetilde{\Theta})$.

6. Let $\sigma \in \widetilde{\Delta}$. Then $\sigma \not\approx \tau$ for any $\tau \in \Delta$. Let $\tau' \in \Delta;\Theta$. Then there is a $\tau \in \Delta$ s.t. $\tau \sqsubseteq \tau'$. As $\not\approx$ is closed under extension, $\sigma \not\approx \tau'$. Then $\sigma \in \widetilde{\Delta;\Theta}$. □

The converse of the fourth item does not hold generally. As for any Δ, $\widetilde{\widetilde{\Delta}}$ is closed under extension, we can get that for any Δ, if Δ is not closed under extension, then $\widetilde{\widetilde{\Delta}} \nsubseteq \Delta$. Here is an example: let $\Pi_0 = \{a, b\}$ and $\Delta = \{aa, ab\}$; then $\widetilde{\Delta} = b\Pi_0^*$ and $\widetilde{\widetilde{\Delta}} = a\Pi_0^*$; then $aaa \in \widetilde{\widetilde{\Delta}}$ but $aaa \notin \Delta$. The converse of the fifth item does not hold either and the reason is that $(\Delta;\widetilde{\Theta}) \subseteq \widetilde{\Delta;\Theta}$ might not hold. What follows is a counter-example: let $\Pi_0 = \{a, b\}$, $\Delta = \{aa, a\}$ and $\Theta = \{ab\}$; then $\widetilde{\Theta} = b\Pi_0^* \cup aa\Pi_0^*$; then $aab \in \Delta;\widetilde{\Theta}$; as $aab \in \Delta;\Theta$, $aab \notin \widetilde{\Delta;\Theta}$. The fifth item has a condition, that is, $\Theta \neq \varnothing$. This item does not hold without the condition. For a counter-example, let $\Pi_0 = \{a, b\}$ and $\Delta = \{ab\}$. Then $\widetilde{\Delta;\Theta} = \mathcal{CS}$, as

$\Delta;\Theta = \varnothing$. We see that $a \notin \widetilde{\Delta}$ and $a \notin \Delta;\widetilde{\Theta}$.

Proposition 4.5 *For any* $\alpha \in \Pi_{PDL}$, *there is a* $\beta \in \Pi_{PDL}$ *s.t.* $CS(\beta) = \widetilde{CS(\alpha)}$.

Proof. As shown in the literature of *automata theory*, a set Δ of seqs is a so called *regular language* if and only if there is a $\alpha \in \Pi_{\text{PDL}}$ s.t. $CS(\alpha) = \Delta$ [1]. Therefore, it suffices to show that $\widetilde{CS(\alpha)}$ is a regular language. As mentioned in **section 4**, $\widetilde{CS(\alpha)} = \overline{CS(\alpha)^T}$ where $CS(\alpha)^T$ is the tree generated from $CS(\alpha)$. Then it suffices to show that $\overline{CS(\alpha)^T}$ is a regular language. Let Θ be the smallest set which contains $CS(\alpha)$ and is closed under extension. It can be seen that $CS(\alpha;(a_1 \cup \cdots \cup a_n)^*) = \Theta$ where $\Pi_0 = \{a_1, \dots, a_n\}$. Then Θ is a regular language. Let Θ' be the smallest set containing Θ which is closed under initial segments. By [5], the closure of a regular language under initial segments is also a regular language. Then Θ' is a regular language. As stated in **section 4**, this Θ' equals to $CS(\alpha)^T$. Then $CS(\alpha)^T$ is a regular language. By [5], the complement of a regular language is also a regular language. Then $\overline{CS(\alpha)^T}$ is a regular language. □

This β is called the *opposite* of α, denoted by $\widetilde{\alpha}$. Here is an example: let $\Pi_0 = \{a, b, c\}$; then $\widetilde{a} = (b \cup c);(a \cup b \cup c)^*$. It can be easily shown that $CS(\widetilde{\alpha}) = \bigcup\{CS(\gamma) \mid \gamma \not\approx \alpha\}$. Hence, $\widetilde{\alpha}$ is the union of all the actions which are something else but α. To refrain to do α is to do something else; to do *anything else* is to do $\widetilde{\alpha}$.

As mentioned in the introduction, it is reasonable to require that anything else but α has empty intersections with α and with $\alpha;\beta$. The following proposition states that this is indeed the case:

Proposition 4.6 $S_{\widetilde{\alpha}} \cap S_\alpha = \varnothing$ *and* $S_{\widetilde{\alpha}} \cap S_{\alpha;\beta} = \varnothing$.

This result can be proved by use of **proposition 4.1 and 4.4**.

In standard relational semantics, an action α is interpreted as a binary relation R_α. Then neither $R_{\widetilde{\alpha}} \cap R_\alpha = \varnothing$ nor $R_{\widetilde{\alpha}} \cap R_{\alpha;\beta} = \varnothing$ is generally the case even if atomic actions are pairwise disjoint. Here is a counter-example for both. Let a, b and c be three atomic actions. Let $R_a = \{(w_1, w_2)\}$, $R_b = \{(w_2, w_3)\}$ and $R_c = \{(w_1, w_3)\}$. We see that the three atomic actions are pairwise disjoint. As c is x-different from $a;b$ and $\widetilde{a;b}$ is the union of all the actions x-different from $a;b$, we know $R_c \subseteq R_{\widetilde{a;b}}$. As $R_c \cap R_{a;b} = \{(w_1, w_3)\}$, $R_{\widetilde{a;b}} \cap R_{a;b} \neq \varnothing$. c is x-different from a, then $R_c \subseteq R_{\widetilde{a}}$. $R_c \cap R_{a;b} = \{(w_1, w_3)\}$, then $R_{\widetilde{a}} \cap R_{a;b} \neq \varnothing$. In usual process logics, atomic actions are viewed as sets of state sequences which might not be binary relations. Then $S_{\widetilde{\alpha}} \cap S_\alpha = \varnothing$ and $S_{\widetilde{\alpha}} \cap S_{\alpha;\beta} = \varnothing$ do not generally hold, given that atomic actions are pairwise disjoint. What follows is a counter-example for both. Let $S_a = \{w_1 w_2\}$, $S_b = \{w_2 w_3\}$ and $S_c = \{w_1 w_2 w_3\}$. a, b and c are pairwise disjoint. c is x-different from $a;b$, then $S_c \subseteq S_{\widetilde{a;b}}$. $S_c \cap S_{a;b} = \{w_1 w_2 w_3\}$, then $S_{\widetilde{a;b}} \cap S_{a;b} \neq \varnothing$. c is x-different from a, then $S_c \subseteq S_{\widetilde{a}}$. $S_c \cap S_{a;b} = \{w_1 w_2 w_3\}$, then $S_{\widetilde{a}} \cap S_{a;b} \neq \varnothing$.

---

[1] Regular languages are defined in terms of *finite deterministic automata*. For details of this, we refer to [5].

By **proposition 4.4** we can get that $S_\alpha \subseteq S_{\widetilde{\widetilde{\alpha}}}$ and $S_{\widetilde{\alpha;\beta}} \subseteq S_{\widetilde{\alpha}} \cup S_{\alpha;\widetilde{\beta}}$. It can be verified that neither of the converses of the two results holds. Considering that opposite is some type of negation, one might wonder about this. However, when restricted to the class of *normatively concise* actions, the two converses hold. What is a normatively concise action? Here we just show its idea by an example and does not give its formal definition. Assume that there are only two atomic actions: a and b. Look at the two sentences: "the agent ought to do $a;a$ or $a;b$" and "the agent ought to do a". The two sentences have the same meaning but the first one is not given concisely. In this sense, we say that the action $(a;a) \cup (a;b)$ is not normatively concise but a is. We leave exploring this issue further as our future work.

5 Validity

By means of *to do anything else*, we now can express obligations. $O\alpha$, α is *obligated*, is defined as $\|\widetilde{\alpha}\|\mathfrak{b}$; it means that no matter what alternative β to α is done, and now matter how β is performed, at some point in the process a bad state will be encountered. The truth condition of $O\alpha$ is as follows:

11. $\mathfrak{M}, w \Vdash O\alpha \Leftrightarrow$ for any trace $w_0 \ldots w_n$, if $w_0 = w$ and $w_0 \ldots w_n \in S_{\widetilde{\alpha}}$, then $w_i \Vdash \mathfrak{b}$ for some i s.t. $1 \leq i \leq n$

By now all the three normative notions are defined and we can illustrate the logic PDDL a bit.

PDDL has the following two features: its semantics does not take the starting point of doing an action as a point of the process of doing this action; whether an action is allowed is totally determined by what happens during the process of doing this action. The two features together imply whether an action is allowed at a state has nothing to do with this state. One may wonder what if the starting point of doing an action counts in the process of doing this action. Suppose so. Then $\phi \to \|\alpha\|\phi$ would be valid for any α and ϕ. Then both $\mathfrak{b} \to F\alpha$ and $\mathfrak{b} \to O\alpha$ would be valid. This means that in bad states, everything is forbidden and everything is obligated. This is of course undesirable. Our present definition at least has the advantage that it is possible to escape from a bad state with a good action.

There is some bonus which we can get from the two features mentioned above. For ease of stating our core points for refraining to do something, we in this work does not introduce the action constructor *test*. A test $\phi?$ in trace semantics is a set of states in which ϕ is true. As the starting point of doing an action does not count in the process of doing this action, the action of testing does not have a process. Then trivially, $\|\phi?\|\psi$ is not satisfiable and $\langle\!\langle\phi?\rangle\!\rangle\psi$ is valid. As a result, $F(\phi?)$ is not satisfiable and $P(\phi?)$ is valid. This means that there is no restriction on testing and testing is always free. Considering that testing is just some mental action and does not directly change the world, we think that this is desirable.

The following valid formulas express some connections between the deontic operators:

1. $P\alpha \leftrightarrow \neg F\alpha$
2. $O\alpha \leftrightarrow F\widetilde{\alpha}$
3. $P\alpha \to \langle\!\langle \alpha \rangle\!\rangle \top$

The first formula says that an action is permitted if and only if it is not forbidden. In addition, we can verify that $P(a \cup b) \to (Pa \wedge Pb)$ is not a valid formula. Putting the two facts together we can get that the operator P introduced in this work is not for the so called *free choice permission* but for *lack-of-prohibition permission*. The second formula tells that an action is obligated if and only if not doing it is forbidden. If an action is permitted, then it is *doable*; this is what the last formula says. *Kant's Law*, whatever should be done can possibly be done, expressed as $O\alpha \to \langle\!\langle \alpha \rangle\!\rangle \top$, does not generally hold in PDDL. To see this, imagine a model with a *dead* state, that is, one from which no transition starts. Then for any atomic action a, a is obligated trivially but not doable at this dead state.

What follows are some valid formulas which essentially involve action constructors:

1. $O\alpha \to O(\alpha \cup \beta)$
2. $F\alpha \to F(\alpha; \beta)$
3. $P(\alpha; \beta) \to P\alpha$
4. $O(\alpha; \beta) \to O\alpha$

The first formula shows that *Ross's Paradox* is not avoided: the agent has the duty to post the letter; therefore, he/she has the duty to post it or burn it. As argued in [8], we do not think that this is a problem. By the second formula, if killing is prohibited, then killing and then surrendering is also prohibited. But note this does not mean that if killing is prohibited, then surrendering is prohibited after killing. Indeed, it can be verified that $Fk \wedge \langle k \rangle Ps$ is satisfiable where k and s represent the actions of killing and surrendering respectively. By the third formula, if smoking and then leaving is permitted, then smoking is permitted. From the fourth formula we can get that the duty of rescuing the injured is implied by the duty of rescuing the injured and then calling an ambulance. These examples show that our logic does not suffer from the problem with [11] that was mentioned in the introduction.

Let's say that a state of a model is an *awkward* state if doing any atomic action at it will end in a bad state. Then at such states, for any atomic action a, a is not allowed. Then at them, nothing is allowed except those actions such as α^* and $\phi?$ which contain *one-element* traces. As a result, neither $O\alpha \to P\alpha$ nor $P\alpha \vee P\widetilde{\alpha}$ is valid.

6 A Variation

We put some constraints on the logic PDDL: in syntax, there are finitely many atomic actions and a special action 0; in semantics, atomic actions are pairwise disjoint. These constraints give PDDL the power to express *to do something else*. This is an implicit way to deal with to do something else. There is a different way to handle it, that is, explicitly introducing an action constructor for it.

Let Π_0 be a *countable* set of atomic actions and Φ_0 a countable set of atomic propositions. Let a range over Π_0 and p over Φ_0. The sets Π_{OPDL} of actions and Φ_{PoDDL} of propositions are defined as follows:

$$\alpha ::= a \mid (\alpha;\alpha) \mid (\alpha \cup \alpha) \mid \alpha^* \mid \widetilde{\alpha}$$
$$\phi ::= p \mid \top \mid \mathfrak{b} \mid \neg\phi \mid (\phi \wedge \phi) \mid \|\alpha\|\phi$$

Here in "Π_{OPDL}" and "Φ_{PoDDL}", "O" is for "opposite". The action $\widetilde{\alpha}$ is called the *opposite* of α; to do $\widetilde{\alpha}$ is to do something else but α. The intuitive reading of this language is as the language Φ_{PDDL} specified in **section 3**. $F\alpha$ and $P\alpha$ are defined as before and $O\alpha$ is directly defined as $\|\widetilde{\alpha}\|\mathfrak{b}$. Compared with Φ_{PDDL}, Φ_{PoDDL} has infinitely many atomic actions and does not have the empty action 0.

$\mathfrak{M} = (W, \{R_a \mid a \in \Pi_0\}, B, V)$ is a model where

1. W is a nonempty set of states
2. for any $a \in \Pi_0$, $R_a \subseteq W \times W$
3. $B \subseteq W$
4. V is a function from Φ_0 to 2^W

Models are understood as before. Here we do not require that atomic actions are pairwise disjoint.

Fix a model $\mathfrak{M} = (W, \{R_a \mid a \in \Pi_0\}, B, V)$. Recall that a sequence $w_0 \ldots w_n$ of states is called a trace if $w_0\mathcal{R}\ldots\mathcal{R}w_n$ where $\mathcal{R} = \bigcup\{R_a \mid a \in \Pi_0\}$. Let $\mathcal{T}$ denote the set of traces as before. In **section 4**, we define a relation *x-different* on $\mathcal{CS}$ which is the set of computation sequences. Here we define it on $\mathcal{T}$ in a similar way: for any traces σ and τ, σ is x-different from τ, $\sigma \not\approx \tau$, if $\sigma \not\trianglelefteq \tau$ and $\tau \not\trianglelefteq \sigma$. By use of the relation x-different, we in **section 4** define a function *opposite* on the power set of $\mathcal{CS}$. We here define it on the power set of $\mathcal{T}$ similarly: for any set Δ of traces, let $\widetilde{\Delta}$, called the opposite of Δ, be the set $\{\tau \in \mathcal{T} \mid \tau \not\approx \sigma \textit{ for any } \sigma \in \Delta\}$. This opposite function also has the properties specified in **proposition 4.4**.

Each $\alpha \in \Pi_{\text{OPDL}}$ is interpreted as a set S_α of traces in the following way:

1. $S_a = R_a$
2. $S_{\beta;\gamma} = S_\beta \otimes S_\gamma$
3. $S_{\beta\cup\gamma} = S_\beta \cup S_\gamma$
4. $S_{\alpha^*} = W \cup S_\alpha \cup S_{\alpha;\alpha} \cup \ldots$
5. $S_{\widetilde{\alpha}} = \widetilde{S_\alpha}$

Here the operation $\otimes$ is defined as in **section 3**. We make a few points in this place. In **section 4**, we assign each α in Π_{PDL} an action $\widetilde{\alpha}$ in Π_{PDL}. The assignment makes use of the relation x-different and the function opposite; the action $\widetilde{\alpha}$ follows our idea for to do something else stated in **section 2**. In this section, $\widetilde{\alpha}$ is directly given in syntax; however, $S_{\widetilde{\alpha}}$, the interpretation of $\widetilde{\alpha}$, uses the relation of x-different and the function of opposite. Here $\widetilde{\alpha}$ also follows our idea for to do something else. $\mathcal{T}$ is the set of state sequences which can be made by performing basic actions. It can be seen that for any α, $S_{\tilde{\alpha}} \subseteq \mathcal{T}$. This means that the action constructor $^{\sim}$ does not essentially introduce new actions in this sense: whichever state can be reached by performing an action with $^{\sim}$ can be reached by performing an action without $^{\sim}$.

$\mathfrak{M}, w \Vdash \phi$, ϕ being true at w in $\mathfrak{M}$, is defined as in **section 3**. The notion of *validity* is defined as usual. This logic is called PoDDL. A check of the formulas from **section 5** shows that the new approach does not make a difference for the validity/invalidity of these formulas.

7 Connections and Future Work

If we accept a state based approach of good and evil, it would be interesting to find out how the two ways of formalizing the notion of refraining to do something are related. Do they have the same expressive power or not? Next, it would be interesting to give complete axiomatisations.

The state based approach to the distinction between good and evil has some inherent limitations that carry over to our proposals above. As mentioned in **section 5**, almost nothing is allowed in the states we called awkward states. In reality, we never stop acting. Even if we are doing *nothing*, we are still doing *something*. There may be cases where, in order to act, we have to violate some prohibition. So what is prudent action in such situations? How should agents act in awkward states? Intuitively, they should transit to those states which are relatively *better* than others. Instead of a black and white division of evil and good states, we need some shades of grey, or even better a relational approach where some states are better than others. This is future work.

Since morality has to do with our interaction with others, another important step to take is from single agent to multiple agent deontic logic. Even more realistic seems an approach where obligations are relational, and where an obligation of some agent A to do something or to refrain from doing something is always an obligation to some other agent B. A proposal for a formalization of this idea in terms of propositional dynamic logic is given in [16]. One of the attractions of this is that it allows us to model conflicts of duty, such as the conflicts between professional obligations and family obligations that we all know so well.

Acknowledgment

Fengkui Ju was supported by the National Social Science Foundation of China (No. 12CZX053) and the Fundamental Research Funds for the Central Universities (No. SKZZY201304). We would like to thank the audience of the Workshop in Memory of Kang Hongkui and the anonymous referees for their useful comments and suggestions.

References

[1] Anderson, A. R., *Some nasty problems in the formal logic of ethics*, Noûs **1** (1967), pp. 345–360.

[2] Broersen, J., *Action negation and alternative reductions for dynamic deontic logics*, Journal of Applied Logic **2** (2004), pp. 153–168.
[3] Harel, D., D. Kozen and R. Parikh, *Process logic: Expressiveness, decidability, completeness*, Journal of computer and system sciences **25** (1982), pp. 144–170.
[4] Hilpinen, R., *Deontic logic*, in: L. Goble, editor, *The Blackwell Guide to Philosophical Logic*, Blackwell Publishing, 2001 pp. 159–182.
[5] Hopcroft, J., R. Motwani and J. Ullman, "Introduction to Automata Theory, Languages, and Computation," Pearson, 2006.
[6] Ju, F. and J. Cai, *Two process modalities in trace semantics* (2015), manuscript.
[7] Ju, F., N. Cui and S. Li, *Trace semantics for IPDL*, in: W. van der Hoek, W. H. Holliday and W. Wang, editors, *Logic, Rationality, and Interaction*, Lecture Notes in Computer Science **9394**, Springer Berlin Heidelberg, 2015 pp. 169–181.
[8] Ju, F. and L. Liang, *A dynamic deontic logic based on histories*, in: D. Grossi, O. Roy and H. Huang, editors, *Logic, Rationality, and Interaction*, Lecture Notes in Computer Science **8196**, Springer Berlin Heidelberg, 2013 pp. 176–189.
[9] Kanger, S., *New foundations for ethical theory*, in: R. Hilpinen, editor, *Deontic Logic: Introductory and Systematic Readings*, Springer Netherlands, 1971 pp. 36–58.
[10] McNamara, P., *Deontic logic*, in: E. N. Zalta, editor, *The Stanford Encyclopedia of Philosophy*, 2014, winter 2014 edition .
[11] Meyer, J.-J. C., *A different approach to deontic logic: deontic logic viewed as a variant of dynamic logic*, Notre Dame Journal of Formal Logic **29** (1988), pp. 109–136.
[12] Pratt, V., *Process logic*, in: *Proceedings of the 6th ACM Symposium on Principles of Programming Languages* (1979), pp. 93–100.
[13] Pucella, R. and V. Weissman, *Reasoning about dynamic policies*, in: I. Walukiewicz, editor, *Proceedings of the 7th International Conference of Foundations of Software Science and Computation Structures* (2004), pp. 453–467.
[14] Sun, X. and H. Dong, *Deontic logic based on a decidable PDL with action negation* (2015), manuscript.
[15] van der Meyden, R., *The dynamic logic of permission*, Journal of Logic and Computation **6** (1996), pp. 465–479.
[16] van Eijck, J. and F. Ju, *Modelling legal relations* (2016), under submission.
[17] Wansing, H., *On the negation of action types: constructive concurrent PDL*, in: P. Hájek, L. Valdes-Villanueva and D. Westerstahl, editors, *Proceedings of the Twelfth International Congress of Logic, Methodology and Philosophy of Science* (2005), pp. 207–225.

Multivalued Logics for Conflicting Norms

Piotr Kulicki

The John Paul II Catholic University of Lublin,
Faculty of Philosophy, Al. Racławickie 14, 20-950 Lublin, Poland
kulicki@kul.pl

Robert Trypuz

The John Paul II Catholic University of Lublin,
Faculty of Philosophy, Al. Racławickie 14, 20-950 Lublin, Poland
trypuz@kul.pl

Abstract

Multivalued setting is quite natural for deontic action logic, where actions are usually treated as obligatory, neutral or forbidden. We apply the ideas of multivalued deontic logic to the phenomenon of a moral dilemma and, broader, to any situation where there are conflicting norms. We formalize three approaches towards normative conflicts. We present matrices for the systems and compare their tautologies. Finally, we present a sound and complete axiomatization of the systems.

Keywords: deontic action logic, multivalued logic, moral dilemma, Belnap-Dunn lattice.

1 Introduction

The need for merging directives coming from different sources is quite common in social life. We may have, for instance, state laws, corporation rules of conduct, religious regulations, orders, requests or expectations coming from different people that apply to the same situation. Usually the different directives can be harmoniously combined. In contemporary European countries state law allows for the freedom of religion, most corporations do not regulate what employees do in their free time so when a person is obliged by the rules of his or her religion to participate in a religious service on Sunday (or another day free from work) he or she can easily comply with such regulations.

However, sooner or later, one can face conflicting regulations, impulses or motivations. It is enough to add to the example the factor that the partner of our agent wants to go hiking for the whole Sunday to have a conflict.

In many cases such a conflict can be quite easily resolved. Several possible ways of solving norm conflicts have been presented, including preferences on norms or norm sources (see e.g. [16]) or Rabbis' decision in the Talmudic system (see e.g. [1]). Applying a game theoretical approach in which an agent gets penalties and payoffs depending on the importance of the norm and the level of violation or compliance would be another one (see e.g. [6]).

Sometimes, however, an agent cannot resolve the conflict. Such situations, especially when they apply to existentially important matters, are recognized in the literature as moral dilemmas and have been extensively discussed in ethics. Moral dilemmas have also been studied in deontic logic. There are many, mutually consistent, definitions of moral dilemmas in the logical literature (see e.g. [15, p.36], [14, p.259], [9, p.283]) [1]. Below we present one of the definitions:

Definition 1.1 By a deontic dilemma I mean a situation in which, in a univocal sense of ought, some state of affairs, A, both ought to be and ought not to be [...] More broadly, a deontic dilemma would be a situation in which there are inconsistent states of affairs, A and B, both of which ought to be [...] More broadly still, a deontic dilemma would be a situation in which it is impossible for both A and B to be realized even though both ought to be, where the sense of impossibility could be anything appropriate to the context of discourse, from some metaphysical impossibility to a more mundane practical incompatibility. [11, p.462]

We will limit ourselves to the situations in which we deal with clearly defined normative systems in which specific actions are obligatory, forbidden or unregulated (indifferent). The systems do not have to be codified, we just assume that there is no doubt how to classify an action within a given system. Loosely speaking we can say that the justification for such norms lies in the fact that actions are regarded, from some point of view, as good, bad and neutral respectively. We will, however, not consider the rationale of norms but accept them as they are.

That allows us to use three/four-valued logic as a technical tool. Multivalued logic has been present in deontic logic from the 1950s [17,10,4], more recent works include [18,7,20]. The biggest advantage of many-valued logic is its conceptual simplicity and efficient decidability. The latter feature is especially important for applications in artificial systems making many-valued logic popular among researchers in computer science.

In the present paper we use a many-valued logic approach for deontic logic focusing on merging norms. The cases of normative conflict, especially dilemmas are most interesting and challenging so we put most of our effort to model these cases. The presented systems, however, can be used also to model merging of non-conflicting normative systems. Finally, we want to obtain the general normative (legal, moral or social) evaluation of actions carried out in a complex

[1] As far as we know ethics provides no definite answer to the question whether moral dilemmas really exist (see e.g., [26,12]). Experiencing conflicting norms in real life we may state following Horty [15, p.37] that: 'even if it does turn out, ultimately, that research in ethics is able to exclude the possibility of conflicts in a correct moral theory, it may be useful all the same to have a logic that allows for conflicting oughts'.

environment consisting of many, possibly inconsistent, normative sub-systems.

As we have mentioned above the idea of multivalued deontic logic is not new. Our contribution lies in providing a new reading of action operators within the logic, making it suitable for dealing with normative conflicts. In the paper we discuss three systems. The first of them is based on the matrices introduced in [17] and complemented with more operators on actions in [10]. The other two systems are original. All of them are presented in a unified way slightly different from the earlier formalizations.

The paper has the following structure. In section 2 we introduce and interpret Antigone's story, a classical example of a moral dilemma. In section 3 we introduce formal tools. We define a language of a deontic action logic (section 3.1), explain the interpretation of its main operator (section 3.2) and define matrix systems (section 3.3). In section 4 we present three logics defining their matrices and axiomatizations. The systems formalize different accounts of assessing decisions in the presence of conflicting norms. In section 5 soundness and completeness of the systems is proved. In section 6 we list tautologies which are common and specific for the three logics. It is important to note that the intuitions are formally reconstructed within the matrix systems and the fact that some formulas are or are not tautologies in a specific system is just a consequence of the application of intuitions on the level of matrices[2].

2 Antigone's example

Let us start our detailed investigations with an example of a moral dilemma from Sophocles' Antigone.

> Creon, as the new ruler of Thebes, has decided that Polyneices will be in public shame and his dead body will not be sanctified by holy rites, but will lie unburied on the battlefield. Polyneices' sister Antigone believes that she should bury his body according to universal laws given by gods. Thus, whatever she does, she is in conflict with one of the directives that she should comply with – as a subject to Creon or as a subject to gods.

Let us analyze this example from the point of view of action theory in a deontic context. Antigone has two options: bury her brother or not. Obviously there is no other possibility. One can look at them at different levels referring to the various intentions or descriptions under which they are carried out (see [3] or [8, essay 3]). At the basic level of crude behavior (bodily movements) *burying Polyneices* is an 'elementary act' done with a basic intention whose content is free from the social context of the situation. The other option is not as simple to interpret at this level but, regardless of any possible and, to some extent justified, criticism, it can be understood as carrying out any other action. This, however, does not help us to understand the essence of the situation. We certainly must look at Antigone's possible acts within their social and

[2] Some results described in the paper were presented in a preliminary form at ESSLLI 2012 workshop 'Trivalent Logics and their applications'.

normative context. Then *burying Polyneices* is no longer elementary. It carries all social saturation. From the point of view of the tragedy it is important that *burying Polyneices* is in defiance of Creon's edict and in accordance with divine law (custom). On the other hand any behavior different from that is in accordance with Creon's edict and in defiance of divine law [3].

Thus, we can think of Antigone's possible acts as determined by their socially grounded interpretations. Those interpretations constitute the characteristics of a behavior that is meaningful from the deontic point of view. Those interpretations have also an essential impact on the agent's choices that in the case of Antigone are: *comply with Creon's edict* and *comply with divine laws (custom)*. As the next step we can attach deontic value (that means a declaration whether the act is obligatory, forbidden or unregulated) to the agent's choices. In our case complying with Creon's edict is obligatory, since he is a king and, on the other hand, complying with divine laws is also obligatory.

Thus, the same action of burying Polyneices is, in the light of the Creon's edict, forbidden and, in the light of customs, obligatory. Analogously, not burying Polyneices is at the same time obligatory and forbidden. Such a situation is in deontic logic treated as (normative) inconsistency. Normally, any action should not be at the same time obligatory and forbidden. The reason of the inconsistency in our case is that, despite the fact that each of the normative systems (Creon's edict and custom) taken separately is consistent, they cannot be harmoniously combined, i.e. it is impossible to fulfill both obligations.

Nonetheless, Antigone herself or any other person may want to judge or assess the behavior in the given situation. Ideally one would like to have a logical system which can help to choose what to do. In the case of a genuine dilemma, as we understand it, it is impossible. Still we can try to define the normative status of the behavior of an agent facing a dilemma. We expect from such a judgment to be consistent and inform us unambiguously whether a given action is obligatory, forbidden or unregulated (neutral).

Thus we have three levels (compare our analysis of Antigone's example illustrated in Figure 1): (1) available actions, (2) normative description – consisting of possibly inconsistent specific norms coming from different systems of norms and (3) final judgment – consistent synthesis of norms applying in a given situation.

The modeling principle accepted in the paper and necessary to use the tools of many-valued logic is that the deontic value of an action can be computed from the deontic values of its basic elements using functions connected with operators in a way analogous to truth values and truth connectives of propositional calculus. We do not claim that this assumption covers all the deontic intuitions, we rather want to explore its possible consequences.

The main problem that remains open in the approach is to decide what

---

[3] There may be of course more socially important, relevant aspects of what Antigone does such as the fact that she is Polyneices' sister, Creon's niece or that she is a woman, but at that point we just focus on the two most important ones.

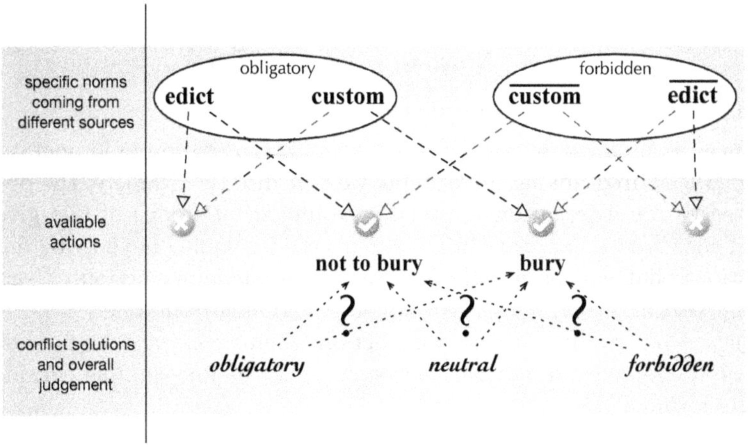

Fig. 1. In the figure **edict** stands for action 'comply with Creon's edict', **custom** – 'comply with divine laws (customs)' and **bury** – 'burying Polyneices'.

should be the deontic value of a combination of obligatory and forbidden action—we will propose and discuss three proposals in section 4. Other combinations like obligatory (forbidden) with obligatory (forbidden) and obligatory (forbidden) with neutral seem to be straightforward.

3 Formalization

Let us start with introducing a formal language we shall use in our considerations.

3.1 Language

The language we shall use can be defined in Backus-Naur notation in the following way:

(1) $$\varphi ::= \mathsf{O}(\alpha) \mid \neg\varphi \mid \varphi \wedge \varphi$$

(2) $$\alpha ::= a_i \mid \overline{\alpha} \mid \alpha \sqcap \alpha$$

where a_i belongs to a finite set of basic actions Act_0, '$\mathsf{O}(\alpha)$' – α is obligatory, '$\alpha \sqcap \beta$' – α and β (aggregation of α and β); '$\overline{\alpha}$' – not α (complement of α). '$\neg$' and '$\wedge$' represent classical negation and conjunction, respectively ('$\vee$', '$\rightarrow$' and '$\equiv$' are the other standard classical operators and are defined in the standard way). Further, for fixed Act_0, by Act we shall understand the set of formulas defined by (2). Let us stress that the language has two kinds of operators: inner ones operating on names – complement and combination, and outer ones operating on propositions – the usual Boolean connectives[4].

We use obligation as the only primitive deontic operator defining weak permission, prohibition and neutrality respectively:

(3) $$\mathsf{P}(\alpha) =_{df} \neg\mathsf{O}(\overline{\alpha})$$

[4] See also [17,22,23,25,21], where the language of deontic logic is built in a similar way.

(4) $$\mathsf{F}(\alpha) =_{df} \mathsf{O}(\overline{\alpha})$$

(5) $$\mathsf{N}(\alpha) =_{df} \neg\mathsf{O}(\alpha) \wedge \neg\mathsf{O}(\overline{\alpha})$$

3.2 The meaning of $\sqcap$ operator

The crucial issue for our formalization is the interpretation of $\sqcap$ operator. Its main idea is shared among all systems we will discuss. Namely, the operator is treated as an aggregation of socially grounded intentions of two given actions. Thus, if $\alpha \sqcap \beta$ appears in a formula, then α and β have to be different descriptions that can be attached to the same particular action. Usually in this context α and β represent types of action coming from different normative systems and $\alpha \sqcap \beta$ refers to the same action when we express its final deontic status after merging the normative systems. Let us for example consider the following formula:

(6) $$\mathsf{O}(\alpha) \wedge \mathsf{N}(\beta) \rightarrow \mathsf{O}(\alpha \sqcap \beta)$$

The intended interpretation of (6) applies to actions that can be called α and β at the same time. α and β are descriptions taken from different normative systems. If in the system using description α any action of type α is obligatory and in the system using description β any action of type β is neutral, then any particular action that is both α and β is obligatory, when both normative systems are taken into account [5].

As a consequence of the accepted interpretation $\sqcap$—as aggregation—should be commutative and associative:

(7) $$\alpha \sqcap \beta = \beta \sqcap \alpha$$

(8) $$(\alpha \sqcap \beta) \sqcap \gamma = \alpha \sqcap (\beta \sqcap \gamma)$$

3.3 Matrices

We shall define our logics by means of matrix semantics. Let us now formally define the principles of matrix systems. Since we are using a 'two level' language we need a slight modification of a usual matrix semantics, which we define below.

Definition 3.1 [Deontic matrix] D-matrix for language L is a tuple $\langle D, \{\mathsf{F}, \mathsf{N}, \mathsf{O}\}, \{\overline{\ }, \sqcap\}\rangle$, where: D is a non-empty set of deontic values; $\{\mathsf{F}, \mathsf{N}, \mathsf{O}\}$ is a set of deontic functions from D to Fregean truth values $\{0, 1\}$; $\{\overline{\ }, \sqcap\}$ consists of functions s.t. $\overline{\ } : D \longrightarrow D$ and $\sqcap : D^2 \longrightarrow D$.

---

[5] To clarify the interpretation let us present a formal notation alternative to the one we use in the paper. Let a refer to a particular action of type $\alpha \sqcap \beta$ (so a is also of type α and of type β), k and l be labels for normative systems and $k \times l$ be a label for a system resulting from merging k and l. Let further the deontic status of actions be recorded using deontic operators (O, F or N) with the label of respective normative system as a subscript and action name as an argument, e.g.: $\mathsf{O}_l(a)$. Now formula (6) takes the form:

$$\mathsf{O}_k(a) \wedge \mathsf{N}_l(a) \rightarrow \mathsf{O}_{k\times l}(a)$$

We prefer our 'main' notation since it is simpler and much closer to the usual language of deontic action logic.

We take into account three matrix systems. In two of them a set of deontic values D will consist of three elements, i.e. $D = \{f, n, o\}$ (forbidden, neutral and obligatory respectively), and in one of them D will include four elements: $b, \top, \bot, g$, where $\top$ and $\bot$ are two special neutrality cases.

Definition 3.2 [Tautology] Formula φ is a tautology of matrix M iff $v(\varphi) = 1$ for every interpretation $\mathcal{I}$ of actions occurring in φ. A set $E(M)$ defined below is a set of tautologies of M.

$$E(M) = \{\varphi \in For : v(\varphi) = 1 \text{ for every interpretation } \mathcal{I}\}$$

4 Different accounts of judgment in the case of conflicting norms

4.1 Pessimistic view on moral dilemmas

Matrices of the pessimistic system As we have mentioned in the introductory section the settings for defining multivalued deontic logic can be found in the classical work of Kalinowski [17]. Its extension presented by M. Fisher [10] employs the pessimistic view on moral dilemmas [6]. In this approach there are three deontic values of actions, i.e., every action is either obligatory (o), forbidden (f) or indifferent/neutral (n). Then deontic operators of permission, obligation, prohibition and neutrality are characterized by referring to deontic and truth values (see table 1). One can see that the deontic operators F, N and O are language counterparts of the deontic values f, n and o, respectively.

α	$\mathsf{F}(\alpha)$	$\mathsf{N}(\alpha)$	$\mathsf{O}(\alpha)$	$\mathsf{P}(\alpha)$
f	1	0	0	0
n	0	1	0	1
o	0	0	1	1

Table 1

The action negation is defined by table 2. A complement of an obligatory action is forbidden, a complement of a forbidden action is obligatory and finally a complement of a neutral action is also neutral.

α	$\overline{\alpha}$
f	o
n	n
o	f

Table 2

$\sqcap$	f	n	o
f	f	f	f
n	f	n	o
o	f	o	o

Table 3

Operation of combination of deontic values is defined by table 3. The view corresponds to the intuition that an action that is from one point of view obligatory and from the other forbidden is here finally regarded as forbidden and

[6] Fisher's system has one more inner connective – alternative, which is defined as a De Morgan dual of the inner conjunction. We do not use that connective because we cannot find a clear intuitive reading of it in the context of our investigations.

therefore judged as bad. That reveals the tragic dimension of moral dilemma. There is no good solution, no matter what an agent does its action is bad in the end.

We can say that in this approach in the conflicting situation both norms that are connected with a compound action are considered in the deontic evaluation of the action but the fact that one of them forbids the action is taken into account as 'more important'.

Now let us try to express the Antigone example in the language. The set of basic actions Act_0 should consist of actions being relevant to the situation; in this case we shall take into account the actions described by socially grounded intentions (as we have defined it above in section 2): *complying with Creon's edict* (let us use *edict* for it) and *complying with divine laws* (*custom*). Thus, *burying Polyneices* can be normatively interpreted as: $\overline{edict} \sqcap custom$, whereas *not burying Polyneices* carries the following normative description[7]: $edict \sqcap \overline{custom}$. Providing $\mathcal{I}(edict) = o$ and $\mathcal{I}(custom) = o$, we obtain that $\mathcal{I}(\overline{edict} \sqcap custom) = f$ and $\mathcal{I}(edict \sqcap \overline{custom}) = f$.

In our formal language we could describe the situation as follows:

(9) $$\mathsf{F}(\overline{edict}) \wedge \mathsf{O}(custom) \rightarrow \mathsf{F}(\overline{edict} \sqcap custom)$$

(10) $$\mathsf{F}(\overline{custom}) \wedge \mathsf{O}(edict) \rightarrow \mathsf{F}(edict \sqcap \overline{custom})$$

Axiomatization of the pessimistic system The matrix representation of the pessimistic system is natural as it is very close to the intuitive investigations. One can, however, ask about the proof theoretic presentation of the systems.

An axiomatisation of the Fishers multivalued deontic logic from [10] with the matrices for negation and conjunction identical to our pessimistic system was presented by L. Aquist in [4][8]. We present an axiomatization that is slightly simpler. Its rules are: Modus Ponens, point substitution (substitution

[7] The Reader may wonder whether $\overline{\alpha} \sqcap \beta$ is a complement of $\alpha \sqcap \overline{\beta}$ and vice versa, i.e., whether $\overline{\overline{\alpha} \sqcap \beta} = \alpha \sqcap \overline{\beta}$. In the systems presented in this paper it is not the case. However, it can be shown that it is the case in the atomic Boolean algebra with two generators α and β and two additional axioms stating that atoms (atomic actions) $\alpha \sqcap \beta$ and $\overline{\alpha} \sqcap \overline{\beta}$ equal 0, i.e., they are impossible. We shall not enter into discussion about algebraical issues here. The interested Reader should consult section 2.3.1 in [24].

[8] Aquist's system F consists of PC laws, Modus Ponens, Substitution rule, Replacement in PC-theorems, and ten axioms: Three of them are mentioned explicitly:

(11) $$\mathsf{O}(a) \rightarrow \mathsf{P}(\overline{a})$$

(12) $$\mathsf{P}(a) \wedge \mathsf{O}(b) \rightarrow \mathsf{O}(a \sqcap b)$$

(13) $$\mathsf{P}(a) \wedge \mathsf{P}(b) \rightarrow \mathsf{P}(a \sqcap b)$$

and the next seven obtained by 'Extensionality (restricted)' rule that turns the seven PC laws chosen by Aquist into deontic formulas, e.g. the rule states that PC law

$$p \wedge q \rightarrow p$$

can be transformed into

(14) $$\mathsf{P}(a \sqcap b) \rightarrow \mathsf{P}(a)$$

by replacing p/a, q/b and putting P in the antecedent and consequent.

for action variables) and extensionality for identity of actions understood as synonymy of action descriptions. Action identity is defined by axioms: (7) and (8) and the following double complement equation:

(15) $$\alpha = \overline{\overline{\alpha}}$$

We assume that all substitutions of PC laws in the system are common axioms of the system. For the specific axioms we use the following formulas:

(16) $$\neg(\mathsf{O}(\alpha) \wedge \mathsf{F}(\alpha))$$

(17) $$\mathsf{F}(\alpha) \rightarrow \mathsf{F}(\alpha \sqcap \beta)$$

(18) $$\mathsf{F}(\alpha \sqcap \beta) \rightarrow \mathsf{F}(\alpha) \vee \mathsf{F}(\beta)$$

(19) $$\mathsf{P}(\alpha) \wedge \mathsf{O}(\beta) \rightarrow \mathsf{O}(\alpha \sqcap \beta)$$

(20) $$\mathsf{O}(\alpha \sqcap \beta) \rightarrow \mathsf{O}(\alpha) \vee \mathsf{O}(\beta)$$

(16) guarantees consistency of norms coming from one source. Formulas (17) and (18) state that a compound action is forbidden if and only it has at least one forbidden component. So since each of Antigone's choices has forbidden aspects, each of them has to be forbidden. (19) states that a compound action is obligatory, providing one of its components is obligatory and the other one permitted (so not forbidden). (20) expresses the fact that an obligatory compound action implies that one of the components is obligatory. Some chosen theses of the system are in Table 8.

4.2 Optimistic view on moral dilemmas

Matrices of the optimistic system In the optimistic view on moral dilemmas the tables for deontic operators and action negation are the same as tables 1 and 2. What differentiates this approach from the previous one is the table for '⊓' (compare tables 3 and 4). With the second solution we have the op-

$\sqcap$	f	n	o
f	f	f	o
n	f	n	o
o	o	o	o

Table 4

posite situation. When we have an act exhibiting elements of something good and something bad the act is always good. Thus providing $\mathcal{I}(edict) = o$ and $\mathcal{I}(custom) = o$, we obtain that $\mathcal{I}(\overline{edict} \sqcap custom) = o$ and $\mathcal{I}(edict \sqcap \overline{custom}) = o$. In this approach Antigone did something good by carrying out the action of *burying Polyneices* (of course if she had chosen otherwise it would have been good as well).

Analogously to the statement saying that in the pessimistic system we have both conflicting norms in force we can say that in the optimistic case only one of the norms is present in a final judgment. Moreover, it is always the norm that was actually followed by the agent.

The optimistic view on moral dilemma, contrary to the approach presented in the previous section, liberates an agent from the burden of guilt in a conflicting situation. Following one obligation is enough to make the decision (whatever it is) good.

Axiomatization of the optimistic system Definitions and rules are as in the pessimistic view. Axioms of the optimistic system consist of (16), (18), (20) and the following ones:

(21) $$\neg\mathsf{O}(\alpha) \wedge \mathsf{F}(\beta) \to \mathsf{F}(\alpha \sqcap \beta)$$

(22) $$\mathsf{O}(\alpha) \to \mathsf{O}(\alpha \sqcap \beta)$$

We can see that in this approach a necessary condition for the compound action to be forbidden is that one of its components has to be forbidden (18). In this approach it is not sufficient that one of the two components of the compound action is forbidden to make the compound action forbidden (formula (17) is not a tautology of the optimistic system); it also has to be guaranteed that the second one is not obligatory (21). Thus obligation in this view is 'stronger' than prohibition in the sense that a compound action is obligatory if and only if one of its components is obligatory – (20) and (22).

Let us observe that axiom (19) of the pessimistic system follows from (22), so it is also a thesis of the optimistic system. On the other hand axiom (21) of the optimistic system follows from (17), so it is a thesis of the pessimistic one. So we may conclude that the formulas (17) and (22) are characteristic for the pessimistic and optimistic systems respectively.

Some chosen theses of the system are in Table 8.

4.3 'In dubio quodlibet' view on moral dilemmas

In this approach the combination of good and bad is treated as neutral, conflicting norms derogate one another. Thus if an action is obligatory for one reason and forbidden for another in the final judgment it is unregulated.

Again, as in the case of the optimistic system, an agent is free from the responsibility for breaking a regulation if it is impossible to follow all of them. The inconsistent norms disappear when the inconsistency is revealed.

In the case of our running example Antigone did something neutral by carrying out the action of *burying Polyneices* and doing opposite would have also been neutral.

That effect, however, cannot be achieved with trivalent matrices with the preservation of associativity, since we would have: $(o \sqcap o) \sqcap f = o \sqcap f = n$ and $o \sqcap (o \sqcap f) = o \sqcap n = o$.

Matrices For that reason we take a structure resembling the Belnap-Dunn [5] construction concerning truth and information, replacing them respectively by moral value and deontic saturation depicted in the following diagram[9]:

[9] The idea of applying Belnap's construction to deontic action logic was first introduced in a preliminary form in [20].

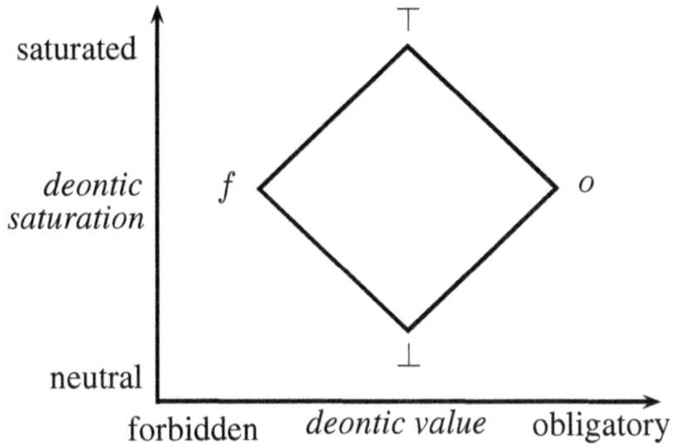

Two new deontic values appear here: $\bot$ and $\top$. The former is attached to actions that are deontically unsaturated (have no deontic value at all, are plainly neutral). The latter is attached to actions that are deontically oversaturated (have obligatory and forbidden components). Both of them are neither 'purely' obligatory nor 'purely' forbidden, and in that sense are neutral.

Formally operator '$\sqcap$' is interpreted as supremum in the structure. Moreover, negation of $\top$ is $\top$ and negation of $\bot$ is $\bot$. The definitions of the operators can be expressed by tables 5 and 6.

α	$\overline{\alpha}$
f	o
$\bot$	$\bot$
$\top$	$\top$
o	f

Table 5

$\sqcap$	f	$\bot$	$\top$	o
f	f	f	$\top$	$\top$
$\bot$	f	$\bot$	$\top$	o
$\top$	$\top$	$\top$	$\top$	$\top$
o	$\top$	o	$\top$	o

Table 6

α	$\mathsf{F}(\alpha)$	$\mathsf{N}_n(\alpha)$	$\mathsf{N}^b(\alpha)$	$\mathsf{N}(\alpha)$	$\mathsf{O}(\alpha)$	$\mathsf{P}(\alpha)$
f	1	0	0	0	0	0
$\bot$	0	1	0	1	0	1
$\top$	0	0	1	1	0	1
o	0	0	0	0	1	1

Table 7

The interpretation of the deontic atoms is defined by Table 7. The table shows that intuitively values $\bot$ and $\top$ are both treated as *neutral*. Thus, in a sense, the system remains trivalent, though formally there are four values that can be attached to actions. We have two new operators in the table: N_n and N^b that are intended to represent deontic values $\bot$ and $\top$ in the language.

The last two views on a moral dilemma, contrary to the approach presented in section 4.1, liberate the agent from the burden of guilt in a conflicting situation. If anybody is to be responsible for the situation, then it is the norm givers.

Axiomatization Definitions and rules are the same as in the pessimistic systems. We have one new definition:

$$\mathsf{N}_n(\alpha) =_{df} \mathsf{P}(\alpha) \wedge \mathsf{P}(\overline{\alpha}) \wedge \neg\mathsf{N}^b(\alpha) \tag{23}$$

Axioms of the system consist of (16), (18), (20) and the following ones:

$$\mathsf{N}^b(\alpha) \rightarrow \mathsf{N}^b(\overline{\alpha}) \tag{24}$$

$$\mathsf{N}^b(\alpha) \rightarrow \mathsf{P}(\alpha) \tag{25}$$

$$\mathsf{N}^b(\alpha) \rightarrow \neg\mathsf{O}(\alpha) \tag{26}$$

$$\mathsf{F}(\alpha) \wedge \mathsf{F}(\beta) \rightarrow \mathsf{F}(\alpha \sqcap \beta) \tag{27}$$

$$\mathsf{O}(\alpha) \wedge \mathsf{O}(\beta) \rightarrow \mathsf{O}(\alpha \sqcap \beta) \tag{28}$$

$$\mathsf{N}^b(\alpha) \vee (\mathsf{O}(\alpha) \wedge \mathsf{F}(\beta)) \rightarrow \mathsf{N}^b(\alpha \sqcap \beta) \tag{29}$$

$$\mathsf{N}^b(\alpha \sqcap \beta) \rightarrow \mathsf{N}^b(\alpha) \vee \mathsf{N}^b(\beta) \vee (\mathsf{O}(\alpha) \wedge \mathsf{F}(\beta)) \tag{30}$$

$$\mathsf{N}_n(\alpha \sqcap \beta) \rightarrow \mathsf{N}_n(\alpha) \wedge \mathsf{N}_n(\beta) \tag{31}$$

$$\mathsf{O}(\alpha) \wedge \mathsf{N}_n(\beta) \rightarrow \mathsf{O}(\alpha \sqcap \beta) \tag{32}$$

$$\mathsf{F}(\alpha) \wedge \mathsf{N}_n(\beta) \rightarrow \mathsf{F}(\alpha \sqcap \beta) \tag{33}$$

(16), (24), (25) and (26) characterize relations between deontic concepts as defined in table 5.

(18) and (27) characterize prohibition and (20) and (28) state the properties of obligation.

(29) and (30) describe deontic saturation. So according to the axioms a compound action is deontically saturated if and only if one of its two components is obligatory and the other one is forbidden or one of them is oversaturated.

Deontic unsaturation is characterized by axioms (31), (32) and (33). (31) (and the implication from right to left which is also a thesis of the system) states that a compound action is deontically unsaturated if and only if all of its components are unsaturated. (32) and (33) express the fact that obligation and prohibition are stronger than deontic unsaturation.

5 Soundness and Completeness of the systems

Theorem 5.1 *The three systems described above are sound and complete with respect to their matrices.*

Proof. We show the complete proofs for the pessimistic and optimistic systems. The proof for the 'in dubio quodlibet' system is straightforwardly analogous. For the completeness part of the proof we use the method of S. Halldén from [13] applied also in [4] for Fisher's deontic logic and in [19] for Kalinowski K_1 system.

To apply the method we have to be able to express the deontic values of actions in the language. Table 1 allows us to do that by connecting uniquely values f, n and o of action α with formulas $\mathsf{F}(\alpha)$, $\mathsf{N}(\alpha)$ and $\mathsf{O}(\alpha)$ respectively.

Thus, we can connect any formula φ and its interpretation $\mathcal{I}$ with a formula describing the interpretation for the action variables occurring in φ – let us use the symbol $\varphi_{\mathcal{I}}$ for that formula. For example, let us consider $\varphi = \mathsf{P}(\alpha \sqcap \beta)$ in which we apply the interpretation $\mathcal{I}_1$ in which we interpret α as f and β as o. Then we have[10]: $\mathsf{P}(\alpha \sqcap \beta)_{\mathcal{I}_1} = \mathsf{F}(\alpha) \wedge \mathsf{O}(\beta)$.

A formula φ is valid in a matrix system if it takes a distinguished value for any interpretation of action expressions.We have to show that in such a case we can prove φ from axioms. We will show that in two steps: (i) for any interpretation $\mathcal{I}$ under which φ takes a distinguished value the formula '$\varphi_{\mathcal{I}} \to \varphi$' is provable and (ii) the disjunction of all $\varphi_{\mathcal{I}}$ for all interpretations is also provable. That will conclude the completeness proof.

For (i) it is enough to show that there exists a thesis of the system corresponding to each 'entry' in the matrices of the system[11] – the straightforward induction on the number of occurrences of action operators completes the proof. The respective formulas are as below.

The matrix of action complement is the same for pessimistic and optimistic systems. It is to be described as follows:

(34) $$\mathsf{F}(\alpha) \to \mathsf{O}(\overline{\alpha})$$

(35) $$\mathsf{N}(\alpha) \to \mathsf{N}(\overline{\alpha})$$

(36) $$\mathsf{O}(\alpha) \to \mathsf{F}(\overline{\alpha})$$

The matrix of accumulated actions for pessimistic system is characterized below:

(37) $$\mathsf{F}(\alpha) \wedge \mathsf{F}(\beta) \to \mathsf{F}(\alpha \sqcap \beta)$$

(38) $$\mathsf{F}(\alpha) \wedge \mathsf{N}(\beta) \to \mathsf{F}(\alpha \sqcap \beta)$$

(39) $$\mathsf{F}(\alpha) \wedge \mathsf{O}(\beta) \to \mathsf{F}(\alpha \sqcap \beta)$$

(40) $$\mathsf{N}(\alpha) \wedge \mathsf{F}(\beta) \to \mathsf{F}(\alpha \sqcap \beta)$$

(41) $$\mathsf{N}(\alpha) \wedge \mathsf{N}(\beta) \to \mathsf{N}(\alpha \sqcap \beta)$$

(42) $$\mathsf{N}(\alpha) \wedge \mathsf{O}(\beta) \to \mathsf{O}(\alpha \sqcap \beta)$$

(43) $$\mathsf{O}(\alpha) \wedge \mathsf{F}(\beta) \to \mathsf{F}(\alpha \sqcap \beta)$$

(44) $$\mathsf{O}(\alpha) \wedge \mathsf{N}(\beta) \to \mathsf{O}(\alpha \sqcap \beta)$$

(45) $$\mathsf{O}(\alpha) \wedge \mathsf{O}(\beta) \to \mathsf{O}(\alpha \sqcap \beta)$$

The matrix of parallel execution for optimistic system is characterized by formulas (37), (38), (40), (41), (42), (44), (45) and two formulas below:

(46) $$\mathsf{F}(\alpha) \wedge \mathsf{O}(\beta) \to \mathsf{O}(\alpha \sqcap \beta)$$

(47) $$\mathsf{O}(\alpha) \wedge \mathsf{F}(\beta) \to \mathsf{O}(\alpha \sqcap \beta)$$

Formulas (34), (35), (36) after the replacement of defined operators with O turn to PC tautologies (in the case of (35) we have to apply the double negation identity axiom (15)).

---

[10] The formula φ is used only to determine the set of variables that are used to construct the formula $\varphi_{\mathcal{I}}$.

[11] The formula is an implication with the conditions defining the position in the matrix as the antecedent and the representation of the value in the matrix as the consequent.

Then for the pessimistic system we prove that formulas (37), (38), (39), (40), (43) follow from (17), formula (41) follows from (20) and (17), formula (42) follows from (19) and formula (44) from (19) and (7), and finally formula (45) follows from (16) and (19).

For the optimistic system we prove that (37) follows from (16) and (21);(38) from (21) and (7); (40) from (21); (41) from (22) and (18) and finally (42), (44), (45) follow from (22). Formulas (46) and (47) follow from (22).

For (ii) we have to notice that the formula in question is a disjunction of formulas representing all possible interpretations for variables in φ. Such a disjunction follows from formula: $\mathsf{F}(\alpha) \vee \mathsf{N}(\alpha) \vee \mathsf{O}(\alpha)$.

For soundness, since we use standard rules of deduction and classical understanding of proposition operators, it is enough to check that identity axioms and the specific axioms of the systems are valid tautologies of the matrices. □

formula / logic in section	4.1	4.2	4.3
$\mathsf{P}(\alpha) \vee \mathsf{P}(\overline{\alpha})$	+	+	+
$\mathsf{O}(\alpha) \to \mathsf{P}(\alpha)$	+	+	+
$\mathsf{F}(\alpha) \equiv \neg\mathsf{P}(\alpha)$	+	+	+
$\mathsf{O}(\alpha) \equiv \neg\mathsf{P}(\overline{\alpha})$	+	+	+
$\neg(\mathsf{O}(\alpha) \wedge \mathsf{O}(\overline{\alpha}))$	+	+	+
$\neg(\mathsf{O}(\alpha) \wedge \mathsf{F}(\alpha))$	+	+	+
$\mathsf{P}(\alpha) \wedge \mathsf{P}(\beta) \to \mathsf{P}(\alpha \sqcap \beta)$	+	+	+
$\mathsf{O}(\alpha) \wedge \mathsf{O}(\beta) \to \mathsf{O}(\alpha \sqcap \beta)$	+	+	+
$\mathsf{F}(\alpha) \wedge \mathsf{F}(\beta) \to \mathsf{F}(\alpha \sqcap \beta)$	+	+	+
$\mathsf{N}(\alpha) \wedge \mathsf{N}(\beta) \to \mathsf{N}(\alpha \sqcap \beta)$	+	+	+
$\mathsf{O}(\alpha \sqcap \beta) \to \mathsf{O}(\alpha) \vee \mathsf{O}(\beta)$	+	+	+
$\mathsf{O}(\alpha) \wedge \mathsf{F}(\beta) \to \mathsf{F}(\alpha \sqcap \beta)$	+	−	−
$\mathsf{F}(\alpha) \to \mathsf{F}(\alpha \sqcap \beta)$	+	−	−
$\mathsf{P}(\alpha \sqcap \beta) \to \mathsf{P}(\alpha) \wedge \mathsf{P}(\beta)$	+	−	−
$\mathsf{O}(\alpha) \wedge \mathsf{F}(\beta) \to \mathsf{O}(\alpha \sqcap \beta)$	−	+	−
$\mathsf{O}(\alpha) \to \mathsf{O}(\alpha \sqcap \beta)$	−	+	−
$\mathsf{O}(\alpha) \wedge \mathsf{F}(\beta) \to \mathsf{N}(\alpha \sqcap \beta)$	−	−	+
$\mathsf{N}(\alpha \sqcap \beta) \to \mathsf{N}(\alpha) \wedge \mathsf{N}(\beta)$	+	+	−
$\mathsf{N}(\alpha) \wedge \mathsf{O}(\beta) \to \mathsf{O}(\alpha \sqcap \beta)$	+	+	−
$\mathsf{O}(\alpha \sqcap \beta) \to \mathsf{O}(\alpha) \wedge \mathsf{O}(\beta)$	−	−	−
$\mathsf{F}(\alpha \sqcap \beta) \to \mathsf{F}(\alpha) \wedge \mathsf{F}(\beta)$	−	−	−

Table 8
'+' indicates that a formula is a tautology of the system; '−' indicated otherwise.

6 Some formulas

In Table 8 formulas being tautologies in the indicated systems are gathered. All the systems have formulas defining the relation among the notions of permission, prohibition and obligation as common tautologies. None of them has a counterpart of Ross formula $\mathsf{O}(\alpha \sqcap \beta) \to \mathsf{O}(\alpha) \wedge \mathsf{O}(\beta)$. Not surprisingly the differences occur when we deal with combinations of actions. In such cases the systems behave according to the content of their deontic tables.

From the content of table 8 it is easy to notice that none of the 3-valued systems either contain or is contained in the 4-valued system. Moreover, pessimistic and optimistic systems are in the same relation – none of them is contained in the other.

7 Conclusions and further work

Moral dilemmas are present in human considerations concerning morality from the very beginning. They are the kind of phenomenon which in principle cannot be solved – we can either look for ways of eliminating them or try to understand them better from various points of view. In the paper we have chosen the second possibility. We have presented three approaches towards assessing an agent facing a dilemma in the framework of multivalued deontic action logic.

One of them, which we call pessimistic, reveals the tragic character of dilemmas. Any action of an agent facing conflicting norms is treated as forbidden. Another one, which we call optimistic, makes us treat any decision in the case of normative conflict as good, provided that at least one obligation is fulfilled. Yet another, which we call 'in dubio quodlibet', reflects the intuition that conflicting norm derogate one another and the situation of normative conflict is treated as unregulated. Thus any decision in the presence of a dilemma is treated as deontically neutral.

Our study shows that all those points of view lead to consistent formal systems that take the form of different deontic action logics. The sound and complete axiomatic systems are presented and their theses are compared. We believe that formalizing them can help in their critical analysis.

We can say that what we have presented consists of different attempts to create a consistent normative system from mutually inconsistent components.

From the technical point of view the third system is the most interesting. It uses a four value structure similar to the Belnap-Dunn lattice used in the epistemic context.

Still a lot of questions that can be formulated within the proposed framework remain untouched in the paper. Let us just mention two of them. The first one is extending the formalization and concerns the notion of permission. In the paper we use the notion of *weak* permission understood as the lack of obligation to do the opposite ([2]). However, in normative systems we also deal with *explicit* or *strong* permission. The question remains how actions explicitly permitted in one normative system and forbidden in another should be judged.

The other problem is of a more intuitive and applicative character. In the present paper we just present three possible formal accounts of judging an agent's actions in the context of normative conflicts. It is a matter of further study to compare them with the actual practice of ethical, legal or psychological assessment of actions in real life situations.

Acknowledgements

This research was supported by the National Science Centre of Poland (grant no. UMO-2014/15/G/HS1/04514). We would also like to thank anonymous reviewers for they valuable remarks. Some of them that we could not address adequately in the present paper will be an inspiration for future works.

References

[1] Abraham, M., D. M. Gabbay and U. J. Schild, *Obligations and prohibitions in talmudic deontic logic.*, in: G. Governatori and G. Sartor, editors, *DEON*, Lecture Notes in Computer Science **6181** (2010), pp. 166–178.

[2] Anglberger, A. J., N. Gratzl and O. Roy, *Obligation, Free Choice, and the Logic of Weakest Permissions*, The Review of Symbolic Logic **8** (2015), pp. 807–827.

[3] Anscombe, E., "Intention," Ithaca, NY:Cornell University Press, 1957.

[4] Aquist, L., *Postulate sets and decision procedures for some systems of deontic logic*, Theoria **29** (1963), pp. 154–175.

[5] Belnap, N., *A useful four-valued logic*, in: J. Dunn and G. Epstein, editors, *Modern uses of multiple-valued logic*, Springer, 1977 pp. 8–37.

[6] Boella, G. and L. W. van der Torre, *A game-theoretic approach to normative multi-agent systems*, in: *Normative Multi-agent Systems*, 2007.

[7] Craven, R., *Policies, Norms and Actions: Groundwork for a Framework*, Technical report, Imperial College, Department of Computing (2011).

[8] Davidson, D., "Essays on Actions and Events," Clarendon Press, Oxford, 1991.

[9] De Haan, J., *The Definition of Moral Dilemmas: A Logical Problem*, Ethical Theory and Moral Practice **4** (2001), pp. 267–284.

[10] Fisher, M., *A three-valued calculus for deontic logic*, Theoria **27** (1961), pp. 107–118.

[11] Goble, L., *A logic for deontic dilemmas*, Journal of Applied Logic **3** (2005), pp. 461–483.

[12] Gowans, C. W., "Innocence Lost: An Examination of Inescapable Moral Wrongdoing," Oxford University Press, 1994.

[13] Halldén, S., "The Logic of Nonsense," Upsala Universitets Arsskrift, 1949.

[14] Holbo, J., *Moral Dilemmas and the Logic of Obligation*, Americal Philosophical Quarterly **39** (2002), pp. 259–274.

[15] Horty, J. F., *Moral dilemmas and nonmonotonic logic*, Journal of Philosophical Logic **23** (1994), pp. 35–65.

[16] Horty, J. F., "Agency and Deontic Logic," Oxford University Press, Oxford, 2001.

[17] Kalinowski, J., *Theorie des propositions normatives*, Studia Logica **1** (1953), pp. 147–182.

[18] Kouznetsov, A., *Quasi-matrix deontic logic*, in: A. Lomuscio and D. Nute, editors, *Deontic Logic in Computer Science*, Lecture Notes in Computer Science **3065**, Springer, 2004 pp. 191–208.

[19] Kulicki, P., *A note on the adequacy of Jerzy Kalinowski K1 logic*, Bulletin of the Section of Logic (2014), pp. 183–190.

[20] Kulicki, P. and R. Trypuz, *How to build a deontic action logic*, in: M. Pelis and V. Punchochar, editors, *The Logica Yearbook 2011*, College Publications, 2012 pp. 107–120.

[21] Kulicki, P. and R. Trypuz, *Completely and partially executable sequences of actions in deontic context*, Synthese **192** (2015), pp. 1117–1138.

[22] Segerberg, K., *A deontic logic of action*, Studia Logica **41** (1982), pp. 269–282.

[23] Trypuz, R. and P. Kulicki, *A systematics of deontic action logics based on boolean algebra*, Logic and Logical Philosophy **18** (2009), pp. 263–279.

[24] Trypuz, R. and P. Kulicki, *Jerzy Kalinowski's logic of normative sentences revisited*, Studia Logica **103** (2015), pp. 389–412.

[25] Trypuz, R. and P. Kulicki, *On deontic action logics based on boolean algebra*, Journal of Logic and Computation **25** (2015), pp. 1241–1260.

[26] Williams, B. A. O. and W. F. Atkinson, *Symposium: Ethical Consistency*, Aristotelian Society Supplementary Volume **39** (1965), pp. 103–138.

Prioritized Norms and Defaults in Formal Argumentation

Beishui Liao

Zhejiang University, China
baiseliao@zju.edu.cn

Nir Oren

University of Aberdeen, UK
n.oren@abdn.ac.uk

Leendert van der Torre

University of Luxembourg, Luxembourg
leon.vandertorre@uni.lu

Serena Villata

CNRS, Laboratoire I3S, France
villata@i3s.unice.fr

Abstract

Deontic logic sentences define what an agent ought to do when faced with a set of norms. These norms may come into conflict such that a priority ordering over them is necessary to resolve these conflicts. Dung's seminal paper raised the — so far open — challenge of how to use formal argumentation to represent non monotonic logics, highlighting argumentation's value in exchanging, communicating and resolving possibly conflicting viewpoints in distributed scenarios. In this paper, we propose a formal framework to study various properties of prioritized non monotonic reasoning in formal argumentation, in line with this idea. More precisely, we show how a version of prioritized default logic and Brewka-Eiter's construction in answer set programming can be obtained in argumentation via the weakest and last link principles. We also show how to represent Hansen's recent construction for prioritized normative reasoning by adding arguments using weak contraposition via permissive norms, and their relationship to Caminada's "hang yourself" arguments.

Keywords: Abstract argumentation theory, prioritized normative reasoning.

1 Introduction

Since the work of Alchourrón and Makinson [1] on hierarchical normative systems, in which a priority or strength is associated with the authority which promulgated a norm, reasoning with priorities of norms has been a central challenge in deontic logic. This has led to a variety of non-monotonic formalisms for prioritized reasoning in deontic logic, including a well known approach from prioritized default logic (PDL) and answer set programming — recently given argumentation semantics [13] (and to which we refer as the *greedy* approach); an approach by Brewka and Eiter [3] (which we refer to as the Brewka-Eiter construction); and a recent approach in hierarchical normative reasoning by Hansen [9], which we refer to as the Hansen construction. Given as input a set of norms with priorities, these approaches may produce different outputs. Consider the following benchmark example introduced by Hansen [9], and which results in the *prioritized triangle.*

Example 1.1 [Prioritized triangle – Hansen [9]]

Imagine you have been invited to a party. Before the event, you receive several imperatives, which we consider as the following set of norms.
- Your mother says: if you drink (p), then don't drive ($\neg x$).
- Your best friend says: if you go to the party (a), then you'll drive (x) us.
- An acquaintance says: if you go to the party (a), then have a drink with me (p).

We assign numerical priorities to these norms, namely '3', '2' and '1' corresponding to the sources 'your mother', 'your best friend' and 'your acquaintance', respectively. Whereas default and answer set programming-based approaches derive p, Hansen [9] argues convincingly that in normative reasoning p should not be derived. Meanwhile, the greedy approach and the Hansen construction return x, but the Brewka-Eiter construction returns $\neg x$.

Given that these different non-monotonic approaches yield different results, and further given Young and colleagues [13] representation result for prioritized default logic in argumentation, we wish to investigate the representation of such prioritized normative systems in formal argumentation. Therefore, the research question we answer in this paper is: *how can Brewka-Eiter's and Hansen's approaches for prioritized non monotonic reasoning be represented in formal argumentation?*

In this paper, we aim to make as few commitments as possible to specific argumentation systems. We therefore build on Tosatto et al. [11]'s abstract normative systems, and a relatively basic structured argumentation framework which admits undercuts and rebuts between arguments, and allows for priorities between rules making up arguments. We show that different approaches to lifting priorities from rules to arguments (based on the weakest and last link principles) allow us to capture the greedy and Brewka-Eiter approaches, while the introduction of additional arguments through the principle of weak contraposition, or through so called *hang yourself arguments*, allows us to obtain the Hansen construction.

A key point of our formal framework is that it addresses the challenge raised by Dung [6] aiming at representing non-monotonic logics through formal argumentation. In particular, argumentation is a way to exchange and communicate viewpoints, thus having an argumentation theory representing a non-monotonic logic is desirable for such a logic, in particular when the argumentation theory is simple and efficient. Note that it is not helpful for the development of non-monotonic logics themselves, but it helps when we want to apply such logics in distributed and multiagent scenarios.

The layout of the paper is as follows. First, we introduce our formal framework, and the three constructions. Second, we present our representation results, and demonstrate the relation between weak contraposition and hang yourself arguments. Finally, in concluding remarks, we discuss the main contributions of our approach, and highlight the future directions to be investigated.

2 Prioritised abstract normative system

In this section, we introduce the notion of prioritized abstract normative system (PANS) and three different approaches to compute what normative conclusions hold (referred to as an *extension*). A PANS captures the context of a system and the normative rules in force in such a system, together with a set of permissive norms which identify exceptions under which the normative rules should not apply. There is an element in the universe called $\top$, contained in every context, and in this paper we consider only a finite universe. A PANS also encodes a ranking function over the normative rules to allow for the resolution of conflicts.

Tosatto et al. [11] introduce a graph based reasoning framework to classify and organize theories of normative reasoning. Roughly, an *abstract normative system* (ANS) is a directed graph, and a context is the set of nodes of the graph containing the universe. In a context, an abstract normative system generates or produces an obligation set, a subset of the universe, reflecting the obligatory elements of the universe.

Based on the notion of abstract normative system defined by Tosatto and colleagues [11], a PANS is defined as follows.

Definition 2.1 [Prioritized abstract normative system] A prioritized abstract normative system PANS is a tuple $\mathcal{P} = \langle L, N, P, A, r\rangle$, where

- $L = E \cup \{\neg e \mid e \in E\} \cup \{\top\}$ is the universe, a set of literals based on some finite set E of atomic elements;
- $N \subseteq L \times L$ is a set of ordinary norms;
- $P \subseteq L \times L$ is a set of permissive norms;
- $A \subseteq L$ is a subset of the universe, called a context, such that for all a in E, $\{a, \neg a\} \not\subseteq A$;
- $r : N \cup P \to I\!N$ is a function from the norms to the natural numbers;

and where $N \cap P = \emptyset$.

Ordinary norms are of the kind "if you go to the party, then you should have a drink with me", whilst permissive norms take the form of statements such as "if you go to the party, then you don't have to have a drink with me". Both ordinary norms and permissive norms are *conditional norms*, requiring some condition to hold (e.g., going to the party) before their conclusion can be drawn. To distinguish the ordinary norms of N from the permissive norms of P, we write (a, x) for the former and $\langle a, x\rangle$ for the latter, where $a, x \in L$ are the antecedent and conclusion of the norm respectively. When no confusion can arise, a permissive norm is also represented as (a, x). Let $u, v \in N \cup P$ be two norms, we say that v is at least as preferred as u (denoted $u \leq v$) if and only if $r(u)$ is no more than $r(v)$ (denoted $r(u) \leq r(v)$), where $r(u)$ is also called a rank of u. We write $u < v$ or $v > u$ iff $u \leq v$ and $v \nleq u$. Given a norm $u = (a, x)$ or $\langle a, x\rangle$, we write $ant(u)$ for a to represent the antecedent of the norm, and $con(u)$ for x to represent the conclusion of the norm. We say that a PANS is totally ordered if and only if the ordering $\leq$ over $N \cup P$ is antisymmetric, transitive and total. We assume that the set of norms is finite. For $a \in L$, we write $\overline{a} = \neg a$ if and only if $a \in E$, and $\overline{a} = e$ for $e \in E$ if and only if $a = \neg e$. Given a set S, we use $S \nvdash \perp$ to denote that $\nexists a, b \in S$ s.t. $a = \overline{b}$, i.e., a and b are not contradictory.

Example 2.2 [Prioritized triangle [9]] In terms of Def. 2.1, the prioritized triangle can be represented as a PANS $\mathcal{P}_1 = \langle L, N, P, A, r\rangle$, where

- $L = \{a, p, x, \neg a, \neg p, \neg x\}$,
- $N = \{(a, p), (p, \neg x), (a, x)\}$,
- $P = \emptyset$, $A = \{a, \top\}$,
- $r((a, p)) = 1$, $r((p, \neg x)) = 3$, and $r((a, x)) = 2$.

Figure 1 visualizes the prioritized triangle, with the crossed line between a and $\neg x$ denoting the norm (a, x).

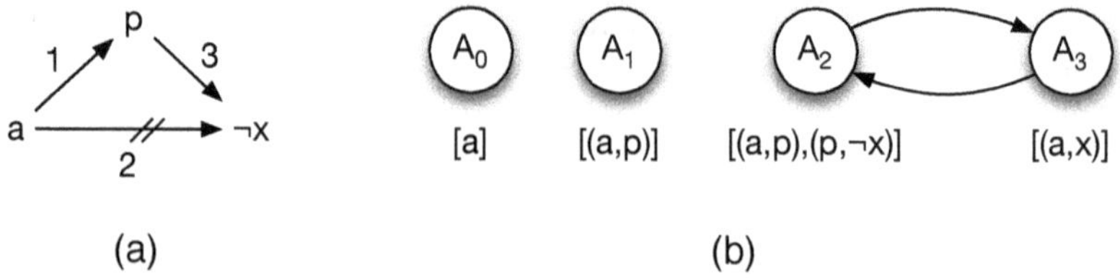

Fig. 1. The prioritized triangle (a), with the related arguments and the attacks among them visualized as directed arrows (b).

Given a totally ordered PANS, existing approaches of reasoning with prioritized norms may give different consequences. We consider three approaches (among others) that give three distinct consequences to the prioritized triangle example: the greedy approach of PDL, the Brewka-Eiter construction and the Hansen construction. Existing approaches consider only PANSs without permissive norms (i.e., $P = \emptyset$). In this paper, we extend these approaches to PANSs with permissive norms. So, the following definitions are applicable for both cases when $P = \emptyset$ and $P \neq \emptyset$.

First, a greedy approach (as used in PDL) always applies the norm with the highest priority among those which can be applied if this does not make the extension inconsistent.

Definition 2.3 [Greedy approach] Given a totally ordered PANS $\mathcal{P} = \langle L, N, P, A, r \rangle$, a norm $u \in N \cup P$ and a set $S \subseteq L$:

- We say that u is acceptable with respect to S, if and only if the following conditions holds:
  - · $ant(u) \in S$,
  - · $S \cup \{con(u)\} \nvdash \bot$, and
  - · $\nexists v \in N \cup P$ such that $v > u$, v has not been previously applied, $ant(v) \in S$, and $S \cup \{con(v)\} \nvdash \bot$.
- Let $G_{\mathcal{P}} : 2^L \rightarrow 2^L$ be a function, such that $G_{\mathcal{P}}(S) = S \cup \{con(u)\}$ if $u \in N \cup P$ is acceptable with respect to S; otherwise, $G_{\mathcal{P}}(S) = S$.
- Given A, $G_{\mathcal{P}}$ has a fixed point (denoted as $G^{\infty}_{\mathcal{P}}(A)$, such that the extension of $\mathcal{P}$ by using the Greedy approach (denoted as *Greedy*($\mathcal{P}$)) is equal to:

$$\begin{aligned}\{a \in G^{\infty}_{\mathcal{P}}(A) \mid \exists \{b_1, \ldots, b_k\} \subseteq G^{\infty}_{\mathcal{P}}(A) : b_1 \in A, \\ \forall i \in \{1, \ldots, k-1\}, (b_i, b_{i+1}) \in N, \\ \text{and } (b_k, a) \in N\}\end{aligned}$$

Note that since $\mathcal{P}$ is totally ordered, using the Greedy approach guarantees that there is a unique extension.

Building on the Greedy approach, Brewka and Eiter [3] defined the following construction.

Definition 2.4 [Brewka-Eiter construction] Given a totally ordered PANS $\mathcal{P} = \langle L, N, P, A, r \rangle$, and a set $X \supseteq A$:

- Let $\mathcal{P}^X = \langle L, N', P', A, r' \rangle$, where
  - · $N' = \{(\top, l_2) \mid (l_1, l_2) \in N, l_1 \in X\}$ is the set of ordinary norms,
  - · $P' = \{\langle \top, l_2 \rangle \mid \langle l_1, l_2 \rangle \in P, l_1 \in X\}$ is the set of permissive norms,
  - · and $r'((\top, l_2)) = r((l_1, l_2))$ for all $(l_1, l_2) \in N \cup P$ are priorities over norms.
- If $X =$ *Greedy*($\mathcal{P}^X$), then X is an extension of $\mathcal{P}$ by using the Brewka-Eiter construction, denoted as $X \in$ *BnE*($\mathcal{P}$).

Our definition (Def. 2.4) and the original formalism of Brewka and Eiter [3] are different, in the sense that in our definition we do not make use of default negation to represent the exceptions, i.e., the defeasibility of a (strict) rule, but we use defeasible rules and the notion of applicability of such rules. This means that the correct translation of the prioritized triangle of Example 2.2 ends up as the following logic program [1]:

[1] Note that in [3] $\mathtt{r}_0 < \mathtt{r}_3$ means that $\mathtt{r}_0$ has higher priority than $\mathtt{r}_3$.

```
r0 : a.
r1 : p :- not ¬p, a.
r2 : x :- not ¬x, a.
r3 : ¬x :- not x, p.
r0 < r3 < r2 < r1
```

If priorities are disregarded, then this logic program has two answer sets: $\{a, p, x\}$ and $\{a, p, \neg x\}$. Thus, considering priorities, the former is the unique preferred answered set, as pointed out in Example 2.6 below.

Similarly, Hansen [9] defined the following construction by building on the Greedy approach.

Definition 2.5 [Hansen construction] Given a totally ordered PANS $\mathcal{P} = \langle L, N, P, A, r\rangle$:

- Let $T = \{u_1, u_2, \ldots, u_n\}$ be a linear order on $N \cup P$ such that $u_1 > u_2 > \cdots > u_n$.
- For all $R \subseteq N \cup P$, let $R(A) = \{x \mid x$ can be derived from A with respect to $R\}$.
- We define a set Φ as $\Phi = \Phi_n$ such that
  - $\Phi_0 = \emptyset$,
  - $\Phi_{i+1} = \Phi_i \cup \{u_i\}$, if $A \cup R(A) \nvdash \perp$ where $R = \Phi_i \cup \{u_i\}$; otherwise, $\Phi_{i+1} = \Phi_i$.
- The extension of $\mathcal{P}$ by using Hansen construction (denoted as *Hansen*($\mathcal{P}$)) is equal to *Greedy*($\mathcal{P}'$), where $\mathcal{P}' = \langle L, N', P', A, r\rangle$, in which $N' = N \cap \Phi$ and $P' = P \cap \Phi$.

Example 2.6 [Prioritised triangle: extensions] Regarding $\mathcal{P}_1$ in Example 2.2, we get three different extensions when using these approaches. For the greedy approach we obtain $S_1 = \{a\}$, $G^1_{\mathcal{P}_1}(S_1) = \{a, x\}$, $G^{\infty}_{\mathcal{P}_1}(S_1) = G^2_{\mathcal{P}_1}(S_1) = \{a, p, x\}$. For the Brewka-Eiter construction, given $X = \{a, p, \neg x\}$, we have $\mathcal{P}_1^X = \langle L, N', P', A, r'\rangle$, where $N' = \{(\top, p), (\top, x), (\top, \neg x)\}$, $P' = \emptyset$, $r'((\top, p)) = 1$, $r'((\top, \neg x)) = 3$ and $r'((\top, x)) = 2$; *Greedy*($\mathcal{P}_1^X$) $= X$. Since no other set could be an extension, *BnE*($\mathcal{P}$) $= \{\{a, p, \neg x\}\}$. Finally, for the Hansen construction, let $u_1 = (p, \neg x)$, $u_2 = (a,\ x)$, and $u_3 = (a, p)$, and $T = \{u_1, u_2, u_3\}$. Then $\Phi_0 = \emptyset$, $\Phi_1 = \{u_1\}$, $\Phi_2 = \{u_1, u_2\}$, and $\Phi = \Phi_3 = \Phi_2 = \{u_1, u_2\}$. So, $\mathcal{P}_1' = \langle L, N', P', A,\ r\rangle$, where $N' = \{u_1, u_2\}$, $P' = \emptyset$. Since *Greedy*($\mathcal{P}_1'$) $= \{a, x\}$, *Hansen*($\mathcal{P}_1$) $= \{a, x\}$.

3 Argumentation theory for a PANS

In this section, we introduce an argumentation theory on prioritised norms. This theory builds on ideas from *ASPIC*$^+$ [10]. Given a PANS, we first define arguments and defeats between them, then compute extensions of arguments in terms of Dung's theory [6], and from these, obtain conclusions.

In a PANS, an argument is an acyclic path in the graph starting in an element of the context. We assume minimal arguments — no norm can be applied twice in an argument and no redundant norm is included in an argument. Permissions are undercutting arguments containing at least one permissive norm. We use $concl(\alpha)$ to denote the conclusion of an argument α, and $concl(E) = \{concl(\alpha) \mid \alpha \in E\}$ for the conclusions of a set of arguments E.

Definition 3.1 [Arguments and sub-arguments] Let $\mathcal{P} = \langle L, N, P, A, r\rangle$ be a PANS.

A context argument in $\mathcal{P}$ is an element $a \in A$, and its conclusion is $concl(a) = a$.

An ordinary argument in $\mathcal{P}$ is an acyclic path $\alpha = [u_1, \ldots, u_n]$, $n \geq 1$, such that:
(i) $\forall i \in \{1, \ldots, n\}$, $u_i \in N$;
(ii) $ant(u_1) \in A$;
(iii) $con(u_i) = ant(u_{i+1})$, $1 \leq i \leq n-1$;
(iv) $\{ant(u_1), \ldots, ant(u_n)\} \nvdash \bot$; and
(v) $\nexists i, j \in \{1, \ldots, n\}$ such that $i \neq j$ and $u_i = u_j$.
Moreover, we have that $concl(\alpha) = con(u_n)$.

An undercutting argument in $\mathcal{P}$ is defined in terms of an ordinary argument, by replacing the first condition with (1') $\exists i \in \{1, \ldots, n\}$ such that $u_i \in P$.

The sub-arguments of argument $[u_1, \ldots, u_n]$ are, for $1 \leq i \leq n$, $[u_1, \ldots, u_i]$. Note that context arguments do not have sub-arguments.

The set of all arguments constructed from $\mathcal{P}$ is denoted as $Arg(\mathcal{P})$. For readability, $[(a_1, a_2), \ldots, (a_{n-1}, a_n)]$ may be written as $(a_1, a_2, \ldots, a_{n-1}, a_n)$. The set of sub-arguments of an argument α is denoted as $sub(\alpha)$.

We follow the tradition in much of preference-based argumentation [2,10], and use *defeat* as the relation among arguments on which the semantics is based, whereas *attack* is used for a relation among arguments which does not take the priorities among arguments into account. To define the defeat relation among prioritized arguments, we assume that *only* the priorities of the norms are used to compare arguments. In other words, we assume a lifting of the ordering on norms to a binary relation on sequences of norms, written as $\alpha \succeq \beta$, where α and β are two arguments, indicating that α is at least as preferred as β.

There is no common agreement about the best way to lift $\geq$ to $\succeq$. In argumentation, two common approaches are the weakest and last link principles, combined with the elitist and democratic ordering [10]. However, Young and colleagues [13] show that elitist weakest link cannot be used to calculate $\succeq$, and proposes a *disjoint elitist order* which ignores shared rules. Based on these ideas we define the orderings between arguments according to the weakest link and last link principles (denoted as $\succeq_w$ and $\succeq_l$ respectively) as follows.

Definition 3.2 [Weakest link and last link] Let $\mathcal{P} = \langle L, N, P, A, r\rangle$ be a PANS, and $\alpha = [u_1, \ldots, u_n]$ and $\beta = [v_1, \ldots, v_m]$ be two arguments in $Arg(\mathcal{P})$. Let $\Phi_1 = \{u_1, \ldots, u_n\}$ and $\Phi_2 = \{v_1, \ldots, v_m\}$. By the weakest link principle,

$\alpha \succeq_w \beta$ iff $\exists v \in \Phi_2 \setminus \Phi_1$ s.t. $\forall u \in \Phi_1 \setminus \Phi_2, v \leq u$. By the last link principle, $\alpha \succeq_l \beta$ iff $u_n \geq v_m$.

When the context is clear, we write $\succeq$ for $\succeq_w$ or $\succeq_l$. We write $\alpha \succ \beta$ for $\alpha \succeq \beta$ without $\beta \succeq \alpha$.

Given a way to lift the ordering on norms to an ordering on arguments, the notion of defeat can be defined.

Definition 3.3 [Defeat among arguments] Let $\mathcal{P} = \langle L, N, P, A, r\rangle$ be a PANS. For all $\alpha, \beta \in Arg(\mathcal{P})$,

α **attacks** β iff β has a sub-argument β' such that
(i) $concl(\alpha) = \overline{concl(\beta')}$

α **defeats** β iff β has a sub-argument β' such that
(i) $concl(\alpha) = \overline{concl(\beta')}$ and
(ii) α is a context argument, or $\beta' \not\succ \alpha$.

The set of defeats between the arguments in $Arg(\mathcal{P})$ is denoted as $Def(\mathcal{P}, \succeq)$.

In what follows, an argument $\alpha = [u_1, \ldots, u_n]$ with ranking on norms is denoted as $u_1 \ldots u_n : r(\alpha)$, where $r(\alpha) = (r(u_1), \ldots, r(u_n))$.

Example 3.4 [Prioritised triangle, continued] Consider the prioritised triangle in Example 2.2. We have the following arguments, visualized in Figure 1.b:

A_0 a (context argument)

A_1 $(a, p) : (1)$ (ordinary argument)

A_2 $(a, p)(p, \neg x) : (1, 3)$ (ordinary argument)

A_3 $(a, x) : (2)$ (ordinary argument)

We have that A_2 attacks A_3 and vice versa, and there are no other attacks among the arguments. Moreover, A_2 defeats A_3 if $(2) \not\succ (1,3)$ (last link), and A_3 defeats A_2 if $(1,3) \not\succ (2)$ (weakest link).

It is worth mentioning that Dung [7] proposes the notion of a *normal attack relation*, which satisfies some desirable properties that cannot be satisfied by the $ASPIC^+$ semantics, i.e., the semantics of structured argumentation with respect to a given ordering of structured arguments (elitist or democratic pre-order) in $ASPIC^+$. In the context of the current paper, this notion could be defined as follows. Let $\alpha = (a_1, \ldots, a_n)$ and $\beta = (b_1, \ldots, b_m)$ be arguments constructed from a PANS. Since we have no Pollock style undercutting argument (as in $ASPIC^+$) and each norm is assumed to be defeasible, it says that α normally attacks argument β iff β has a sub-argument β' s.t. $concl(\alpha) = \overline{concl(\beta')}$, and $r((a_{n-1}, a_n)) \geq r((b_{m-1}, b_m))$. According to Def. 3.2 and 3.3, the normal defeat relation is equivalent to the defeat relation using the last link principle in this paper.

Given a set of arguments $\mathcal{A} = Arg(\mathcal{P})$ and a set of defeats $\mathcal{R} = Def(\mathcal{P}, \succeq)$, we get an argumentation framework (AF) $\mathcal{F} = (\mathcal{A}, \mathcal{R})$. For a set $B \subseteq \mathcal{A}$, B is conflict-free iff $\nexists \alpha, \beta \in B$ s.t. $(\alpha, \beta) \in \mathcal{R}$. B defends an argument α iff $\forall(\beta, \alpha) \in \mathcal{R}$, $\exists \gamma \in B$ s.t. $(\gamma, \beta) \in \mathcal{R}$. The set of arguments defended by B

in $\mathcal{F}$ is denoted as $\mathcal{D}_{\mathcal{F}}(B)$. A set of B is a complete extension of $\mathcal{F}$, iff B is conflict-free and $B = \mathcal{D}_{\mathcal{F}}(B)$. B is a preferred (grounded) extension iff B is a maximal (resp. minimal) complete extension. B is a stable extension, iff B is conflict-free, and $\forall \alpha \in \mathcal{A} \setminus B$, $\exists \beta \in B$ s.t. $(\beta, \alpha) \in \mathcal{R}$. We use $sem \in \{cmp, prf, grd, stb\}$ to denote complete, preferred, grounded, or stable semantics. A set of argument extensions of $\mathcal{F} = (\mathcal{A}, \mathcal{R})$ is denoted as $sem(\mathcal{F})$. Then, we write *Outfamily* for the set of conclusions from the extensions of the argumentation theory, as in [12].

Definition 3.5 [Conclusion extensions] Given a prioritised abstract normative system $\mathcal{P} = \langle L, N, P, A, r \rangle$, let $\mathcal{F} = (Arg(\mathcal{P}), Def(\mathcal{P}, \succeq))$ be the AF constructed from $\mathcal{P}$. The conclusion extensions, written as $Outfamily(\mathcal{P}, \succeq, sem)$, are the conclusions of the ordinary and context arguments in argument extensions.

$$\{\{concl(\alpha) \mid \alpha \in S, \alpha \text{ is an ordinary or context argument}\} \mid S \in sem(\mathcal{F})\}$$

Multi-extension semantics can yield different conclusions when norms may yield multiple most preferred results. Additionally, it is important to note that conclusions of a PANS are drawn only from ordinary and context arguments.

Example 3.6 [Prioritized triangle, continued] According to Example 3.4, let $\mathcal{A} = \{A_0, \ldots, A_3\}$. We have $\mathcal{F}_1 = (\mathcal{A}, \{(A_2, A_3)\})$ where $A_2 \succeq_l A_3$, and $\mathcal{F}_2 = (\mathcal{A}, \{(A_3, A_2)\})$ where $A_3 \succeq_w A_2$. For all $sem \in \{cmp, prf, grd, stb\}$, $Outfamily(\mathcal{P}, \succeq_l, sem) = \{\{a, p, \neg x\}\}$, and $Outfamily(\mathcal{P}, \succeq_w, sem) = \{\{a, p, x\}\}$.

We now turn our attention to the properties of the argumentation theory for a PANS. Since all norms in a PANS are defeasible, it is obvious that our theory maps to the framework of $ASPIC^+$. According to the corresponding properties in [10], the following three propositions follow directly.

Proposition 3.7 *Let $\mathcal{F} = (\mathcal{A}, \mathcal{R})$ be an AF constructed from a PANS. For all $\alpha, \beta \in \mathcal{A}$: if α attacks β, then α attacks arguments that have β as a sub-argument; if α defeats β, then α defeats arguments that have β as a sub-argument.*

Proposition 3.8 (Closure under sub-arguments) *Let $\mathcal{F} = (\mathcal{A}, \mathcal{R})$ be an AF constructed from a PANS. For all $sem \in \{cmp, prf, grd, stb\}$, $\forall E \in sem(\mathcal{F})$, if an argument $\alpha \in E$, then $sub(\alpha) \subseteq E$.*

Proposition 3.9 (Consistency) *Elements of Outfamily are conflict free.*

The following two properties formulate the relations between non-argument-based and argument-based approaches for reasoning with a totally ordered PANS without permissive norms.

Proposition 3.10 (Greedy is weakest link) *Given a totally ordered PANS $\mathcal{P} = \langle L, N, P, A, r \rangle$ where $P = \emptyset$, and $\mathcal{F} = (Arg(\mathcal{P}), Def(\mathcal{P}, \succeq_w))$. It holds that $\mathcal{F}$ is acyclic, and $Greedy(\mathcal{P}) = concl(E)$ where E is the unique complete extension of $\mathcal{F}$.*

Proof. First, since $\mathcal{P}$ is totally ordered, under $\succeq_w$, the relation $\succeq_w$ among arguments is acyclic. Hence, $\mathcal{F}$ is acyclic, and therefore has a unique extension under all argumentation semantics mentioned above.

Second, given *Greedy*($\mathcal{P}$), let $E = \{(a_1, \ldots, a_n) \in Arg(\mathcal{P}) \mid \{a_1, \ldots, a_n\} \subseteq Greedy(\mathcal{P})\}$. According to Def. 2.3, it holds that $concl(E) = Greedy(\mathcal{P})$. Now, we verify that E is a stable extension of $\mathcal{F}$:

(1) Since all premises and the conclusion of each argument of E are contained in *Greedy*($\mathcal{P}$) which is conflict-free, it holds that E is conflict-free.

(2) $\forall \beta = (b_1, \ldots, b_m) \in Arg(\mathcal{P}) \setminus E$, $b_m \notin Greedy(\mathcal{P})$ (otherwise, if $b_m \in Greedy(\mathcal{P})$, then $(b_1, \ldots, b_{m-1}) \subseteq Greedy(\mathcal{P})$, and thus $\beta \in E$, contradicting to $\beta \notin E$). Then $\exists \alpha = (a_1, \ldots, a_n) \in E$, s.t. $a_n = \overline{b_j}$, $2 \leq j < m$. Then, we have the following two possible cases:

- (a_{n-1}, a_n) and (b_{j-1}, b_j) are applicable at the same time: in this case, since $a_n \in Greedy(\mathcal{P})$, $r((a_{n-1}, a_n)) \geq r((b_{j-1}, b_j))$. It follows that $(a_1, \ldots, a_n) \succeq_w (b_1, \ldots, b_j)$. So, β is defeated by α.
- (a_{n-1}, a_n) is applicable, (b_{j-1}, b_j) is not applicable: in this case, there are in turn two possibilities:
  · $(a_1, \ldots, a_n) \succeq_w (b_1, \ldots, b_j)$: β is defeated by α.
  · $(b_1, \ldots, b_j) \succ_w (a_1, \ldots, a_n)$: in this case, $\exists \gamma = (c_1, \ldots, c_k) \in E$ s.t.: $c_k = \overline{b_i}$, $(c_1, \ldots, c_k) \succeq_w (b_1, \ldots, b_i)$, $2 \leq i < j$. Then, β is defeated by γ.

Since E is conflict-free and for all $\beta \in Arg(\mathcal{P}) \setminus E$, β is defeated by an argument in E, E is a stable extension. Since $\mathcal{F}$ is acyclic, E is the unique complete extension of $\mathcal{F}$. □

Proposition 3.11 (Brewka-Eiter is last link) *Given a totally ordered PANS* $\mathcal{P} = \langle L, N, P, A, r\rangle$ *where* $P = \emptyset$, *and* $\mathcal{F} = (Arg(\mathcal{P}), Def(\mathcal{P}, \succeq_l))$. *It holds that* $BnE(\mathcal{P}) = \{concl(E) \mid E \in stb(\mathcal{F})\}$.

Proof. ($\Rightarrow$:) $\forall H \in BnE(\mathcal{P})$, let $E = \{(a_1, \ldots, a_n) \in Arg(\mathcal{P}) \mid \{a_1, \ldots, a_n\} \subseteq H\}$. According to the Brewka-Eiter construction [3], $H = concl(E)$, because $\forall a \in H$, there exists at least one argument $(a_1, \ldots, a_n)$ s.t. $a_n = a$ and $\{a_1, \ldots, a_{n-1}\} \subseteq H$, which is in turn because if $a_n \in H$, then (a_{n-1}, a_n) is applicable w.r.t. H, and hence $a_{n-1} \in H$; recursively, we have $a_i \in H$ for all $i \in \{1, \ldots, n-1\}$.

Let $(Args_0, Defeats_0)$ be an AF, in which $Args_0 = \{\alpha \mid sub(\alpha) \subseteq E\}$, $Defeats_0 \subseteq Args_0 \times Args_0$ that is constructed in terms of the last link principle. It holds that $Defeats_0 \subseteq Def(\mathcal{P}, \succeq_l)$. For all $\alpha \in Args_0 \setminus E$, $concl(\alpha) \notin H$. Then, $\exists \beta \in E$ s.t. $concl(\alpha) = \overline{concl(\beta)}$ and β defeats α by using the last link principle. It follows that E is a stable extension of $(Args_0, defeats_0)$. Now, let us prove that E is a stable extension of $\mathcal{F}$.

We need only to verify that for all $\alpha \in Arg(\mathcal{P}) \setminus Args_0$, α is defeated by E. It follows that α has at least one sub-argument (otherwise, it should be included in E, contradicting $\alpha \notin Args_0$). Let β be a sub-argument of α such that β has no sub-argument. It follows that β is in $Args_0$. Then we have the following two possible cases:

- β is defeated by E: In this case, α is defeated by E.
- β is not defeated by E: In this case, β is in E (since E is a stable extension). Then, according to the definition of $Args_0$, the direct super argument of β (say β') is in $Args_0$. We in turn have two possible cases similar to the cases with respect to β. Recursively, we may conclude that α is defeated by E or, α is in E (this case does not exist).

($\Leftarrow$:) For all $E \in stb(\mathcal{F})$, let $\mathcal{P}' = \langle L, N', P', A, r' \rangle$ where $P' = \emptyset$, $N' = \{(\top, b) \mid (a, b) \in N \text{ and } a \in concl(E)\}$, and $r'(\top, b) = r(a, b)$ for all $(a, b) \in N$ and $a \in concl(E)$.

Let $E' = \{(\top, a_n) \mid (a_1, \ldots, a_n) \in E\}$.

In order to prove that $concl(E)$ is an extension of $\mathcal{P}$ in terms of the Brewka-Eiter construction, according to Proposition 3.10, we only need to verify that E' is a stable extension of $(Arg(\mathcal{P}'), Def(\mathcal{P}', \succeq'_w))$ which is an AF of $\mathcal{P}'$ by using the weakest link principle. This is true because:

- Since E is conflict-free, E' is conflict-free.
- For all $\beta' \in Arg(\mathcal{P}') \setminus E'$, let β be a corresponding argument in $Arg(\mathcal{P}) \setminus E$ s.t. $\beta = (b_1, \ldots, b_n)$, $\beta' = (\top, b_n)$, and all sub-arguments of β are in E. Since β is not in E, it is defeated by E. Since all sub-arguments of β are not defeated by E, there exists an argument in E whose conclusion is in conflict with $concl(\beta) = concl(\beta')$. So, β' is defeated by E'.

□

4 Weak contraposition

Geffner and Pearl [8] introduces conditional entailment, combining extensional and conditional approaches to default reasoning. Conditional entailment determines a prioritization of default knowledge bases. A distinguishing property of conditional entailment is what we can call *weak contraposition*, which inspires our weak contraposition property.

Output under weak contraposition is obtained by adding the contrapositives of the norms to the permissive norms. The priorities of the permissive norms are the same as the priorities of the original norms.

Definition 4.1 [Weak contraposition] Let $wcp(N) = \{\langle \overline{x}, \overline{a} \rangle \mid (a, x) \in N\}$. $Outfamily_{wcp}(\langle L, N, P, A, r \rangle, \succeq, sem) = Outfamily(\langle L, N, P \cup wcp(N), A, r' \rangle, \succeq, sem)$, where $r'(\langle \overline{x}, \overline{a} \rangle) = r((a, x))$, and $r'((a, x)) = r((a, x))$ otherwise.

In the running example we add three contrapositives. Given a contextual argument a, the undercutting arguments for $\neg a$ do not affect the result, as they are always defeated by the contextual argument. So the only additional argument to be considered is the undercutting argument for $\neg p$. This can block the argument for p, as required.

Example 4.2 [Prioritized triangle, continued] Consider $\mathcal{P}_1$ in Example 2.2, visualized in Figure 2.a. We have $wcp(N) = \{\langle \neg p, \neg a \rangle, \langle x, \neg p \rangle, \langle \neg x, \neg a \rangle\}$, and assume that contrapositives have the same priority as the original norms, i.e., $r(wcp(N)) = (1, 3, 2)$. We have the following arguments:

A_0 a (context argument)

A_1 $(a, p) : (1)$ (ordinary argument)

A_2 $(a, p)(p, \neg x) : (1, 3)$ (ordinary argument)

A_3 $(a, x) : (2)$ (ordinary argument)

A_4 $(a, x)\langle x, \neg p\rangle : (2, 3)$ (undercutting argument)

A_5 $(a, x)\langle x, \neg p\rangle\langle \neg p, \neg a\rangle : (2, 3, 1)$ (undercutting arg.)

A_6 $(a, p)(p, \neg x)\langle \neg x, \neg a\rangle : (1, 3, 2)$ (undercutting arg.)

Argument A_0 is not defeated by any argument, and defeats A_5 and A_6. We therefore consider only arguments A_1 to A_4.

As before, A_2 attacks A_3 and vice versa. In addition, A_4 attacks both A_1 and A_2, A_1 attacks A_4, and A_2 attacks A_4.

By using the last link principle, we have that A_4 defeats A_1 and thus A_2; and that A_2 defeats A_3 and thus A_4. In this case, under the stable and preferred semantics, there are two extensions $\{A_0, A_1, A_2\}$ and $\{A_0, A_3, A_4\}$. So, $Outfamily_{wcp}(\mathcal{P}_1, \succeq_l, sem) = \{\{a, p, \neg x\}, \{a, x\}\}$, where $sem \in \{prf, stb\}$.

By using the weakest link principle, we have that A_4 defeats A_1 and thus A_2; and that A_3 defeats A_2. In this case, for all $sem \in \{cmp, grd, prf, stb\}$, $\{A_0, A_3, A_4\}$ is the only extension. So, $Outfamily_{wcp}(\mathcal{P}_1, \succeq_w, sem) = \{\{a, x\}\}$.

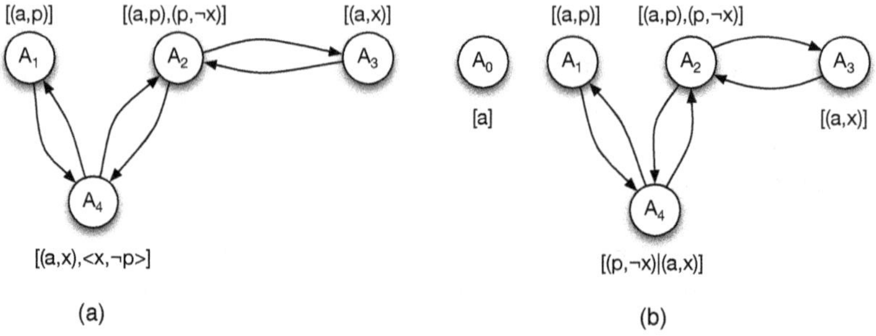

Fig. 2. The prioritized triangle of Example 4.2 (a) and of Example 5.3 (b).

The following proposition shows that the Hansen construction can be represented in formal argumentation by weakest link, if the set of permissive norms is extended with the contrapositions of the norms in N. Note that to capture Hansen's reading of the prioritized triangle, we need to add more structure to the example. The proof is along the lines of the proof of Proposition 3.10.

Proposition 4.3 (Hansen is weakest link plus wcp) *Given a totally ordered PANS* $\mathcal{P} = \langle L, N, P, A, r\rangle$ *where* $P = \emptyset$, $\mathcal{P}' = \langle L, N, P', A, r'\rangle$ *with* $P' = wcp(N)$ *and* $r'(\langle \overline{x}, \overline{a}\rangle) = r((a, x))$, *and* $r'((a, x)) = r((a, x))$ *otherwise, and* $\mathcal{F} = (Arg(\mathcal{P}'), Def(\mathcal{P}', \succeq_w))$. *It holds that* $Hansen(\mathcal{P}) = concl(E)$ *where* E *is the set of ordinary arguments of the unique complete extension of* $\mathcal{F}$.

Proof. [Sketch] First, closure under sub-arguments and consistency follow from Proposition 2 and 3, because we reuse the definitions of weakest link.

Second, $\mathcal{P}$ does not have to be totally ordered, as there may be permissive norms with the same rank as one of the ordinary norms. Thus, it no longer holds that $Arg(\mathcal{P})$ must be totally ordered under $\succeq_w$, and thus $\mathcal{F}$ is not necessarily acyclic. Nevertheless, thanks to the properties we imposed on arguments, there is still only one unique extension under all the argumentation semantics mentioned above.

Third, let $E = \{(a_1, \ldots, a_n) \in Arg(\mathcal{P}) \mid \{a_1, \ldots, a_n\} \subseteq Hansen(\mathcal{P})$ or $\exists i < n$ such that $a_i \notin Hansen(\mathcal{P})\}$. According to Definition 4, it holds that $concl(E) = Hansen(\mathcal{P})$. Now, we first prove that E is a stable extension of $\mathcal{F}$:

(1) Since all premises and the conclusion of each argument of E are contained in $Hansen(\mathcal{P})$ which is conflict-free, or one of the premises is not in $Hansen(\mathcal{P})$, it holds that E is conflict-free.

(2) $\forall \beta = (b_1, \ldots, b_m) \in Arg(\mathcal{P}) \setminus E$, $b_m \notin Hansen(\mathcal{P})$ (otherwise, if $b_m \in Hansen(\mathcal{P})$, then $(b_1, \ldots, b_{m-1}) \subseteq Hansen(\mathcal{P})$, and thus $\beta \in E$, contradicting the requirement that $\beta \notin E$). Then $\exists \alpha = (a_1, \ldots, a_n) \in E$, such that $a_n = \overline{b_j}$, $2 \leq j < m$. The two cases are analogous to the two cases in the proof of Proposition 3.10.

Since E is conflict-free and for all $\beta \in Arg(\mathcal{P}) \setminus E$, β is defeated by an argument in E, E is a stable extension. E is thus the unique complete extension of $\mathcal{F}$. □

5 Hang Yourself Arguments

We now introduce another type of argument, the *hang yourself argument* (abbreviated HYA) for prioritized normative systems. HYAs were introduced in a non-prioritized setting by [4,5] [2]. A HYA is made up of a *hypothetical argument* α, and an ordinary argument β, with contradictory conclusions. A third argument, γ, serves as the premise for α. If argument $\gamma;\alpha$ (where ; denotes concatenation of arguments to obtain a super-argument) is an ordinary argument which conflicts with β, then a contradiction exists, meaning that either γ or the HYA is invalid.

Definition 5.1 [Hang yourself arguments] Given a prioritized abstract normative system PANS $\mathcal{P} = \langle L, N, P, A, r\rangle$.

A hypothetical argument in $\mathcal{P}$ is similar to an ordinary argument in Definition 3.1. The only difference is that in a hypothetical argument, $ant(u_1) \notin A$.

A hang yourself argument in $\mathcal{P}$, written $\alpha|\beta$ consists of a hypothetical argument α and an ordinary argument β with opposite conclusions, such that for sub-arguments α', β' of α and β respectively, we have that if α' and β' have opposite conclusions, then $\alpha = \alpha'$ and $\beta = \beta'$.

[2] They are also called Socratic-style arguments due to their connection with Socratic style argumentation.

For convenience, given an argument $\beta;\alpha$ where $\beta = [(a_1, a_2), \dots, (a_{i-1}, a_i)]$ and $\alpha = [(a_i, a_{i+1}), \dots, (a_{n-1}, a_n)]$, where $(a_j, a_j + 1)$ has rank r_j, we write $r(\beta;\alpha^{-1})$ to denote the priority obtained from the sequence of ranks $r_1, \dots, r_i, r_{n-1}, \dots, r_{i+1}$.

Definition 5.2 [Defeat for HYAs] A HYA $\alpha|\beta$ defeats an argument γ iff there is a sub argument γ' of γ such that $\gamma';\alpha$ is an argument, and $r(\beta;\alpha^{-1}) \not\prec r(\gamma')$. A HYA $\alpha|\beta$ is defeated by an argument γ if and only if

(i) γ defeats β; or

(ii) there is a sub argument γ' of γ such that $\gamma';\alpha$ is an argument, and $r(\gamma') \not\prec r(\beta;\alpha^{-1})$;

Example 5.3 [Prioritized triangle, continued] Consider $\mathcal{P}_1$ in Example 2.2, visualized in Figure 2.b. The only relevant hang yourself argument is $(p,x)|(a,\neg x)$ which defeats (a,p) depending on the ranking of $(p,x),(a,\neg x)$. We thus have the following arguments:

A_0	a	(context argument)
A_1	$(a,p) : (1)$	(ordinary argument)
A_2	$(a,p)(p,\neg x) : (1,3)$	(ordinary argument)
A_3	$(a,x) : (2)$	(ordinary argument)
A_4	$(p,\neg x)|(a,x) : (3),(2)$	(hang yourself arg.)

A_1 and A_2 each defeats A_4 if $(2,3) \not\succ (1)$. A_3 defeats A_2 if $(1,3) \not\succ (2)$. A_4 defeats A_1 and A_2 if (1) and $(1,3) \not\succ (2,3)$ respectively.

For weakest link, A_4 defeats A_1 and A_2, and A_3 defeats A_2. We therefore have *Outfamily*$(\mathcal{P}_1, \succeq_w, sem) = \{\{a,x\}\}$ for all complete semantics.

An argument containing weak contrapositives may be seen as a kind of HYA. More precisely, consider an argument $A = [(a_1,a_2),\dots,(a_{n-1},a_n)]$, and another argument $B = [(b_1,b_2),\dots,(b_m,\overline{a_n})]$. These two arguments result in a sequence of weak contrapositive arguments:

$B; [\langle\overline{a_n}, \overline{a_{n-1}}\rangle]$
$B; [\langle\overline{a_n}, \overline{a_{n-1}}\rangle, \langle\overline{a_{n-1}}, \overline{a_{n-2}}\rangle]$
...
$B; [\dots; \langle\overline{a_2}, \overline{a_1}\rangle]$

Note that the last argument in the sequence is always defeated by the context argument. The remaining arguments attack (and may defeat) the different sub-arguments of A.

We now prove that the hang yourself argument is equivalent to the weak contrapositive argument.

Proposition 5.4 *The HYA $\delta = [(a_i,a_{i+1}),\dots,(a_{n-1},a_n)]|\beta$ is equivalent to the weak contrapositive argument $\omega = \beta;\dots;(\overline{a_{i+1}},\overline{a_i})$ in the sense that δ defeats a subargument α' of α if and only if ω defeats α'.*

Proof. Without loss of generality, assume that ω attacks α' on its last argument. Then the rank of ω is $r(\omega) = r(\beta), r((a_{n-1}, a_n)), \ldots, r((a_i, a_{i+1}))$. From Definition 5.2, the HYA defeats α if $r(\alpha') \not\succ r(\omega)$. Similarly, α defeats the HYA if $r(\omega) \not\succ r(\alpha')$. The final situation in which the weak contraposition is defeated holds if α defeats β. In such a situation, the HYA is also defeated. Thus, the situation where the weak contraposition defeats (is defeated by) α is identical to when the HYA defeats (is defeated by) α. □

6 Conclusions

In this paper, we provide a step towards studying non-monotonic logics through formal argumentation theory. Here, we begin addressing this challenge by considering three distinct systems for prioritized nonmonotonic reasoning, showing that they are different forms of our theory of argumentation. In particular, we showed how the Greedy approach of prioritized default logic can be represented by the weakest link principle; the Brewka-Eiter approach of answer set programming by the last link principle; and the Hansen approach of deontic logic using the weakest link principle extended with weak contraposition. We also showed that for weakest link, weak contraposition is a special case of hang yourself arguments.

While most work in formal argumentation uses very general frameworks to study argumentation systems, we use a very simple argument system to study the links between argumentation and prioritized norms. In particular, we utilised prioritized abstract normative systems, where norms are represented by a binary relation on literals, priorities are represented by natural numbers, and all norms have a distinct priority.

The main lessons that can be learned from our results are as follows. The weakest link principle corresponds to the greedy approach which is computationally attractive, but conceptually flawed. It should be adopted only when computational efficiency is the most important property. Thus, to get a more balanced result, the last link approach seems to be better for a wide number of potential applications, e.g., multiagent systems. This means that the pros and cons of both solutions have to be considered, and the decision regarding which to use depends on the application scenario of interest. Finally, Hansen's approach is a sophisticated way to deal with prioritized rules, and can be modeled using weakest link to handle conflicts, as we have shown. Our results are relevant not only when modelling normative systems, but also potentially when a developer must make a choice regarding which link principles to use when developing an argumentation system.

Acknowledgements

B. Liao, L. van der Torre and S. Villata have received funding from the European Union's Horizon 2020 research and innovation programme under the Marie Skodowska-Curie grant agreement No. 690974 for the project "MIREL: MIning and REasoning with Legal texts". B. Liao was partially supported by Zhejiang Provincial Natural Science Foundation of China (No. LY14F030014).

References

[1] Alchourron, C. E. and D. Makinson, *Hierarchies of regulations and their logic*, in: R. Hilpinen, editor, *New studies in deontic logic*, 1981 pp. 125–148.

[2] Amgoud, L. and C. Cayrol, *Integrating preference orderings into argument-based reasoning*, in: *ECSQARU-FAPR*, 1997, pp. 159–170.

[3] Brewka, G. and T. Eiter, *Preferred answer sets for extended logic programs*, Artificial Intelligence **109** (1999), pp. 297–356.

[4] Caminada, M., "For the sake of the argument: explorations into argument-based reasoning," Ph.D. thesis, Vrije Universiteit Amsterdam (2004).

[5] Caminada, M., *A formal account of socratic-style argumentation*, Journal of Applied Logic **6** (2008), pp. 109–132.
URL `http://dx.doi.org/10.1016/j.jal.2006.04.001`

[6] Dung, P. M., *On the acceptability of arguments and its fundamental role in nonmonotonic reasoning, logic programming and n-person games*, Artificial Intelligence **77** (1995), pp. 321–358.

[7] Dung, P. M., *An axiomatic analysis of structured argumentation for prioritized default reasoning*, in: *ECAI 2014*, 2014, pp. 267–272.

[8] Geffner, H. and J. Pearl, *Conditional entailment: Bridging two approaches to default reasoning*, Artificial Intelligence **53** (1992), pp. 209–244.
URL `http://dx.doi.org/10.1016/0004-3702(92)90071-5`

[9] Hansen, J., *Prioritized conditional imperatives: problems and a new proposal*, Autonomous Agents and Multi-Agent Systems **17** (2008), pp. 11–35.
URL `http://dx.doi.org/10.1007/s10458-007-9016-7`

[10] Modgil, S. and H. Prakken, *A general account of argumentation with preferences*, Artificial Intelligence **195** (2013), pp. 361–397.
URL `http://dx.doi.org/10.1016/j.artint.2012.10.008`

[11] Tosatto, S. C., G. Boella, L. van der Torre and S. Villata, *Abstract normative systems: Semantics and proof theory*, in: *KR 2012*, 2012.

[12] van der Torre, L. and S. Villata, *An aspic-based legal argumentation framework for deontic reasoning*, in: *COMMA 2014*, 2014, pp. 421–432.

[13] Young, A. P., S. Modgil and O. Rodrigues, *Argumentation semantics for prioritised default logic*, Technical report (2015).

Reasons to Believe in a Social Environment

Fenrong Liu

Department of Philosophy, Tsinghua University
Beijing, China

Emiliano Lorini [1]

IRIT-CNRS, Toulouse University
Toulouse, France

Abstract

We present a logic which supports reasoning about an agent's belief formation and belief change due to evidence provided by other agents in the society. We call this logic DEL-ES which stands for "Dynamic Epistemic Logic of Evidence Sources". The term 'evidence source' refers to an agent in the society who provides evidence to believe something to another agent. According to DEL-ES, if an agent has gathered a sufficient amount of evidence in support a given fact φ then, as a consequence, she should start to believe that φ is true. A sound and complete axiomatization for DEL-ES is given. We discuss some of its interesting properties and illustrate it in a concrete example from legal contexts.

Keywords: Modal logic, Epistemic Logic, Evidences, Reasons.

1 Introduction

A. J. Ayer put in his book [2, p. 3] :

> A rational man is one who makes a proper use of reason: and this implies, among other things, that he correctly estimates the strength of evidence.

In the same vain, this paper attempts to look into the ways that a rational agent handles evidence, as *reasons*, to support or reject her beliefs. Notions of evidence and justification pervade in legal contexts, in that various parties that are involved would collect evidence, and then use them to support their own claims or beliefs, or reject those of the opposing parties. The following three aspects suddenly become relevant, namely, evidence collection, belief formation

[1] We would like to thank the three anonymous reviewers of the DEON conference for their useful comments and suggestions. This work has been presented at the Workshop on Logical Dynamics of Social Influence and Information Change, March 22, 2016 in Amsterdam, and at the Chinese Conference on Logic and Argumentation, April 2-3, 2016 in Hangzhou, we thank the audience and their inspiring questions.

and belief revision. We will pursue the problems that arise when we pay more attention to the social enviroment in which evidence is not simply given, but provided by certain sources through communication.

Logical investigation of evidence and justification is not something new to us. Here we only mention two research areas that have inspired our work. [1] proposed so-called justication logic which explicitly expresses evidence as a term and possible manipulations of evidence as operations over terms. This framework has been further connected to the notion of beliefs and belief revision in [5]. The notion of evidence, understood as a proposition,[2] and its relationships with belief, are studied in recent papers [7,8]. The latter work has a single-agent perspective and clearly states that it left open the issues of evidence sources (i.e., where does the evidence come from?). This is where we started our journey.

Motivated differently, social influence in terms of individual's belief change has caught a lot of attention in recent years. [16] presented a finite state automata model with a threshold to deal with social influence. As a simple case, agent i would change her belief from p to $\neg p$ if all her neighbors believe $\neg p$. This model can successfully explain social phenomena, like peer pressure, and behavior adoption. [3] further enriches this model by introducing quantitative measurement on trust between agents and strength of evidence, and stipulates how these parameters influence one's valuation of evidence. [18] studies the phenonemon of trust-based belief change, that is, belief change that depends on the degree of trust the receiver has in the source of information. Viewed in line of social choice theory, one can also think of belief formation or change as a process of aggregating opinions from different reliable information sources, as e.g. [13].

In this paper we would like to combine ideas from the above two research areas, taking evidence source into account, and consider its roles in formation and dynamics of beliefs. A rational agent is someone who is aware of reasons for her beliefs, and who is willing to change her beliefs when facing new evidence.

The contribution of the paper is a new logic which supports reasoning about an agent's belief formation and belief change due to evidence provided by other agents in the society. We call this logic DEL-ES which stands for "Dynamic Epistemic Logic of Evidence Sources". It is assumed that the evidence source is social, that is, it is an agent in the society who provides evidence to believe something to another agent. The central idea of the logic DEL-ES is that an agent accumulates evidence in support of a given proposition φ from other agents in the society and the body of evidence in support of φ can become a reason to believe φ. This is reminiscient of Keynes's well-known concept of "weight of evidence" (or "weight of argument"), as clearly defined in the following paragraph from the famous treatise on probability [14, p. 77]:

> As the relevant evidence at our disposal increases, the magnitude of the probability of the argument may either decrease or increase, according as the

[2] Semantically, as a set of possible worlds.

new knowledge strengthens the unfavourable or the favourable evidence; but something seems to have increased in either case, - we have more substantial basis upon which to rest our conclusion. I express this by saying that an accession of new evidence increases the weight of argument.

Using Keynes' terminology, we assume that an agent has a reason to believe that φ is true if the weight of evidence supporting φ is considered by her *sufficient* to believe φ. In this paper we take the notion of weight in a qualitative sense, namely, the amount of evidence matters. We will leave the quantitative reading for future investigation.

The paper is organized as follows. In Section 2 we present the syntax and semantics of the logic DEL-ES, while in Section 3 we discuss some of its general properties. The semantics of DEL-ES combines a relational semantics for the concepts of knowledge and belief and a neighbourhood semantics for the concept of evidence. A sound and complete axiomatization for the logic is given in Section 4. The completeness proof is non-standard, given the interrelation between the concepts of belief and knowledge, on the one hand, and the concept of evidence, on the other hand. In Section 5 we illustrate the logic DEL-ES in a concrete example. Finally, in Section 6 we conclude.

2 Dynamic epistemic logic of evidence sources

In this section, we present the syntax and the semantics of the logic DEL-ES. DEL-ES has a static component, called EL-ES, which includes modal operators for beliefs, knowledge and evidence sources *plus* special atomic formulas that allow us to represent an agent's disposition to form beliefs based on evidence, namely, how much evidence the agent needs to collect in support of a fact before starting to believe that the fact is true. DEL-ES extends EL-ES with two kinds of dynamic operators: (i) operators for describing the consequences on an agent's epistemic state of the operation of receiving some new evidence, and (ii) operators for describing the consequences on an agent's epistemic state of the operation of restoring belief consistency.

On the technical level, DEL-ES combine methods and techniques from Dynamic Epistemic Logic (DEL), that has been developed in the past decades (cf. [4,11,6]), with methods and techniques from neighbourhood semantics for modal logic (cf. [10]).

2.1 Syntax

Assume a countable set of atomic propositions $Atm = \{p, q, \dots\}$ and a finite set of agents $Agt = \{1, \dots, n\}$. The set of groups (or coalitions) is defined to be $2^{Agt*} = 2^{Agt} \setminus \{\emptyset\}$. Elements of 2^{Agt*} are denoted by $J, J', \dots$ For every $J \in 2^{Agt*}$, $card(J)$ denotes the cardinality of J.

The language of DEL-ES, denoted by $\mathcal{L}_{\mathsf{DEL\text{-}ES}}$, is defined by the following grammar in Backus-Naur Form (BNF):

$$\begin{array}{lll} \alpha & ::= & \varphi!_{i \hookleftarrow j} \mid \circ_i \\ \varphi & ::= & p \mid \mathsf{trs}(i,x) \mid \neg\varphi \mid \varphi \wedge \psi \mid \mathsf{K}_i\varphi \mid \mathsf{B}_i\varphi \mid \mathsf{E}_{i,j}\varphi \mid [\alpha]\varphi \end{array}$$

where p ranges over Atm, i, j over Agt and $1 \leq x \leq card(Agt)$. The other Boolean constructions $\top$, $\bot$, $\vee$, $\rightarrow$ and $\leftrightarrow$ are defined from p, $\neg$ and $\wedge$ in the standard way.

The language of EL-ES (Epistemic Logic of Evidence Sources), denoted by $\mathcal{L}_{\textsf{EL-ES}}$, the fragment of DEL-ES without dynamic operators, is defined by the following:

$$\varphi \quad ::= \quad p \mid \mathsf{trs}(i, x) \mid \neg\varphi \mid \varphi \wedge \psi \mid \mathsf{K}_i\varphi \mid \mathsf{B}_i\varphi \mid \mathsf{E}_{i,j}\varphi$$

K_i is the standard modal operator of knowledge and $\mathsf{K}_i\varphi$ has to be read "agent i knows that φ is true". B_i is an operator for belief and $\mathsf{B}_i\varphi$ has to be read "agent i believes that φ is true". The dual of the knowledge operator is defined as follows:

$$\widehat{\mathsf{K}}_i\varphi \stackrel{\mathsf{def}}{=} \neg\mathsf{K}_i\neg\varphi$$

while the dual of the belief operator is defined as follows:

$$\widehat{\mathsf{B}}_i\varphi \stackrel{\mathsf{def}}{=} \neg\mathsf{B}_i\neg\varphi$$

$\mathsf{E}_{i,j}\varphi$ has to be read "agent i has evidence in support of φ based on the information provided by agent j".

$\mathsf{trs}(i, x)$ is a constant that has to be read "agent i has a level of epistemic cautiousness equal to x". Agent i's level of epistemic cautiousness corresponds to the amount of evidence in support of a given fact that agent i needs to collect before forming the belief that the fact is true.

We distinguish two types of events denoted by α: $\varphi!_{i \hookleftarrow j}$ and $\circ_i$. The symbol $\varphi!_{i \hookleftarrow j}$ denotes the event which consists in agent j providing evidence to agent i in support of φ, whereas the symbol $\circ_i$ denotes the event which consists in agent i restoring the consistency of her beliefs. The formula $[\alpha]\varphi$ has to be read "φ will hold after the occurrence of the event α". The dual of the dynamic operator $[\alpha]$ is defined as follows:

$$\langle\alpha\rangle\varphi \stackrel{\mathsf{def}}{=} \neg[\alpha]\neg\varphi$$

where $\langle\alpha\rangle\varphi$ has to be read "it is possible that the event α occurs and, if it occurs, φ will hold afterwards". Clearly, $\langle\alpha\rangle\top$ and $[\alpha]\bot$ have to read, respectively, "it is possible that the event α occurs" (or α is executable) and "it is impossible that the event α occurs" (or α is inexecutable).

The reason why we introduce events of the form $\circ_i$ and corresponding dynamic operators $[\circ_i]$ is that the process of accumulating new evidence may lead to inconsistency of beliefs. In such a situation, an agent may want to restore consistency of her beliefs and start the accumulation of new evidence to discover new truths. This issue will be clearly illustrated in Section 3.

Let us immediately define the following abbreviations for every $i \in Agt$ and

$1 \leq x \leq card(Agt)$:

$$\mathsf{E}_i^{\geq x}\varphi \overset{\text{def}}{=} \bigvee_{J \in 2^{Agt*}: card(J)=x} \bigwedge_{j \in J} \mathsf{E}_{i,j}\varphi$$

$$\mathsf{E}_i\varphi \overset{\text{def}}{=} \bigvee_{1 \leq x \leq card(Agt)} (\mathsf{E}_i^{\geq x}\varphi \wedge \neg \mathsf{E}_i^{\geq x}\neg\varphi)$$

$$\mathsf{R}_i\varphi \overset{\text{def}}{=} \mathsf{E}_i\varphi \wedge \bigvee_{1 \leq x \leq card(Agt)} (\mathsf{E}_i^{\geq x}\varphi \wedge \mathsf{trs}(i,x))$$

$$\mathsf{QR}_i\varphi \overset{\text{def}}{=} \neg\mathsf{E}_i\neg\varphi \wedge \bigvee_{1 \leq x \leq card(Agt)} (\mathsf{E}_i^{\geq x-1}\varphi \wedge \mathsf{trs}(i,x))$$

We use the convention $\mathsf{E}_i^{\geq 0}\varphi \overset{\text{def}}{=} \top$ and $\mathsf{E}_i^{\geq card(Agt)+1}\varphi \overset{\text{def}}{=} \bot$.

$\mathsf{E}_i^{\geq x}\varphi$ has to be read "agent i has at least x evidence in support of φ". $\mathsf{E}_i\varphi$ has to be read "agent i has a *decisive* evidence for φ" in the sense that she has more evidence in support of φ than evidence in support of $\neg\varphi$. $\mathsf{R}_i\varphi$ has to be read "agent i has a *sufficient* reason to believe that φ is true". According to our definition, an agent has a sufficient reason to believe that φ is true if and only if:

(i) she has a decisive evidence for φ,

(ii) the amount of evidence in support of φ is equal to or above her threshold of epistemic cautiousness.

As we will highlight in Section 4, a sufficient reason to believe that φ is true ensures that the agent will form the corresponding belief that φ is true. The last abbreviation QR_i defines the concept of quasi-reason: an agent has a *quasi-sufficient reason* to believe that φ is true, denoted by $\mathsf{QR}_i\varphi$, if and only if an additional evidence in support of φ will provide to the agent a sufficient reason to believe that φ is true.

2.2 Semantics

The main notion in semantics is given by the following definition of evidence source model which provides the basic components for the interpretation of the logic DEL-ES:

Definition 2.1 [Evidence Source Model] An evidence source model (ESM) is a tuple $M = (W, E, D, S, T, V)$ where:

- W is a set of worlds or situations;
- $E : Agt \longrightarrow 2^{W \times W}$ s.t. for all $i \in Agt$, $E(i)$ is an epistemic relation on W;
- $D : Agt \longrightarrow 2^{W \times W}$ s.t. for all $i \in Agt$, $D(i)$ is a doxastic relation on W;
- $S : Agt \times Agt \times W \longrightarrow 2^{2^W}$ is an evidence source function;
- $T : Agt \times W \longrightarrow \{k \in \mathbb{N} : 0 \leq k \leq card(Agt)\}$ is an epistemic threshold function;
- $V : W \longrightarrow 2^{Atm}$ is a valuation function;

and which satisfies the following conditions for all $i, j \in Agt$ and $w, v \in W$:

(C1) every $E(i)$ is an equivalence relation;

(C2) $D(i) \subseteq E(i)$;

(C3) if $wE(i)v$ then $D(i)(w) = D(i)(v)$;

(C4) if $wE(i)v$ then $S(i, j, w) = S(i, j, v)$;

(C5) if $X \in S(i, j, v)$ then $X \subseteq E(i)(w)$;

(C6) $\emptyset \notin S(i, j, v)$;

(C7) if $wE(i)v$ then $T(i, w) = T(i, v)$;

(C8) if $card(Agt_{i,X,M,w}) > card(Agt_{i,W \setminus X,M,w})$ and $card(Agt_{i,X,M,w}) \geq T(i, w)$
then $D(i)(w) \subseteq X$;

where, for any binary relation R on W, $R(w) = \{v \in W : wRv\}$ and for all $X \subseteq W$:

$$Agt_{i,X,M,w} = \{j \in Agt : X \in S(i, j, w)\}.$$

For notational convenience, we write E_i instead of $E(i)$ and D_i instead of $D(i)$. For every $w \in W$, $E_i(w)$ and $D_i(w)$ are called, respectively, agent i's information set and belief set at w. Agent i's information set at w is the set of worlds that agent i envisages at world w, while agent i's belief set at w is the set of worlds that agent i thinks to be possible at world w.

Constraint C1 ensures that the epistemic relation $E(i)$ is nothing but the indistinguishability relation traditionally used to model a fully introspective and truthful notion of knowledge.

Constraint C2 ensures that the set of possible worlds is included in the set of envisaged worlds. Indeed, following [15], a ESM requires that an agent is capable of assessing whether an envisaged situation is *possible* or not.[3]

Constraint C3 just means that if two worlds are in the same information set of agent i, then agent i has the same belief set at these two worlds. In other words, an agent knows her beliefs.

$S(i, j, w)$ is the set of evidence that agent j has provided to agent i where, following [7], a piece of evidence is identified with a set of worlds. Constraint C4 means that if two worlds are in the same information set of agent i, then agent i has the same evidence at these two worlds. In other words, an agent knows her evidence. Constraint C5 just means that an agent can have evidence only about facts which are compatible with her current information set. Constraint C6 means that an agent cannot have evidence about inconsistent facts.

[3] Here we take the term "envisaged" to be synonymous of the term "imagined". Clearly, there are situations that one can imagine that she considers impossible. For example, a person can imagine a situation in which she is the president of French republic and, at the same time, considers this situation impossible.

$T(i,w)$ corresponds to agent i's level of epistemic cautiousness at world w. Constraint C7 just means that if two worlds are in the same information set of agent i, then agent i has the same level of epistemic cautiousness at these two worlds. In other words, an agent knows her level of epistemic cautiousness.

Constraint C8 relates evidence with belief. Suppose that the amount of evidence in support of a given fact is: (i) equal or higher than my level of epistemic cautiousness and (ii) is higher than the amount of evidence in support of its negation. Then, I should start to believe this fact. Specifically, conditions (i) and (ii) together provide a sufficient reason to believe the fact in question.

Truth conditions of DEL-ES formulas are inductively defined as follows.

Definition 2.2 [Truth conditions] Let $M = (W, E, D, S, T, V)$ be a ESM and let $w \in W$. Then:

$$
\begin{aligned}
M, w \models p &\Longleftrightarrow p \in V(w)\\
M, w \models \mathsf{trs}(i, x) &\Longleftrightarrow T(i, w) = x\\
M, w \models \neg\varphi &\Longleftrightarrow M, w \not\models \varphi\\
M, w \models \varphi \wedge \psi &\Longleftrightarrow M, w \models \varphi \text{ and } M, w \models \psi\\
M, w \models \mathsf{K}_i\varphi &\Longleftrightarrow \forall v \in E_i(w) : M, v \models \varphi\\
M, w \models \mathsf{B}_i\varphi &\Longleftrightarrow \forall v \in D_i(w) : M, v \models \varphi\\
M, w \models \mathsf{E}_{i,j}\varphi &\Longleftrightarrow ||\varphi||^M_{i,w} \in S(i, j, w)\\
M, w \models [\varphi!_{i\leftarrow j}]\psi &\Longleftrightarrow \text{ if } M, w \models \widehat{\mathsf{K}}_i\varphi \text{ then } M^{\varphi!_{i\leftarrow j}}, w \models \psi\\
M, w \models [\circ_i]\psi &\Longleftrightarrow \text{ if } M, w \models \mathsf{B}_i\bot \text{ then } M^{\circ_i}, w \models \psi
\end{aligned}
$$

where

$$||\varphi||^M_{i,w} = \{v \in W : M, v \models \varphi\} \cap E_i(w),$$

$M^{\varphi!_{i\leftarrow j}}$ and M°_i} are updated models defined according to the following Definitions 2.3 and 2.4.

According to the truth conditions: agent i knows that φ at world w if and only if φ is true in all worlds that at w agent i envisages, and agent i believes that φ at world w if and only if φ is true in all worlds that at w agent i considers possible. Moreover, at world w agent j has provided evidence in support of φ to agent i if and only if, at w, agent i has the fact corresponding to the formula φ (i.e., $||\varphi||^M_{i,w}$) included in her evidence set $S(i,j,w)$. In what follows, we define the updated models triggered by the two kinds of events:

Definition 2.3 [Update via $\varphi!_{i\leftarrow j}$] Let $M = (W, E, D, S, T, V)$ be a ESM. Then, $M^{\varphi!_{i\leftarrow j}}$ is the tuple $(W, E, D^{\varphi!_{i\leftarrow j}}, S^{\varphi!_{i\leftarrow j}}, T, V)$ such that, for all $k, l \in$

Agt and $w \in W$:

$$D_k^{\varphi!_{i \leftarrow j}}(w) = \begin{cases} D_k(w) \cap ||\varphi||_{k,w}^M & \text{if } k = i \text{ and } M, w \models \mathsf{QR}_i\varphi \\ D_k(w) & \text{otherwise} \end{cases}$$

$$S^{\varphi!_{i \leftarrow j}}(k, l, w) = \begin{cases} S(k, l, w) \cup \{||\varphi||_{k,w}^M\} & \text{if } k = i \text{ and } l = j \\ S(k, l, w) & \text{otherwise} \end{cases}$$

Definition 2.4 [Update via $\circ_i$] Let $M = (W, E, D, S, T, V)$ be a SSM. Then, M°_i} is the tuple $(W, E, D^{\circ_i}, S^{\circ_i}, T, V)$ such that, for all $j, k \in Agt$ and $w \in W$:

$$D_j^{\circ_i}(w) = \begin{cases} E_j(w) & \text{if } j = i \\ D_j(w) & \text{otherwise} \end{cases}$$

$$S^{\circ_i}(j, k, w) = \begin{cases} \emptyset & \text{if } j = i \\ S(j, k, w) & \text{otherwise} \end{cases}$$

As highlighted in Definition 2.3, the event consisting in agent j providing evidence to agent i in support of φ, has two consequences: (i) a new evidence in support of φ is added to the set of evidence provided by agent j to agent i, and (ii) if before getting the new information agent i has a quasi-sufficient reason to believe φ then, after getting it, the agent will start to believe φ.

Again in Definition 2.4, the event consisting in restoring the consistency of agent i's beliefs has two consequences: (i) all agent i's sets of evidence become empty, and (ii) agent i starts to consider possible all situations that she envisages. More concisely, the operation of restoring belief consistency makes an agent to forget everything she has in her mind except her knowledge. This includes the agent's evidence as well as her beliefs. In other words, by restoring belief consistency, an agent "cleans up" her mind in order to start the accumulation of new evidence and the discovery of new truths.

Notice that the event $\varphi!_{i \leftarrow j}$ is executable, denoted by $\langle \varphi!_{i \leftarrow j} \rangle \top$, if and only if $\widehat{\mathsf{K}}_i\varphi$ holds and the event $\circ_i$ is executable, denoted by $\langle \circ_i \rangle \top$, if and only if $\mathsf{B}_i\bot$ holds. This means that an agent cannot provide to another agent evidence in support of φ if this evidence conflicts with her knowledge, and an agent will not restore consistency of her beliefs unless her beliefs are inconsistent.

For every $\varphi \in \mathcal{L}_{\mathsf{DEL\text{-}ES}}$, we write $\models \varphi$ to mean that φ is valid w.r.t. the class of ESMs, that is, for every $M = (W, E, D, S, T, V)$ and for every $w \in W$ we have $M, w \models \varphi$. We say that φ is satisfiable w.r.t. the class of ESMs if and only if $\neg\varphi$ is not valid w.r.t. the class of ESMs.

3 Properties

In this section we want to focus on some interesting properties of the logic DEL-ES. We start with the following static properties about the relationship between sufficient reason and quasi-sufficient reason:

$$\models \mathsf{R}_i\varphi \rightarrow \mathsf{QR}_i\varphi \tag{1}$$

$$\models (\mathsf{R}_i\varphi \wedge \mathsf{R}_i\neg\varphi) \rightarrow \mathsf{B}_i\bot \tag{2}$$

The validity (1) highlights that sufficient reason is stronger than quasi-sufficient reason, while, according to the validity (2), two conflicting reasons lead to belief inconsistency.

Let us now consider some properties that only apply to the propositional fragment of the logic DEL-ES. Let $\mathcal{L}_{Atm}$ be the propositional language build out of the set of atoms Atm. Then, we have the following validities for $\varphi, \psi \in \mathcal{L}_{Atm}$:

$$\models (\neg\mathsf{E}_{i,j}\varphi \wedge \mathsf{QR}_i\varphi) \rightarrow [\varphi!_{i\leftrightarrow j}]\mathsf{R}_i\varphi \tag{3}$$

$$\models (\neg\mathsf{E}_{i,j}\varphi \wedge \mathsf{QR}_i\varphi) \rightarrow [\varphi!_{i\leftrightarrow j}]\mathsf{B}_i\varphi \tag{4}$$

$$\models (\neg\mathsf{E}_{i,j_1}\varphi \wedge \ldots \wedge \neg\mathsf{E}_{i,j_x}\varphi \wedge \mathsf{trs}(i,x)) \rightarrow [\varphi!_{i\leftrightarrow j_1}]\ldots[\varphi!_{i\leftrightarrow j_x}]\mathsf{R}_i\varphi \tag{5}$$

$$\models (\neg\mathsf{E}_{i,j_1}\varphi \wedge \ldots \wedge \neg\mathsf{E}_{i,j_x}\varphi \wedge \mathsf{trs}(i,x)) \rightarrow [\varphi!_{i\leftrightarrow j_1}]\ldots[\varphi!_{i\leftrightarrow j_x}]\mathsf{B}_i\varphi \tag{6}$$

$$[\chi!_{i\leftrightarrow j}]B_i\varphi \rightarrow [\chi!_{i\leftrightarrow j}][\theta!_{i\leftrightarrow k}]B_i\varphi \tag{7}$$

$$\models (\neg\mathsf{E}_{i,j}\varphi \wedge \mathsf{QR}_i\varphi \wedge \mathsf{K}_i(\varphi \rightarrow \psi)) \rightarrow [\varphi!_{i\leftrightarrow j}]\mathsf{B}_i\psi \tag{8}$$

According to (3), if agent i has a quasi-sufficient reason to believe that the propositional formula φ is true and agent j has not provided to agent i evidence in support of φ then, after j does that, i will have a sufficient reason to believe φ. According to (4), if agent i has a quasi-sufficient reason to believe that the propositional formula φ is true and agent j has not provided to agent i evidence in support of φ then, after j does that, i will start to believe φ. The validities (5) and (6) are similar properties for sequences of informative events: if agent i has a level of epistemic cautiousness equal to x and there are x agents who have not provided to agent i evidence in support of the propositional formula φ then, after they do that, i will have a sufficient reason to believe φ and, as a consequence, i will start to believe φ. The validity (7) highlights that, by getting more evidence, an agent decreases her uncertainty about the truth of propositional formulas. The validity (8) highlights the relationship between knowledge and belief from a dynamic point of view.

The reason why we need to impose that φ and ψ are propositional formulas is that there are DEL-ES-formulas such as the Moore-like formula $p \wedge \neg\mathsf{B}_i p$ for which the previous validities (3)-(8) do not hold. For instance, the following formula is not valid:

$$(\neg\mathsf{E}_{i,j}(p \wedge \neg\mathsf{B}_i p) \wedge \mathsf{QR}_i(p \wedge \neg\mathsf{B}_i p)) \rightarrow [(p \wedge \neg\mathsf{B}_i p)!_{i\leftrightarrow j}]\mathsf{B}_i(p \wedge \neg\mathsf{B}_i p)$$

This is intuitive since if I think that my uncertainty about p could be unjustified since p is possibly true and someone gives me a decisive evidence in support of this fact then, as a consequence, I should start to believe that p is true and that I believe this (since I have introspection over my beliefs).

The following two validities apply to any formula of the language $\mathcal{L}_{\mathsf{DEL\text{-}ES}}$:

$$\models (\mathsf{B}_i\varphi \wedge \neg\mathsf{E}_{i,j}\neg\varphi \wedge \mathsf{QR}_i\neg\varphi) \rightarrow [\neg\varphi!_{i\leftrightarrow j}]\mathsf{B}_i\bot \tag{9}$$

$$\models [\circ_i]\neg\mathsf{B}_i\bot \tag{10}$$

According to the validity (9) if agent i believes that φ is true, has a quasi-sufficient reason to believe that φ is false and agent j has not provided to agent i evidence in support of the fact that φ is false then, after j does that, i's beliefs will become inconsistent. Validity (10) highlights the role of the event o_i in restoring consistency of i's beliefs.

4 Axiomatization

Let us now present sound and complete axiomatizations for the logic EL-ES and its dynamic extension DEL-ES. As we will show, the completeness proof of the logic EL-ES is non-standard, given the interrelation between the concepts of belief and knowledge, on the one hand, and the concept of evidence, on the other hand. The completeness proof of EL-ES is based on a canonical model construnction. All axioms of EL-ES, except one, are used in the usual way to prove that the canonical model so constructed is a ESM. There is a special axiom of the logic EL-ES, about the interrelation between knowledge and evidence, that is used in an unusual way to prove the truth lemma.

Definition 4.1 [EL-ES] We define EL-ES to be the extension of classical propositional logic given by the following rules and axioms:

$$(\mathsf{K}_i\varphi \wedge \mathsf{K}_i(\varphi \rightarrow \psi)) \rightarrow \mathsf{K}_i\psi \qquad (\mathbf{K}_{\mathsf{K}_i})$$
$$\mathsf{K}_i\varphi \rightarrow \varphi \qquad (\mathbf{T}_{\mathsf{K}_i})$$
$$\mathsf{K}_i\varphi \rightarrow \mathsf{K}_i\mathsf{K}_i\varphi \qquad (\mathbf{4}_{\mathsf{K}_i})$$
$$\neg\mathsf{K}_i\varphi \rightarrow \mathsf{K}_i\neg\mathsf{K}_i\varphi \qquad (\mathbf{5}_{\mathsf{K}_i})$$
$$(\mathsf{B}_i\varphi \wedge \mathsf{B}_i(\varphi \rightarrow \psi)) \rightarrow \mathsf{B}_i\psi \qquad (\mathbf{K}_{\mathsf{B}_i})$$
$$\bigvee_{0 \leq x \leq card(Agt)} \mathsf{trs}(i,x) \qquad (\mathbf{AtLeast}_{\mathsf{trs}(i,x)})$$
$$\mathsf{trs}(i,x) \rightarrow \neg\mathsf{trs}(i,y) \text{ if } x \neq y \qquad (\mathbf{AtMost}_{\mathsf{trs}(i,x)})$$
$$\mathsf{K}_i\varphi \rightarrow \mathsf{B}_i\varphi \qquad (\mathbf{Mix1}_{\mathsf{K}_i,\mathsf{B}_i})$$
$$\mathsf{B}_i\varphi \rightarrow \mathsf{K}_i\mathsf{B}_i\varphi \qquad (\mathbf{Mix2}_{\mathsf{K}_i,\mathsf{B}_i})$$
$$\mathsf{trs}(i,x) \rightarrow \mathsf{K}_i\mathsf{trs}(i,x) \qquad (\mathbf{Mix}_{\mathsf{K}_i,\mathsf{trs}(i,x)})$$
$$\mathsf{E}_{i,j}\varphi \rightarrow \mathsf{K}_i\mathsf{E}_{i,j}\varphi \qquad (\mathbf{Mix1}_{\mathsf{K}_i,\mathsf{E}_{i,j}})$$
$$\neg\mathsf{E}_{i,j}\bot \qquad (\mathbf{Cons}_{\mathsf{E}_{i,j}})$$
$$(\mathsf{E}_{i,j}\varphi \wedge \mathsf{K}_i(\varphi \leftrightarrow \psi)) \rightarrow \mathsf{E}_{i,j}\psi \qquad (\mathbf{Mix2}_{\mathsf{K}_i,\mathsf{E}_{i,j}})$$
$$\mathsf{R}_i\varphi \rightarrow \mathsf{B}_i\varphi \qquad (\mathbf{SuffReas})$$
$$\frac{\varphi}{\mathsf{K}_i\varphi} \qquad (\mathbf{Nec}_{\mathsf{K}_i})$$

Notice that the rule of necessitation for the belief operator is provable by means of the rule of inference ($\mathbf{Nec}_{\mathsf{K}_i}$) and Axiom ($\mathbf{Mix1}_{\mathsf{K}_i,\mathsf{B}_i}$). Moreover, Axiom 4 for the belief operator is provable by means of Axioms ($\mathbf{Mix1}_{\mathsf{K}_i,\mathsf{B}_i}$) and ($\mathbf{Mix2}_{\mathsf{K}_i,\mathsf{B}_i}$). Axiom 5 for the belief operator is provable by means of Axioms ($\mathbf{Mix1}_{\mathsf{K}_i,\mathsf{B}_i}$), ($\mathbf{Mix2}_{\mathsf{K}_i,\mathsf{B}_i}$), $\mathbf{K}_{\mathsf{K}_i}$, $\mathbf{T}_{\mathsf{K}_i}$, $\mathbf{4}_{\mathsf{K}_i}$ and $\mathbf{5}_{\mathsf{K}_i}$. A syntactic proof can be found in [17].

Theorem 4.2 *The logic EL-ES is sound and complete for the class of ESMs.*

Proof. It is routine to check that the axioms of EL-ES are all valid w.r.t. the class of ESMs and that the rule of inference ($\mathbf{Nec}_{\mathsf{K}_i}$) preserves validity.

To prove completeness, we use a canonical model argument.

We consider maximally consistent sets of formulas in $\mathcal{L}_{\text{EL-ES}}$ (MCSs). The following proposition specifies some usual properties of MCSs.

Proposition 4.3 *Let* Γ *be a MCS and let* $\varphi, \psi \in \mathcal{L}_{\text{EL-ES}}$. *Then:*

- *if* $\varphi, \varphi \to \psi \in \Gamma$ *then* $\psi \in \Gamma$;
- $\varphi \in \Gamma$ *or* $\neg\varphi \in \Gamma$;
- $\varphi \vee \psi \in \Gamma$ *iff* $\varphi \in \Gamma$ *or* $\psi \in \Gamma$.

The following is the Lindenbaum's lemma for our logic. As the proof is standard (cf. [9, Lemma 4.17]) we omit it here.

Lemma 4.4 *Let* Δ *be a EL-ES-consistent set of formulas. Then, there exists a MCS* Γ *such that* $\Delta \subseteq \Gamma$.

Let the canonical ESM model be the tuple $M^c = (W^c, E^c, D^c, S^c, T^c, V^c)$ such that:

- W^c is set of all MCSs;
- for all $w, v \in W^c$ and $i \in Agt$, $wE_i^c v$ iff, for all $\varphi \in \mathcal{L}_{\text{EL-ES}}$, if $\mathsf{K}_i\varphi \in w$ then $\varphi \in v$;
- for all $w, v \in W^c$ and $i \in Agt$, $wD_i^c v$ iff, for all $\varphi \in \mathcal{L}_{\text{EL-ES}}$, if $\mathsf{B}_i\varphi \in w$ then $\varphi \in v$;
- for all $w \in W^c$ and $i, j \in Agt$, $S^c(i, j, w) = \{A_\varphi(i, j, w) : \mathsf{E}_{i,j}\varphi \in w\}$;
- for all $w \in W^c$ and $i \in Agt$, $T^c(i, w) = x$ iff $\mathsf{trs}(i, x) \in w$;
- for all $w \in W^c$ and $p \in Atm$, $p \in V^c(w)$ iff $p \in w$;

where $A_\varphi(i, j, w) = \{v \in E_i^c(w) : \varphi \in v\}$.

Thanks to Axioms $\mathbf{AtLeast}_{\mathsf{trs}(i,x)}$ and $\mathbf{AtMost}_{\mathsf{trs}(i,x)}$, it is easy to check that the model M^c is well-defined as the function T^c exists.

We have to prove that M^c is a ESM by showing that it satisfies conditions C1-C8 in Definition 2.1. The proof is a routine exercise and uses of Proposition 4.3: Condition C1 is satisfied because of Axioms $\mathbf{T}_{\mathsf{K}_i}$, $\mathbf{4}_{\mathsf{K}_i}$ and $\mathbf{5}_{\mathsf{K}_i}$; Condition C2 is satisfied because of Axiom $\mathbf{Mix1}_{\mathsf{K}_i,\mathsf{B}_i}$; Condition C3 is satisfied because of Axiom $\mathbf{Mix2}_{\mathsf{K}_i,\mathsf{B}_i}$; Condition C4 is satisfied because of Axiom $\mathbf{Mix1}_{\mathsf{K}_i,\mathsf{E}_{i,j}}$; Condition C6 is satisfied because of Axiom $\mathbf{Cons}_{\mathsf{E}_{i,j}}$; Condition C7 is satisfied because of Axiom $\mathbf{Mix}_{\mathsf{K}_i,\mathsf{trs}(i,x)}$; Condition C8 is satisfied because of Axiom **SuffReas**; Condition C5 is satisfied by construction of the model M^c and, in particular, by definition of $A_\varphi(i, j, w)$. Here we only show that M^c satisfies Conditions C4 and C5.

As for C4, suppose that $wE_i^c v$ and $X \in S^c(i, j, w)$. The latter means that $X = \{u \in E_i^c(w) : \varphi \in u\}$ and $\mathsf{E}_{i,j}\varphi \in w$ for some φ. Hence, by Proposition 4.3 and Axiom $\mathbf{Mix1}_{\mathsf{K}_i,\mathsf{E}_{i,j}}$, we have $\mathsf{K}_i\mathsf{E}_{i,j}\varphi \in w$. By $wE_i^c v$ and the definition of

E_i^c, from the latter it follows that $\mathsf{E}_{i,j}\varphi \in v$. Hence, by the definition of S^c, we have $Y = \{u \in E_i^c(v) : \varphi \in u\} \in S^c(i,j,v)$. Since E_i^c is an equivalence relation and wE_i^cv, we have $E_i^c(w) = E_i^c(v)$. Thus, $X = Y$. Hence, $X \in S^c(i,j,v)$.

As for C5, suppose that $X \in S^c(i,j,w)$. The latter means that $X = \{u \in E_i^c(w) : \varphi \in u\}$ and $\mathsf{E}_{i,j}\varphi \in w$ for some φ. Thus, clearly, $X \subseteq E_i^c(w)$.

The next step in the proof consists in stating the following existence lemma. The proof is again standard (cf. [9, Lemma 4.20]) and we omit it.

Lemma 4.5 *Let $\varphi \in \mathcal{L}_{\mathsf{EL\text{-}ES}}$ and $w \in W^c$. Then:*

- *if $\widehat{\mathsf{K}}_i\varphi \in w$ then there exists $v \in W^c$ such that wE_i^cv and $\varphi \in v$;*
- *if $\widehat{\mathsf{B}}_i\varphi \in w$ then there exists $v \in W^c$ such that wD_i^cv and $\varphi \in v$.*

Finally, we can prove the following truth lemma.

Lemma 4.6 *Let $\varphi \in \mathcal{L}_{\mathsf{EL\text{-}ES}}$ and $w \in W^c$. Then, $M^c, w \models \varphi$ iff $\varphi \in w$.*

Proof. The proof is by induction on the structure of the formula. Here we only prove the case $\varphi = \mathsf{E}_{i,j}\psi$ which is the most interesting one as it uses a non-standard technique. The other cases are provable in the standard way (cf. [9, Lemma 4.21]).

($\Rightarrow$) Suppose $M^c, w \models \mathsf{E}_{i,j}\psi$. Thus, $\{u \in E_i^c(w) : M^c, u \models \psi\} \in S^c(i,j,w)$. Hence, by definition of S^c, there exists χ such that $\mathsf{E}_{i,j}\chi \in w$ and $\{u \in E_i^c(w) : \chi \in u\} = \{u \in E_i^c(w) : M^c, u \models \psi\}$. Thus, by induction hypothesis, $\{u \in E_i^c(w) : \chi \in u\} = \{u \in E_i^c(w) : \psi \in u\}$. Now, suppose that $\mathsf{K}_i(\chi \leftrightarrow \psi) \notin w$. By Proposition 4.3, it follows that $\neg\mathsf{K}_i(\chi \leftrightarrow \psi) \in w$. This means that $\widehat{\mathsf{K}}_i((\chi \wedge \neg\psi) \vee (\neg\chi \wedge \psi)) \in w$. By Lemma 4.5, the latter implies that there exists $v \in W^c$ such that wE_i^cv and $((\chi \wedge \neg\psi) \vee (\neg\chi \wedge \psi)) \in v$ which is in contradiction with $\{u \in E_i^c(w) : \chi \in u\} = \{u \in E_i^c(w) : \psi \in u\}$. Thus, we have $\mathsf{K}_i(\chi \leftrightarrow \psi) \in w$. From $\mathsf{E}_{i,j}\chi \in w$ and $\mathsf{K}_i(\chi \leftrightarrow \psi) \in w$, by Proposition 4.3 and Axiom $\mathbf{Mix2}_{\mathsf{K}_i,\mathsf{E}_{i,j}}$, it follows that $\mathsf{E}_{i,j}\psi \in w$.

($\Leftarrow$) Suppose $\mathsf{E}_{i,j}\psi \in w$. Thus, by the definition of S^c, $A_\psi(i,j,w) = \{v \in E_i^c(w) : \psi \in v\} \in S^c(i,j,w)$. Hence, by induction hypothesis, $\{v \in E_i^c(w) : M^c, v \models \psi\} \in S^c(i,j,w)$. The latter means that $M^c, w \models \mathsf{E}_{i,j}\psi$. □

To conclude the proof, suppose that φ is a EL-ES-consistent formula in $\mathcal{L}_{\mathsf{EL\text{-}ES}}$. By Lemma 4.4, there exists $w \in W^c$ such that $\varphi \in w$. Hence, by Lemma 4.6, there exists $w \in W^c$ such that $M^c, w \models \varphi$.

□

The axiomatics of the logic DEL-ES includes all principles of the logic EL-ES *plus* a set of reduction axioms and the rule of replacement of equivalents.

Definition 4.7 We define DEL-ES to be the extension of EL-ES generated by the following reduction axioms for the dynamic operators $[\varphi!_{i\hookleftarrow j}]$:

$$[\varphi!_{i\hookleftarrow j}]p \leftrightarrow (\widehat{\mathsf{K}}_i\varphi \to p) \quad (\mathbf{Red}_{\varphi!_{i\hookleftarrow j},p})$$

$$[\varphi!_{i\hookleftarrow j}]\mathsf{trs}(k,x) \leftrightarrow (\widehat{\mathsf{K}}_i\varphi \to \mathsf{trs}(k,x)) \quad (\mathbf{Red}_{\varphi!_{i\hookleftarrow j},\mathsf{trs}(l,x)})$$

$$[\varphi!_{i\hookleftarrow j}]\neg\psi \leftrightarrow (\widehat{\mathsf{K}}_i\varphi \to \neg[\varphi!_{i\hookleftarrow j}]\psi) \quad (\mathbf{Red}_{\varphi!_{i\hookleftarrow j},\neg})$$

$$[\varphi!_{i\hookleftarrow j}](\psi \wedge \chi) \leftrightarrow ([\varphi!_{i\hookleftarrow j}]\psi \wedge [\varphi!_{i\hookleftarrow j}]\chi) \quad (\mathbf{Red}_{\varphi!_{i\hookleftarrow j},\wedge})$$

$$[\varphi!_{i\hookleftarrow j}]\mathsf{K}_k\varphi \leftrightarrow (\widehat{\mathsf{K}}_i\varphi \to \mathsf{K}_k[\varphi!_{i\hookleftarrow j}]\varphi) \quad (\mathbf{Red}_{\varphi!_{i\hookleftarrow j},\mathsf{K}_k})$$

$$[\varphi!_{i\hookleftarrow j}]\mathsf{B}_k\psi \leftrightarrow (\widehat{\mathsf{K}}_i\varphi \to \mathsf{B}_k[\varphi!_{i\hookleftarrow j}]\varphi) \text{ if } k \neq i \quad (\mathbf{Red}_{\varphi!_{i\hookleftarrow j},\mathsf{B}_k})$$

$$[\varphi!_{i\hookleftarrow j}]\mathsf{B}_i\psi \leftrightarrow \Big(\widehat{\mathsf{K}}_i\varphi \to ((\neg\mathsf{QR}_i\varphi \wedge \mathsf{B}_i[\varphi!_{i\hookleftarrow j}]\psi)\vee (\mathsf{QR}_i\varphi \wedge \mathsf{B}_i(\varphi \to [\varphi!_{i\hookleftarrow j}]\psi)))\Big) \quad (\mathbf{Red}_{\varphi!_{i\hookleftarrow j},\mathsf{B}_i})$$

$$[\varphi!_{i\hookleftarrow j}]\mathsf{E}_{k,l}\psi \leftrightarrow (\widehat{\mathsf{K}}_i\varphi \to \mathsf{E}_{k,l}[\varphi!_{i\hookleftarrow j}]\psi) \text{ if } k \neq i \text{ or } l \neq j \quad (\mathbf{Red}_{\varphi!_{i\hookleftarrow j},\mathsf{E}_{k,l}})$$

$$[\varphi!_{i\hookleftarrow j}]\mathsf{E}_{i,j}\psi \leftrightarrow \Big(\widehat{\mathsf{K}}_i\varphi \to ((\mathsf{E}_{i,j}[\varphi!_{i\hookleftarrow j}]\psi\vee \mathsf{K}_i([\varphi!_{i\hookleftarrow j}]\psi \leftrightarrow \varphi))\Big) \quad (\mathbf{Red}_{\varphi!_{i\hookleftarrow j},\mathsf{E}_{i,j}})$$

the following ones for the dynamic operators $[\circ_i]$:

$$[\circ_i]p \leftrightarrow (\mathsf{B}_i\bot \to p) \quad (\mathbf{Red}_{\circ_i,p})$$

$$[\circ_i]\mathsf{trs}(k,x) \leftrightarrow (\mathsf{B}_i\bot \to \mathsf{trs}(k,x)) \quad (\mathbf{Red}_{\circ_i,\mathsf{trs}(k,x)})$$

$$[\circ_i]\neg\varphi \leftrightarrow (\mathsf{B}_i\bot \to \neg[\circ_i]\varphi) \quad (\mathbf{Red}_{\circ_i,\neg})$$

$$[\circ_i](\varphi \wedge \psi) \leftrightarrow ([\circ_i]\varphi \wedge [\circ_i]\psi) \quad (\mathbf{Red}_{\circ_i,\wedge})$$

$$[\circ_i]\mathsf{K}_j\varphi \leftrightarrow (\mathsf{B}_i\bot \to \mathsf{K}_j[\circ_i]\varphi) \quad (\mathbf{Red}_{\circ_i,\mathsf{K}_j})$$

$$[\circ_i]\mathsf{B}_j\varphi \leftrightarrow (\mathsf{B}_i\bot \to \mathsf{B}_j[\circ_i]\varphi) \text{ if } j \neq i \quad (\mathbf{Red}_{\circ_i,\mathsf{B}_j})$$

$$[\circ_i]\mathsf{B}_i\varphi \leftrightarrow (\mathsf{B}_i\bot \to \mathsf{K}_i[\circ_i]\varphi) \quad (\mathbf{Red}_{\circ_i,\mathsf{B}_i})$$

$$[\circ_i]\mathsf{E}_{j,k}\varphi \leftrightarrow (\mathsf{B}_i\bot \to \mathsf{E}_{j,k}[\circ_i]\psi) \text{ if } j \neq i \quad (\mathbf{Red}_{\circ_i,\mathsf{E}_{j,k}})$$

$$[\circ_i]\mathsf{E}_{i,j}\varphi \leftrightarrow \neg\mathsf{B}_i\bot \quad (\mathbf{Red}_{\circ_i,\mathsf{E}_{i,j}})$$

and the following rule of inference:

$$\frac{\psi_1 \leftrightarrow \psi_2}{\varphi \leftrightarrow \varphi[\psi_1/\psi_2]} \quad (\mathbf{RRE})$$

We write $\vdash_{\mathsf{DEL\text{-}ES}} \varphi$ to denote the fact that φ is a theorem of DEL-ES.

The completeness of DEL-ES follows from Theorem 4.2, in view of the fact that the reduction axioms may be used to find, for any DEL-ES formula, a provably equivalent EL-ES formula.

Lemma 4.8 *If φ is any formula of $\mathcal{L}_{\mathsf{DEL\text{-}ES}}$, there is a formula $red(\varphi)$ in $\mathcal{L}_{\mathsf{EL\text{-}ES}}$ such that $\vdash_{\mathsf{DEL\text{-}ES}} \varphi \leftrightarrow red(\varphi)$.*

Proof. This follows by a routine induction on φ using the reduction axioms and the rule of replacement of equivalents (**RRE**) from Definition 4.7. □

As a corollary, we get the following:

Theorem 4.9 *DEL-ES is sound and complete complete for the class of ESMs.*

Proof. It is a routine exercise to check that all principles in Definition 4.7 are valid and that the rule of inference (**RRE**) preserves validity. As for completeness, if Γ is a consistent set of $\mathcal{L}_{\mathsf{DEL\text{-}ES}}$ formulas, then $red(\Gamma) = \bigwedge\{red(\varphi) : \varphi \in \Gamma\}$ is a consistent set of $\mathcal{L}_{\mathsf{EL\text{-}ES}}$ formulas (since DEL-ES is an extension of EL-ES), and hence by Theorem 4.2 there is a model M with a world w such that $M, w \models red(\Gamma)$. But, since DEL-ES is sound and for each $\varphi \in \Gamma$, $\vdash_{\mathsf{DEL\text{-}ES}} \varphi \leftrightarrow red(\varphi)$, it follows that $M, w \models red(\Gamma)$.

□

5 Illustration: is Peterson guilty?

In this section we want to illustrate how the concepts and framework we proposed in the paper can be used in understanding issues from our real life. Again, take the legal case, a judge is someone whom we trust as a rational agent. Her decision has to be made on the basis of reasons, or rather evidence. Let us consider a recent case in the US, and a small part of the timeline from online Fox News ([12]), with our notes italic in brackets. We use g to denote the proposition that "Scott Peterson is guilty" and we single out some events along the timeline that provide evidence for g or $\neg g$.

Dec. 24, 2002: Laci Peterson, while 8-months pregnant, is reported missing from her home in Modesto, Calif., by husband Scott Peterson. He claimed to have returned from a fishing trip and was unable to find his wife.
Jan. 24, 2003: Amber Frey, a massage therapist from Fresno, confirms she had a romantic relationship with Scott Peterson. [*evidence, at least, in favor of g*]
Aug. 22, 2003: ... Later that day, sources tell Fox News that Scott Peterson had admitted – then denied – involvement in his wife's disappearance in a wiretapped telephone conversation with his then-girlfriend Amber Frey. [*evidence supporting g*]
Oct. 15, 2003: Sources tell Fox News that telephone logs show that Scott Peterson called Frey hundreds of times after his wife's disappearance, contradicting prior claims that Frey pursued him. [*evidence supporting g*]
Nov. 3, 2003: A defense expert testifies that mitochondrial DNA tests, which cannot link evidence to a specific individual, are scientifically flawed. [evidence supporting $\neg g$]
Nov. 6, 2003: A police detective testifies that Scott Peterson told Frey he was a recent widower on Dec. 9, 2002, two weeks before his wife disappeared. [*evidence supporting g*]
March 16, 2005: Judge Alfred Delucchi formally sentences Peterson to death, calling the murder of his wife "cruel, uncaring, heartless, and callous." [*final decision made*]

As we can see, while time goes, the evidence is accumulated. Some are supporting g, some are not. This dynamic process leads the judge to form the belief that Scott Peterson is guilty ($\mathsf{B}_{judge}g$). Let us assume that the judge's level of cautiousness is equal to 4. Then, the above example can be expressed by the following formula:

$$\mathsf{trs}(judge, 4) \wedge [g!_{judge \hookleftarrow source1}][g!_{judge \hookleftarrow source2}]$$
$$[g!_{judge \hookleftarrow source3}][\neg g!_{judge \hookleftarrow source4}][g!_{judge \hookleftarrow source5}]\mathsf{B}_{judge} g$$

Here the judge's epistemic cautiousness, as well as the amount of evidence that have been collected determines the final decision. This might look too simple. However, we hope to have shown the potential of connecting our work to real legal practice. We believe that evidence-based analysis of legal texts can facilitate the justice system.

6 Conclusion and future directions

In this paper we have proposed a new logic, called "Dynamic Epistemic Logic of Evidence Sources", which enables us to reason about an agent's evidence-based belief formation and belief change, triggered by social communication. We have provided a complete axiomatization for both the static Epistemic Logic of Evidence Sources and its dynamic extension. We have discussed several interesting concepts that we can use in talking about evidence or reasons. For instance, having decisive evidence, and having sufficient reasons to believe. In particular, we have explicitly introduced evidence sources into our language. The new logic can be adopted to analyze. issues in legal contexts. For future directions, we identify a few. (i) We want to further study the relation between the evidence sources and the sources themselves. The same evidence provided by different sources who are situated in various communities should carry different strength. For instance, evidence from independent sources may be treated heavily than that from an internally closed community. (ii) We have emphasized that the accumulation of evidence leads to an agent's belief change and that the amount of evidence plays a role in relation with the level of epistemic cautiousness. However, it is sometimes the case that one piece of evidence counts much more than many other pieces all together. We would like to deal with such situations. (iii) Finally, an agent obtains information by social communication, and forms her beliefs on the basis of reasons. In this paper, we have investigated epistemic reasons. We plan to extend our logical framework with agents' preferences and choices in order to incorporate practical reasons in our analysis and to study their connection with epistemic reasons.

References

[1] Artemov, S. N., *The logic of justication*, The Review of Symbolic Logic **1** (2001), pp. 477–513.

[2] Ayer, A. J., "Probability and Evidence," Columbia University Press, 1972.

[3] Baltag, A., F. Liu and S. Smets, *Reason-based belief revision in social networks*, Slides, KNAW-Workshp on The Logical Dynamics of Information, Agency and Interaction, Amsterdam (2014).

[4] Baltag, A., L. Moss and S. Solecki, *The logic of common knowledge, public announcements, and private suspicions*, in: I. Gilboa, editor, *Proceedings of the 7th Conference on Theoretical Aspects of Rationality and Knowledge (TARK 98)*, 1998, pp. 43–56.

[5] Baltag, A., B. Renne and S. Smets, *The logic of justified belief, explicit knowledge, and conclusive evidence*, Annals of Pure and Applied Logic **165** (2014), pp. 49–81.

[6] Benthem, J., "Logical Dynamics of Information and Interaction," Cambridge: Cambridge University Press, 2011.

[7] Benthem, J., D. Fernández-Duque and E. Pacuit, *Evidence logic: A new look at neighborhood structures*, in: T. Bolander, T. Braüner, S. Ghilardi and L. Moss, editors, *Proceedings of Advances in Modal Logic Volume 9* (2012), pp. 97 – 118.

[8] Benthem, J. and E. Pacuit, *Dynamic logics of evidence-based beliefs*, Studia Logica **99** (2011), pp. 61 – 92.

[9] Blackburn, P., M. de Rijke and Y. Venema, "Modal Logic," Cambridge: Cambridge University Press, 2001.

[10] Chellas, B. F., "Modal logic: an introduction," Cambridge University Press, 1980.

[11] Ditmarsch, H., W. van der Hoek and B. Kooi, "Dynamic Epistemic Logic," Berlin: Springer, 2007.

[12] Foxnews, *Timeline: The scott peterson case*, online (2012).

[13] Grandi, U., E. Lorini and L. Perrussel, *Propositional opinion diffusion*, in: *Proceedings of the 14th International Conference on Autonomous Agents and Multiagent Systems (AAMAS 2015)* (2015), pp. 989–997.

[14] Keynes, J. M., "A Treatise on Probability," The Collected Writings of John Maynard Keynes **8**, Macmillan, 1973.

[15] Kraus, S. and D. J. Lehmann, *Knowledge, belief and time*, Theoretical Computer Science **58** (1988), pp. 155–174.

[16] Liu, F., J. Seligman and P. Girard, *Logical dynamics of belief change in the community*, Synthese **191** (2014), pp. 2403–2431.

[17] Lorini, E., *A minimal logic for interactive epistemology*, Synthese **193** (2016), pp. 725–755.

[18] Lorini, E., G. Jiang and L. Perrussel, *Trust-based belief change*, in: *Proceedings of the 21st European Conference on Artificial Intelligence (ECAI 2014)* (2014), pp. 549–554.

Objective Oughts and a Puzzle about Futurity

Alessandra Marra

Tilburg Center for Logic, Ethics and Philosophy of Science
Tilburg University
Tilburg, The Netherlands

Abstract

The paper argues that the standard definition of objective *oughts* (as *oughts* in light of the relevant facts) leaves room for a puzzle about futurity: Can objective *oughts* depend on what will happen in the future? If yes, how to account for those objective *oughts* that are future-dependent? Two main options are investigated: (i) treating the future as a fact and, hence, committing to the inevitability of the future, or (ii) adopting a branching-time account of the future, and weighting future possibilities in terms of their objective probability. I argue that (ii) is more promising. Given an appropriate account of objective probability, option (ii) allows for a univocal definition of the meaning of objective *oughts* without endorsing (i)'s commitment to the inevitability of the future. Finally, it is shown that, if (ii) is adopted, it is possible to construct examples like the Miners' Paradox involving exclusively objective oughts.

Keywords: Objective Oughts, Branching Time, Objective Probabilities, Miners' Paradox.

1 Introduction

Suppose that, unbeknownst to you, a friend of yours broke her leg yesterday. Some moral theorists, linguists and logicians would say that there exists a sense of *ought* according to which the sentence "You ought to visit your friend at the hospital" is true. It is called objective *ought*, and constitutes the focus of this paper.

To a first approximation, objective *oughts* are meant to indicate what is right, what is the best action to perform regardless of an agent's beliefs or information. So, in our hospital example, the sentence "You ought to visit your friend at the hospital" indicates what is the right thing to do, the best course of action given what *is* the case (namely, that your friend is at the hospital with a broken leg). But, of course, you might object that you have no clue that your friend is at the hospital. Indeed, under a subjective sense of *ought*, the sentence "You ought to visit your friend at the hospital" seems false. In light of *your* information, it might not be true that the best action for you to perform is going to the hospital.

Contrary to subjective *oughts*, objective *oughts* seem to have little role in contexts of deliberation, i.e., when it comes to decide what we ought to do.[1] Moreover, it is even open to debate whether objective *oughts* exist in natural language.[2] And, if they exist, whether they deserve a separate semantic treatment.[3] This paper does not aim to take side on those issues. Rather, I put them aside, and simply start by assuming the notion of objective *ought* as it is standardly presented in the literature. Starting point of the paper is therefore the following:

Definition 1.1 [Objective Oughts] An agent α ought objectively to do X if and only if X is the best thing to do in light of all the relevant facts.

Despite all the difficulties mentioned above, objective *oughts* as defined in Def.1.1 still feature in moral theories, as well as in several logical and linguistic frameworks.[4] Some of the best-known frameworks of deontic logic are indeed based on such a notion of objective *ought*. Those frameworks typically come with a state (or modal base) of possible worlds describing a body of contextually relevant facts, and an ideality function which indicates the deontically best worlds within the state. Kratzer's ([11]) semantics is an example.[5]

It seems then that such a notion of objective *ought* is worth a closer inspection. In the paper, I focus on a particular aspect of Def.1.1 which, to the best of my knowledge, has been overlooked so far. In particular, I argue that Def.1.1, by linking the meaning of objective *ought* to the "facts", leaves room for a **puzzle about futurity**: What if X is the best thing to do in light of what will happen in the future? Such "future-dependent" *oughts* are certainly conceivable, yet what will happen in the future does not immediately strike us as a "fact". Is it possible to have future-dependent objective *oughts*? If yes, how should we account for them?

The paper considers some possible ways of making sense of future-dependent objective *oughts*, without postulating truth-value gaps. In particular, two main

---

1 To the extent that Kolodny and MacFarlane [10] deny that objective *oughts* have any role in deliberation. Given that we always find ourselves in making decision under partial information, we are never in a position to determine what we objectively *ought* to do. Carr [4], on the other end, rejects Kolodny and MacFarlane's conclusion. Cf. Kolodny and MacFarlane [10], p.117 and Carr [4], p. 703.

2 For instance, one could argue that, in contexts of advice, subjective *oughts* are actually playing the role that is generally attributed to objective *oughts*. When seeking for advice or giving advice, we strive for finding out what is the case, and then advice on what to do on the basis of that. However, there is no need of objective *oughts* for that: *oughts* in advice could simply be subjective *oughts*, just relative to a more informed agent. See Gibbard [6] on that point. For a defense of the existence of objective *oughts* in natural language, see Carr [4].

3 For instance, Kolodny and MacFarlane [10] argue in favor of a general semantic treatment of informational *oughts*, rather than having separate semantic entries for the different senses of *ought*.

4 Cf. Gibbard [6], Broome [2] (even though he ultimately focuses on another sense of *ought*), Wedgwood [17], Silk [15], Carr [4].

5 In Kratzer [11], deontic modals such as *ought, must, might* quantify indeed over a *circumstantial* modal base (and not over an epistemic modal base).

options are available: (i) treating the future as a fact and, hence, committing to the inevitability of the future, or (ii) adopting a branching-time account of the future, and weighting future possibilities in terms of their objective probabilities. In Sec.3, I present (i). While in Sec.4, I discuss (ii). Option (ii), I argue, allows to account for future-dependent objective *oughts*, and to provide a univocal definition of objective *oughts* while avoiding to commit to the inevitability of the future. Finally, Sec.4.2 is a brief excursus on a well-known deontic puzzle involving reasoning by cases: the Miners' Paradox. While the original Miners' Paradox has been thought to be circumscribed to subjective *oughts*, I show that a similar paradox could emerge also for objective *oughts*. The logics underlying subjective and objective *oughts* might be less different than expected.

Let me now discuss a bit more in detail what being a "fact" means.

2 The "Facts"

What are the "facts" mentioned in Def.1.1? In absence of an explicit answer to that question,[6] looking at the way objective *oughts* are presented in the literature might be of some help. Let us consider the following examples:

> Late for an important meeting, I approach a blind intersection. In fact, nothing is coming on the crossroad, and so in light of all the facts, **I ought to drive on through** without slowing down. I have no way of knowing this, however, until I slow down and look, and so in light of my information, I ought to slow down and look, and proceed only if I see that nothing is coming. Standardly in moral theory, we distinguish what a person ought to do in the *objective* sense and what she ought to do in the *subjective* sense.
>
> Gibbard [6], p. 340 [emphasis added]

> Suppose you buy three rubber duckies for your child, and later learn that one out of every hundred rubber duckies from this manufacturer leaches out toxic chemicals. What should you do? Clearly, **you ought to throw away all three duckies. But that is almost certainly not what you ought to do in the objective sense**. An omniscient being who knew all the facts would know which (if any) of your duckies were toxic, and would counsel you to discard only those, keeping the rest.
>
> MacFarlane [13], p. 282 [emphasis added]

From the quotes above, we can safely assume that the predicate "... is a fact" has (at least) the following characteristics:

(i) It is *non-rigid*. It is a fact that there is no car coming. However, if there is a car coming, it is not a fact that there is no car coming.[7]

(ii) It is *time-connected*, that is, a certain temporal dimension is involved.

---

[6] To the best of my knowledge, the metaphysics of facts underlying Def.1.1 remains, at best, implicit in the literature of objective *oughts*. One of the main aims of the paper is indeed to shed some light on that.

[7] For convenience, I write "It is a fact that φ" in place of "φ is a fact".

For instance, we can say that "It is a fact that there *is* no car", "It is a fact that this ducky *is* toxic", "It is a fact that this ducky *was* produced with toxic material". In that sense, I will talk of **present facts** and **past facts**.[8] It seems then that the account of facts in place here is quite intuitive, and not significantly different from the expression "matter of fact" in ordinary English. I will not assume anything more than (i)-(ii) in the paper.

I take as a *desideratum* for a semantic theory to provide a univocal definition of objective *oughts*, be they dependent on the past, present or future. If we strive for that univocality, as I am trying to do here, then the *puzzle about futurity* mentioned in Sec.1 emerges mainly as a puzzle about the way we can account for the future. Two main options are open.

3 Future Facts and the Inevitability of the Future

One first, straightforward way of accounting for future-dependent objective *oughts* is to treat what will happen in the future as a fact. There would be future facts (e.g., "It is a fact that this coin *will* land tails"), and their status would be on a par with present and past facts.

It is clear that such an account would allow for a univocal definition of objective *oughts*, without even requiring any modification of Def.1.1. In particular, the semantic clause for objective *oughts* would look as follows:

Definition 3.1 [Objective Oughts with Future Facts] At time t it is true that agent α ought objectively to do X if and only if X is the best thing to do in light of all the relevant facts.

There are three main formal ways of representing the conception of time underlying Def.3.1: assuming a linear representation of time, a "Peircean" approach to branching time or the so called "thin red line" approach.

3.1 Linear Time

Future facts and Def.3.1 can be easily formalized against a linear representation of time. That is, we can imagine time as a non-empty set of moments $t, t', t'', ...$ together with an *irreflexive*, *transitive* and *connected* ordering $\prec$ on those moments:[9] If t_0 is the present, our *now*, then what comes before t_0 counts as the past and what comes after t_0 counts as the future. There is one future, as there is only one present and one past. In linear time, the talk about future facts becomes intelligible in the following sense: given a certain past and present, i.e., a linear order up to t_0, what will happen in the future is a fact since it is already uniquely fixed at t_0. At t_0 it is a fact that the coin will land

[8] Gibbard's sentence "there is no car coming" can be also interpreted as future progressive. That alternative interpretation does not undermine the analysis presented in the paper. I will come back to that in Sec.3.3.

[9] Where: *irreflexivity:* $\forall t, \neg t \prec t$; *transitivity:* $\forall t_1, t_2, t_3((t_1 \prec t_2 \wedge t_2 \prec t_3) \rightarrow t_1 \prec t_3)$; *connectedness:* $\forall t_1, t_2(t_1 \prec t_2 \vee t_2 \prec t_1 \vee t_1 = t_2)$. Additionally, one could also assume *no beginning* and *no end*, making time an infinite line of moments without endpoints. That does not matter for the purposes of the paper.

tails because it is already fixed that the coin will land tails, i.e., there exists a certain future moment t_1, $t_0 \prec t_1$, in which the coin lands tails.

The linear time approach provides a formally simple framework to account for future facts and Def.3.1, but does so at considerable costs. First of all, such an account ultimately implies that the future, in all its aspects, is *inevitable*. Since the future is uniquely fixed, if the coin will land tails, then it is inevitable that the coin will land tails, and if the coin will not land tails, then it is inevitable that the coin will not land tails. That clearly clashes with the intuition that there exist future contingents, and that some aspects of the future are genuinely open. Second, and more strongly, in the linear approach it holds that if the coin will land tails, that the coin will land tails is inevitable not only now, but also in all moments of the past. That is, it is fixed now and was fixed in the past that the coin will land tails. The future has never been open.

3.2 Peircean Approach and Branching Time

Adopting a linear representation of time is not the unique way to make sense of future facts and Def.3.1. An alternative representation, widespread in temporal logics, consists in considering the temporal evolution of the world as having the shape of a branching tree: The future is open, and branches represent many possible continuations of the world. [10] For the purposes of this paper, it is crucial to understand the branches as genuine *objective possibilities*. The future is open not only in an epistemic sense, but *in re*. [11]

Formally, branching trees can be defined as follows:

Definition 3.2 [Trees and Histories] Let T be a non-empty set of moments. A *tree* is an irreflexive ordered set $\mathbf{T}=\langle T, \prec\rangle$ in which the set of the $\prec$-predecessors of any moment t of T is linearly ordered by $\prec$. A *history* is a maximal linearly ordered subset of $\mathbf{T}$.

Trees are structurally a generalization of the linear representation of time discussed above; the difference being that, while in linear time there is only one future, trees can be forward branching. The dispensability of the linear representation emerges also with respect to Def.3.1: by adopting a *Peircean interpretation* of the future, [12] according to which at t it is the case that $\ulcorner Will\varphi \urcorner$ if φ is the case in all future branches of t, trees too can provide an adequate formal background for future facts and Def.3.1. In particular, we can say that at t_0, now, it is a fact that φ will happen if φ happens in all possible futures. Or, to use our classic example of the coin: If the coin lands tails in all the future possibilities that are open to us now, then it is already a fact that the coin will land tails.

Branching time together with the Peircean interpretation described above is more attractive than the linear representation of time. Formally, it is a

[10] See Belnap and Green [1], Horty and Belnap [9], and MacFarlane [12].

[11] See Belnap and Green [1], p. 365.

[12] See Prior [14], pp.128-134, and Thomason [16], pp.141-143.

generalization of the latter, and still provides a way to make sense of future facts and, hence, of Def.3.1. Moreover, conceptually, it avoids some of the strong commitments of linear time: First, in the Peircean interpretation, if it does not hold that the coin will land tails, then it might be the case that in some possible futures the coin lands tails and in some other possible futures the coin does not land tails.[13] Second, even if it holds that the coin will land tails, it does not follow that that held always in the past. There might have been a future, possible in the past but not possible anymore, in which it did not hold that the coin would have landed tails.[14]

The Peircean interpretation, however, inherits some of the problematic characteristics of linear time. Despite adopting a forward branching representation of time, the approach treats some relevant aspect of the future as inevitable: If $\ulcorner Will\varphi \urcorner$ is the case at t_0, then φ is inevitable. No matter how the world will develop, φ will be the case. Hence the Peircean approach still implies a form, albeit weaker, of inevitability of the future.

3.3 The Thin Red Line and Branching Time

There is one further approach that could provide a proper formal background for Def.3.1, and it is the so called *thin red line.*[15] We have seen that, according to the standard interpretation of branching time, the world could evolve in the future through different histories. Amongst all those histories, however, there is only one which corresponds to how the world will actually be. The thin red line marks out precisely such actual future.

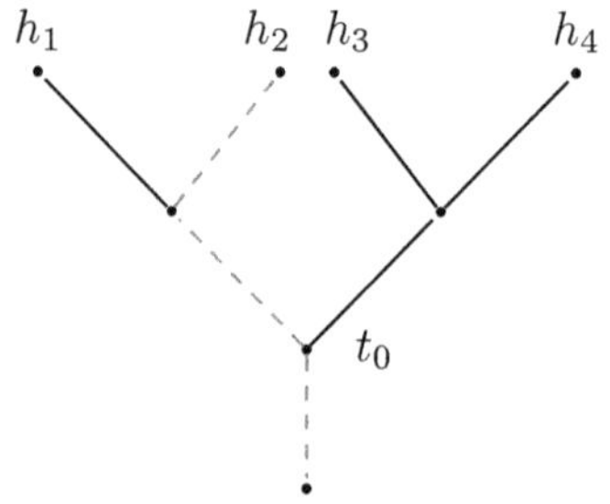

Fig. 1. The Thin Red Line

Consider again our coin example. There is a sense in which at t_0, the moment in which the coin is tossed, the future is open. It is possible that the coin lands tails and it is possible that the coin lands heads. Amongst those two possibilities, however, only one will be realized. That is the one picked up by the thin red line. So, at t_0 it is a fact that the coin will land tails, if

[13] On the Peircean interpretation, $\ulcorner \neg Will\varphi \urcorner$ differs from $\ulcorner Will\neg\varphi \urcorner$: the latter implies the former, but not viceversa. See Prior [14], p.129

[14] See Thomason [16], p.143.

[15] For a critical presentation of the thin red line approach, see Belnap and Green [1] and MacFarlane [12]-[13].

that is what will actually happen, i.e., if the coin lands tails at a certain future moment t within the thin-red line. Postulating a "thin red line" seems to allow to have it both: Future facts and different future possibilities. We can say that it is a fact that the coin will land tails without committing to saying that that is inevitable; after all, the coin could land heads.

The thin red line provides an attractive background for Def.3.1. I suspect that this is indeed the approach implicitly assumed by some moral theorists when dealing with objective *oughts*. It is said that the thin red line marks the evolution of the world from the God's eye perspective, and the same metaphor of the God's eye perspective is often used to characterize objective *oughts* too.[16]

However, the thin red line has some unfortunate shortcomings. Belnap, Green and MacFarlane have extensively argued against such an approach,[17] but I think two specific problems are particularly crucial for the purposes of this paper. The first, standard objection is that the stipulation that there exists one, privileged thin red line amongst the possible histories makes unclear in which sense the other possible histories represent objective, ontological possibilities. If now it is a fact that the coin will land tails, then, in a certain sense, that the coin will land tails is already determined. Hence, the alternative histories might at most be epistemically possible (it is a fact that coin will land tails, but we cannot know it yet), but not possible *in re*. The second objection concerns more closely the semantics for objective *oughts*. According to the thin red line, as described here, $\ulcorner Will\varphi \urcorner$ is true at t_0 if φ holds at some later t along the thin red line, and false otherwise. Suppose it is true that the coin will land tails. Then, it is not clear how to make sense of the difference between the following two indicative conditionals involving a false antecedent: "If the coin will land heads, then I ought to bet on heads" and "If the coin will land heads, then I ought to bet on tails". A distinction that seems important, at least in everyday life.

Clearly what I have said so far does not count as a proof that Def.3.1 is wrong, nor that Def.3.1 is bound to remain ungrounded. Rather, this section was meant to clarify the implications of the main approaches which support Def.3.1. In particular, I have shown that the talk about future facts, which Def.3.1 relies on, implies a certain commitment to the inevitability of the future. A commitment that, even if it might be welcomed by some, is worth to make explicit.

4 Objective Probabilities to the Rescue?

There is an alternative way to make sense of objective *oughts*, and in particular of future-dependent objective *oughts*, which does not require us to decide on

[16] See, e.g., MacFarlane [13], p.282 and Carr [4], p.678. It is worth remarking that the alternative, "future progressing" reading of Gibbard's quote in Sec.2 is compatible with the observation presented here. Cf. Footnote 8.

[17] See Belnap and Green [1], pp. 379-381, MacFarlane [12], p.325-326

whether there exist future facts, nor to take side on whether the future is inevitable or contingent. The following modification of Def.1.1, I argue, does the job:

Definition 4.1 [Objective Oughts with Probabilities] At time t it is true that agent α ought objectively to do X if and only if X is the best thing to do in light of the *objective chances* of all the relevant circumstances.

Where the term "circumstances" is meant to indicate, in a rather neutral way, whatever counts as contextually relevant for establishing the truth of an objective *ought*: A present or a past fact (as already accounted for in Def.1.1) or what will happen in the future. [18]

To see why Def.4.1 can provide a univocal definition of objective *oughts*, it is first necessary to introduce some formal background. Let me start by defining probability trees and future branching frames. [19]

Definition 4.2 [Probability Tree] A probability tree is a tuple $\mathcal{E}=\langle \mathbf{T}, S, Pr^{t_0}\rangle$ where $\mathbf{T}=\langle T, \prec\rangle$ is a finite tree, S is an algebra over $\mathbf{T}$, and Pr^{t_0} is a function assigning to each maximal subtree of $\mathbf{T}$ in S a number [0,1] satisfying the following:

- $Pr^{t_0}(\mathbf{T})=1$
- $Pr^{t_0}(\mathbf{T}' \cup \mathbf{T}'')= Pr^{t_0}(\mathbf{T}')+Pr^{t_0}(\mathbf{T}'')$ if $\mathbf{T}'$ and $\mathbf{T}''$ are disjoint elements of S.

Definition 4.3 [Future Branching Frames] A future branching frame is a tuple $\mathcal{F}=\langle \mathcal{E}, t_0\rangle$ where $\mathcal{E}=\langle \mathbf{T}, S, Pr^{t_0}\rangle$ is a probability tree, and $t_0 \in T$ such that for all histories $h, h' \subset \mathbf{T}$, $t_0 \in h \cap h'$.

As stressed above, it is crucial to interpret the tree $\mathbf{T}$ (and S) *in re*, as describing the objective possible temporal evolution of the world. Moreover, also the Pr^{t_0} function should be interpreted objectively. It represents the *objective chance* of the various histories, not an agent's beliefs about them. Finally, it is worth noticing that, in our frames, the function Pr^{t_0} is defined over the tree $\mathbf{T}$ whose current time is t_0. With the passing of time, the probability tree will change as well: Some future of t_0 will be realized, some of the possibilities will be ruled out, and the new tree of reference will differ from $\mathbf{T}$ and so will do the probability function. Therefore, in Def.4.3, the Pr^{t_0} function should be

[18] Probabilistic analyses of *oughts* can be also found in Carr [4], Wedgwood [18] and Finlay [5]. The approach presented in this paper significantly differs from those previous works. Contrary to Carr [4] and Wedgwood [18], this paper defends a deontic semantics in which probabilities play a significant, non trivial role for (future-dependent) objective *oughts*. Contrary to Finlay [5], the paper is not committed to an end-relational theory of *ought*, and develops a formal analysis of objective *oughts* which links deontic logic with branching time logic.

[19] To keep the presentation simpler, I am limiting myself to the case of $\mathbf{T}$ finite. Def.4.2 above can be easily extended for $\mathbf{T}$ infinite by considering σ-algebras, closed under *countable* union, instead of algebras. Cf. Halpern [7], p. 15.

intended as expressing the objective chance assigned from the perspective of the moment t_0.

One main advantage of the approach described in this section is that it does not force us to assume future facts, nor to explain what it means for a sentence like $\ulcorner Will\varphi \urcorner$ to hold at t_0. I will therefore adopt the "double indices" semantics presented by Prior and then developed by Thomason, Horty and Belnap, according to which sentences are evaluated not only at a moment but also relative to a history:[20]

Definition 4.4 [Future Branching Models and Basic Clauses] A future branching model is a tuple $\mathcal{M} = \langle \mathcal{F}, v \rangle$ with $\mathcal{F}=\langle \mathcal{E}, t_0 \rangle$ a future branching frame. The valuation function v maps each sentence letter from the background language into a set of t/h pairs from T. Truth at t/h is defined as follows:

- $\mathcal{M}, t/h \models \mathrm{p}$ iff $t/h \in v(\mathrm{p})$ for p atomic
- $\mathcal{M}, t/h \models \neg\varphi$ iff $\mathcal{M}, t/h \not\models \varphi$
- $\mathcal{M}, t/h \models \varphi \wedge \psi$ iff $\mathcal{M}, t/h \models \varphi$ and $\mathcal{M}, t/h \models \psi$
- $\mathcal{M}, t/h \models Will\varphi$ iff there is a $t' \in h$ such that $t \prec t'$ and $\mathcal{M}, t'/h \models \varphi$
- $\mathcal{M}, t/h \models Past\varphi$ iff there is a $t'' \in h$ such that $t'' \prec t$ and $\mathcal{M}, t''/h \models \varphi$

Hence the "double indices" semantics establishes that a sentence like "The coin will land tails" is true now with respect to a particular evolution of the world if and only if, according to that particular evolution, it is true that the coin lands tails at a later moment. On the other hand, the semantics remains agnostic on the meaning of "The coin will land tails" at t_0 *simpliciter.*

As it stands, however, the "double indices" semantics has certain features that make it not appropriate for providing a background for Def.4.1. In particular, the semantics does not allow to express what holds in the present and in the past *simpliciter.* In principle, the same atomic formula p can be both true and false at t_0, if different histories are taken into account. To fix that, I adopt the following constraint for overlapping histories:

Definition 4.5 [Overlapping Histories and Constraint on v] Two histories, h and h', overlap if $h \cap h' \neq \emptyset$, and they overlap at t if $t \in h \cap h'$, i.e., if both histories pass through t. If h and h' overlap at t, then, for every atom p: $t/h \in v(\mathrm{p})$ iff $t/h' \in v(\mathrm{p})$

Since in the frames I am considering, **T** is a tree forward branching (at most) at t_0, now, the above constraint guarantees that all histories agree on the present and the past. We can then talk of present facts and past facts, in the sense described in Sec.2. Moreover, the definition of the Pr^{t_0} function provided above makes the objective chance of present and past facts trivial. To see it, let me first define the proposition espressed by a sentence φ:

[20] See Prior [14], Thomason [16], Horty and Belnap [9], and Horty [8].

Definition 4.6 [Proposition] Consider a future branching model $\mathcal{M} = \langle \mathcal{F}, v \rangle$. Let φ be a sentence of the background language. The proposition expressed by φ in $\mathcal{M}$ at t, written $\mathbf{T}_{\varphi}^{\mathcal{M},t}$, is defined as follows:

- $\mathbf{T}_{\varphi}^{\mathcal{M},t} = \bigcup \{h | \mathcal{M}, t/h \models \varphi\}$

Hence the proposition expressed by φ at t in $\mathcal{M}$ corresponds to a subtree $\mathbf{T}_{\varphi}^{\mathcal{M},t} \subseteq \mathbf{T}$ given only by the histories in which φ is true at t. In what follows, I drop the superscript $\mathcal{M}$, and simply write $\mathbf{T}_{\varphi}^{t}$. From Deff.4.2-4.6, it follows that talking about the objective chance of present and past facts is trivial. In particular, for every true φ referring to the present or the past,[21] $\mathbf{T}_{\varphi}^{t_0} = \mathbf{T}$ and, therefore, $Pr^{t_0}(\mathbf{T}_{\varphi}^{t_0}) = 1$. That is a welcome result: It means that with respect to present and past facts, Def.4.1 gives the same predictions as Def.1.1, which was our starting point.

Finally, here is the semantic clause for objective *oughts* as described in Def.4.1:[22]

Definition 4.7 [Clause for Objective Oughts with Probabilities] Let $\mathcal{D} = \langle \mathcal{M}, d \rangle$ be a deontic model such that:

- $\mathcal{M} = \langle \mathcal{E}, t_0, v \rangle$ is a future branching model
- d is a deontic selection function that maps each probability tree $\mathcal{E} = \langle \mathbf{T}, S, Pr^{t_0} \rangle$ to a set of histories of $\mathbf{T}$ that are deontically best.

The evaluation clause for the objective ought $\mathcal{O}$ is the following:

- $\mathcal{D}, t/h \models \mathcal{O}\varphi$ iff $\mathcal{D}', t/h' \models \varphi$ for every $h' \in d(\mathcal{E}')$
  where:
  - $\mathcal{D}' = \langle \mathcal{E}', t_0, v, d \rangle$
  - $\mathcal{E}' = \langle \mathbf{T}, S, Pr^{t} \rangle$
  - $Pr^{t} = Pr^{t_0}(\cdot | \mathbf{T}^{t})$
  - $\mathbf{T}^{t}$ is the subtree given by all histories passing through t.

Crucial component in the semantics of $\mathcal{O}$ is therefore the deontic selection function d, which in turn depends on the probability function Pr^{t_0} (conditional on the histories open at the time of evaluation).

The above semantic clause could be simplified for t_0, given that $\mathbf{T}^{t_0} = \mathbf{T}$. Informally, at t_0 it is true that $\ulcorner \mathcal{O}\varphi \urcorner$ if and only if, given the tree $\mathbf{T}$ and the objective probability function Pr^{t_0}, φ holds in all the deontically best histories selected by the function d.[23] In other words, as stated in Def.4.1, $\ulcorner \mathcal{O}\varphi \urcorner$ is true

[21] E.g., given our point of evaluation t_0, for φ not containing tense operators or of the form $\ulcorner Past\psi \urcorner$ with ψ containing, at most, only past tense operators.

[22] For ease of exposition, I consider all the histories in $\mathbf{T}$ to be contextually relevant. Such assumption could be weakened by adopting a modal base function which indicates, *à la* Kratzer, the contextually relevant histories in $\mathbf{T}$. Cf. Kratzer [11].

[23] A fully fleshed out account of the deontic function d would require to make it relative also to a $Value$ function and some decision norm. In particular, $Value$ would be a function mapping every history in $\mathbf{T}$ into a set of values partially ordered by a relation $\leq$. On the $Value$ function, Cf. Horty [8], p.37. Moreover, I would be happy to take on board

now, if φ is the best thing to do in light of the objective chance of the relevant circumstances (i.e., the histories in $\mathbf{T}$).[24]

To sum up, the approach presented in this section has the following advantages:

I It does not assume the existence of future facts, and therefore avoids the commitments that follow from such an assumption;

II It provides a univocal definition of the meaning of objective *oughts*, be they dependent on the present, past or future;

III It is conservative with respect to Def.1.1 when the relevant circumstances are present or past facts.

I think that I, II and III, taken together, provide good ground to prefer the probabilistic account of objective *oughts* described here to the one based on future facts and discussed in Sec.3.

Before moving over, I give one last definition that will turn out useful in Sec.4.2: The deontic conditional. I follow the standard, Kratzerian analysis of the *if*-clause as restriction over the contexts in which the antecedent is true.[25] I assume that the *if*-clause restricts $\mathbf{T}$ to the set of histories that satisfy the antecedent, shifts the deontic function d and, following Yalcin [20], that it also shifts the Pr^{t_0} function as follows:

Definition 4.8 [Deontic Conditional] Let $\mathcal{D} = \langle \mathcal{M}, d \rangle$ be a deontic model. Then:

- $\mathcal{D}, t/h \models if\psi, \mathcal{O}\varphi$ iff $\mathcal{D}', t/h' \models \varphi$ for every $h' \in d(\mathcal{E}')$
  where:
  · $\mathcal{D}' = \langle \mathcal{E}', t_0, v, d \rangle$
  · $\mathcal{E}' = \langle \mathbf{T}^t_\psi, S, Pr^t_\psi \rangle$
  · $Pr^t_\psi = Pr^{t_0}(\cdot | \mathbf{T}^t_\psi)$

Roughly speaking, a deontic conditional $\ulcorner if\psi, \mathcal{O}\varphi \urcorner$ is true if $\ulcorner \mathcal{O}\varphi \urcorner$ is true with respect to the subtree generated by assuming ψ and the objective chances conditionalized to such subtree.

4.1 A Note on Reasons, Justification and Truth

I would like to draw the attention to one particular feature of the approach presented in this section, and specifically of Deff.4.1-4.7. By providing an analysis of objective *oughts* in terms of objective probabilities I am vulnerable to the following objection: Suppose that at t_0, given the objective chance of the

Carr's suggestion to have a placeholder for decision norms rather than anchoring the deontic semantics to a particular decision rule (e.g., MaxMax, MaxiMin). In such a way what counts as "best" would be determined by the Pr^{t_0} function, the $Value$ function and a contextually chosen decision rule. For a defense of such approach, see Carr [4], pp.703-707.

[24] Notice that the second index, h, in Def.4.7 can be omitted. If $\mathcal{D}, t_0/h \models \mathcal{O}\varphi$ then for all h'' of $\mathbf{T}$: $\mathcal{D}, t_0/h'' \models \mathcal{O}\varphi$.

[25] See also Kolodny and MacFarlane [10], Willer [19], Yalcin [20] and Carr [4].

relevant future circumstances C, I ought objectively to do X. However, in the same way as a biased coin could once land with the heavier side on top, even a high objective chance of C does not guarantee that such C will be actually the case in the future. Therefore it could be the case that, given the objective chance that C will happen tomorrow, now I objectively ought to do X, but that tomorrow it is false that I objectively ought to have done X.

Does the objection imply that Deff.4.1-4.7 provide at most a justification, a reason in support of the claim that now I objectively ought to do X? No, I would say. Because while the objection holds, it does not follow that Deff.4.1-4.7 do not provide a definition of what it means for "I objectively ought to do X" to be *true* now. First of all, it should be noticed that, in the approach presented here, objective chances are temporally dependent. The function Pr^{t_0} is indeed dependent on the tree whose current time is t_0. With the passing of time, the tree loses some branches: some of the future possibilities that were open at t_0 are ruled out. That means that there is no contradiction between having "I objectively ought to do X" true at t_0 in $\mathbf{T}$ and having "I objectively ought to have done X" false at t_1 in $\mathbf{T}'$. The context has changed.

Moreover, I am inclined to say that it is not a requirement for objective *oughts* to be persistent retrospectively, so to speak. Having "I objectively ought to do X" and "I objectively ought to have done X" both true is clearly welcome, but it does not figure as necessary requirement in the meaning objective *oughts*. As the standard Def.1.1 shows, objective *oughts* are rather characterized by being insensitive from any epistemic dimension.

Finally, whoever considered that objective *oughts* need to be persistent retrospectively would simply prefer Def.3.1 to Def. 4.1, and adopt any of the approaches described in Sec.3.

4.2 A Miners-like Scenario

> Ten miners are trapped either in shaft A or in shaft B, but we do not know which. Flood waters threaten to flood the shafts. We have enough sandbags to block one shaft, but not both. If we block one shaft, all the water will go into the other shaft, killing any miners inside it. If we block neither shaft, both shafts will fill halfway with water, and just one miner, the lowest in the shaft, will be killed.
>
> Kolodny and MacFarlane [10], p. 115

The above scenario is known in the literature as the *Miners' Paradox* (or the *Miners' Puzzle*). Why is it a "paradox"? Because while, given the context, the following sentences seem all true:

(i) We ought to block neither shaft

(ii) The miners are in shaft A or the miners are in shaft B

(iii) If the miners are in shaft A, then we ought to block shaft A

(iv) If the miners are in shaft B, then we ought to block shaft B

by Reasoning by Cases[26] from the premises (ii), (iii), (iv), one can derive:

(v) We ought to block shaft A or we ought to block shaft B

which, in turn, contradicts (i). The puzzle has been taken to show that, at least *prima facie*, Reasoning by Cases fails for indicative conditionals with deontic sentences in the consequent.

For the purposes of the present paper, two observations are particularly relevant. First of all, in order to show that Reasoning by Cases fails in the Miners' scenario we do not need a premise as strong as (i). A weaker sentence like " It is not the case that we ought to block shaft A and it is not the case that we ought to block shaft B" would do the job as well (as it corresponds to the negation of v). Premise (i) says something more: we do not simply lack the obligation to block either shaft, we have the obligation to block neither. Such a strong premise may emerge from the application of a particular decision norm: MaxiMin. The following table depicts the decision problem faced in the Miners' scenario:

	miners are in A	miners are in B
Block A	10	0
Block B	0	10
Block neither shaft	9	9

Table 1
Decision problem for the Miners' Paradox

where the numerical values are derived from the number of miners saved, and the states "miners are in A" and "miners are in B" are equally likely (epistemically speaking). By MaxiMin, we ought to block neither shaft as that action guarantees the highest amongst the worst possible outcomes.[27]

Second observation: the Miners' scenario emerges as a puzzle because it involves a certain epistemic uncertainty. Such an uncertainty is relevant for subjective *oughts* or, more generally, for informational *oughts* that are sensitive to the beliefs/information that an agent possesses.[28] However, it clearly does not affect objective *oughts*. The position of the miners is already determined: It is a fact that the miners are in shaft A or it is fact that the miners are in shaft B. In either case, it follows that the best thing to do is blocking the shaft the miners are in. Hence, (i) is false under the objective reading of *ought*, and the consequence (v) is correctly derived from (ii)-(iv) by Reasoning by Cases.

Does that show that objective *oughts* always validate Reasoning by Cases?

[26] Reasoning by Cases is a deductive principle from the premises $\ulcorner\phi$ or $\psi\urcorner$, $\ulcorner$if ϕ, then $\chi\urcorner$ and $\ulcorner$if ψ, then $\sigma\urcorner$ to the conclusion $\ulcorner\chi$ or $\sigma\urcorner$.

[27] See Cariani, Kaufmann and Kaufmann [3] for an analysis of the Miners' Paradox in terms of the decision rule MaxiMin, and Carr [4] for a general overview of the various approaches adopted to solve the Miners' Paradox.

[28] See Carr [4] for an analysis of subjective *oughts*, and MacFarlane [13], Kolodny and MacFarlane [10] and Silk [15] for a defense of informational *oughts*.

No, at least if Def.4.1 and its formal equivalent Def.4.7 are adopted. Consider the following **betting scenario**:

> I offer you a bet. I will toss a fair coin (hence 0.5 objective chance that the coin will land tails and 0.5 objective chance that the coin will land heads) and ask you to guess on which side the coin will land. If your guess is correct, you gain 150$. However, entering in the bet costs you 90$.

The tree in Fig.2 is a model for the betting scenario, and Table 2 represents the decision problem at t_0.

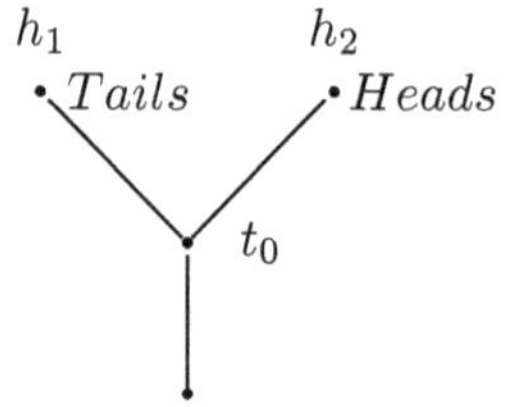

Fig. 2. The Betting Model

	Will Tails (h_1)	Will Heads (h_2)
Bet Tails	60	-90
Bet Heads	-90	60
Bet nothing	0	0

Table 2
Decision Problem for the Betting Scenario

Given the model presented in Fig.2, we get that $\mathcal{D}, t_0/h_1 \models$ *Will Tails* and therefore $\mathcal{D}, t_0/h_1 \models$ *Will Tails*$\vee$ *Will Heads*. Similarly, we get that $\mathcal{D}, t_0/h_2 \models$ *Will Tails*$\vee$ *Will Heads*.

Now assume you are considering whether to bet on tails, bet on heads or refrain from betting at t_0. Given the values depicted in Table 2 and the decision rule MaxiMin, we can define our deontic selection function as follows:

For $\mathcal{E}' = \langle \mathbf{T}_{\text{will tails}}, S, Pr^{t_0}_{\text{will tails}} \rangle$, $d(\mathcal{E}') = \{h | D', t_0/h \models \textit{Bet Tails}\}$. If we look only at the left side of Table 2, we can indeed see that *Bet Tails* is the best action (given the worst possible outcomes).

Similarly, for $\mathcal{E}'' = \langle \mathbf{T}_{\text{will heads}}, S, Pr^{t_0}_{\text{will heads}} \rangle$, $d(\mathcal{E}'') = \{h | D'', t_0/h \models \textit{Bet Heads}\}$. If we look only at the right side of Table 2, *Bet Heads* is the best action (given the worst possible outcomes).

For $\mathcal{E} = \langle \mathbf{T}, S, Pr^{t_0} \rangle$, $d(\mathcal{E}) = \{h | D, t_0/h \models \textit{Bet Nothing}\}$. If we look at the global Table 2, *Bet Nothing* is the best action (since it guarantees the highest amongst the worst possible outcomes).

From Deff. 4.7-4.8, it follows that:

$\mathcal{D}, t_0/h_1 \models if WillTails, \mathcal{O}BetTails,$
$\mathcal{D}, t_0/h_2 \models if WillTails, \mathcal{O}BetTails;$
$\mathcal{D}, t_0/h_1 \models if WillHeads, \mathcal{O}BetHeads;$

$\mathcal{D}, t_0/h_2 \models if WillHeads, \mathcal{O} BetHeads.$

However, we also get that:

$\mathcal{D}, t_0/h_1 \models \mathcal{O} Bet\ Nothing$

$\mathcal{D}, t_0/h_2 \models \mathcal{O} Bet\ Nothing$

To say it in English, there is a sense in which the following sentences are all true:

(vi) The coin will land tails or the coin will land heads

(vii) If the coin will land tails, you ought to bet on tails

(viii) If the coin will land heads, you ought to bet on heads

(ix) You ought not to bet

While, by Reasoning by Cases, it would follow:

(x) You ought to bet on tails or you ought to bet on heads

which contradicts (ix). Hence if the Miners' Paradox is taken to show that Reasoning by Cases may fail for belief/information dependent *oughts*, the same does the Betting Scenario with respect to the objective *oughts* of Def 4.1.[29] The relevant difference between the two, the Miners' Paradox and the Betting Scenario, is that the former turns around uncertainty, while the latter involves indeterminacy.

5 Conclusion

The standard definition of objective *oughts* as *oughts* dependent on the relevant facts leaves room for a puzzle about futurity: While it is clear how to account for *oughts* that depend on present or past facts, "future-dependent" *oughts* remain mysterious. In the paper I have investigated two main strategies to account for future-dependent objective *oughts*. The first one stipulates the existence of future facts, and ultimately is committed to the inevitability of the future. The second one remains agnostic on the existence of future facts, and allows to discriminate between different future possibilities in terms of their objective probability. Following that second strategy, I have reformulated the standard definition of objective *oughts* and proposed an alternative definition as *oughts* dependent on the objective chance of the relevant circumstances. This latter definition, I have argued, meets some important *desiderata*, as it provides a univocal account of the meaning of objective *oughts* without committing to the inevitability of the future. Finally, I have showed that it is possible to construct a Miners-like scenario involving exclusively objective *oughts*. Contrary to what is generally assumed, the Miners' Paradox is not strictly an artifact of uncertainty.

[29] A scenario similar to the Miners' Puzzle is discussed by Horty [8] in the context of Stit-models and utilitarian *oughts*. The counterexample to Reasoning by Cases presented here is more general, as meant to hold for objective *oughts*, and not specifically for Stit utilitarian *oughts*. Moreover, contrary to Horty, here I focus on sentences like (ix) which, it has been noticed, are stronger than the simple negation of (x). Cf. Horty [8], p.110.

References

[1] Belnap, N. and M. Green, *Indeterminism and the thin red line*, Philosophical Perspectives **8** (1994), pp. 365–388.
[2] Broome, J., *Rationality through reasoning* (2013).
[3] Cariani, F., S. Kaufmann and M. Kaufmann, *Deliberative modality under epistemic uncertainty*, Linguistics and Philosophy **36** (2013), pp. 225–259.
[4] Carr, J., *Subjective ought*, Ergo **2** (2015), pp. 678–710.
[5] Finaly, S., *Confusion of tongues* (2014).
[6] Gibbard, A., *Truth and correct belief*, Philosophical Issues **15** (2005), pp. 338–350.
[7] Halpern, J., *Reasoning about uncertainty* (2003).
[8] Horty, J., *Agency and deontic logic* (2001).
[9] Horty, J. and N. Belnap, *The deliberative stit: A study of action, omission, ability and obligation*, Journal of Philosophical Logic **24** (1995), pp. 583–644.
[10] Kolodny, N. and J. MacFarlane, *Ifs and oughts*, Journal of Philosophy **107** (2010), pp. 115–143.
[11] Kratzer, A., *The notional category of modality*, in: *Words, Worlds, and Context. New Approaches in Word Semantics* (1981), pp. 38–74.
[12] MacFarlane, J., *Future contingents and relative truth*, The Philosophical Quarterly **53** (2003), pp. 321–336.
[13] MacFarlane, J., *Assessment sensitivity: Relative truth and its applications* (2014).
[14] Prior, A., *Past, present and future* (1967).
[15] Silk, A., *Evidence sensitivity in weak necessity deontic modals*, Journal of Philosophical Logic **43** (2014), pp. 691–723.
[16] Thomason, R., *Combinations of tense and modality*, in: *Handbook of Philosophical Logic* (1984), pp. 135–165.
[17] Wedgwood, R., *Choosing rationally and choosing correctly*, in: *Weakness of Will and Practical Irrationality* (2003), pp. 201–230.
[18] Wedgwood, R., *Objective and subjective 'ought'*, in: *Deontic Modality* (forthcoming).
[19] Willer, M., *A remark on iffy oughts*, Journal of Philosophy **109** (2012), pp. 449–461.
[20] Yalcin, S., *Probability operators*, Philosophy Compass **5** (2010), pp. 916–937.

Rights in Default Logic

Robert Mullins [1]

Faculty of Law, Oxford University
United Kingdom

Abstract

The paper proposes a defeasible treatment of rights reasoning. First, I introduce a basic Hohfeldian account of rights within the framework of a multi-modal deontic logic. Second, I offer a number of examples of rights reasoning that cannot be appropriately captured within this Hohfeldian framework. The classical Hohfeldian framework is unable to accommodate reasoning processes involving the balancing of rights in order of priority, the disabling or cancellation of rights, and strong permissive rights. I then develop an account of rights within an alternative framework. The particular framework used is the version of default logic developed by Horty in [11]. Rights are embedded within this logic as the premises of default rules. The account is meant to capture Raz's informal characterisation of rights as 'intermediate conclusions' capable of justifying a variety of duties [31]. I argue that this logic brings the role of rights in practical reasoning into greater clarity. I conclude by considering a number of problems with the proposed logic and suggesting areas for further research.

Keywords: rights, Hohfeld, default logic, conditional rights, permissive rights, strong permission

1 Introduction

According to the traditional logical account of rights, the language of rights is used to state the conclusions of our practical deliberation in terms of a set of normative positions that hold between pairs of agents. In John Finnis's words, the language of rights is used as a 'precise instrument for sorting out and expressing the demands of justice' [5, p. 210]. But this is not the only way in which the language of rights is used. Rights often feature within the premises of practical arguments, and may defeasibly entail a variety of different practical conclusions. They may be defeated or undercut by other considerations. A number of authors have noted these features of rights reasoning without seeing the need to adopt an alternative logical framework. [2] In this paper I propose an alternative logical account that embeds rights within default logic.

---

[1] robert.mullins@law.ox.ac.uk. I would like to thank the anonymous referees from the DEON2016 program committee for their many helpful comments on this paper.

[2] Authors who have offered a defeasible treatment of rights within the traditional framework include Kamm [12, chapter 9], Kramer [17], [18, chapter 8], and Thomson [40].

I first introduce a semantic account of the traditional deontic logic of rights, as it was developed by Wesley Hohfeld [9]. I interpret Hohfeld's informal logic as a multi-modal variant of standard deontic logic (SDL) in which the obligation operators are relativised to ordered bearer-counterparty pairings. I then point out a number of problems with this variant of SDL—particularly focussing on problems that arise from the apparent defeasibility of rights reasoning. I follow this criticism by introducing the rudiments of a logical approach to rights that utilises the default theoretic approach to reasoning that originates in [33], and has recently been adapted by John Horty to model various aspects of practical reasoning [11]. The logic follows Joseph Raz's suggestion that rights are 'intermediate conclusions' in practical reasoning [31]. Rights are premises in default rules defeasibly connecting the existence of the right with certain outcomes. I argue that the default theoretic approach better captures the dynamics of rights reasoning.

I rely on the default theoretic approach because it offers an attractive and plausible formalisation of practical reasoning. It would be interesting to explore the connections between this approach and other non-monotonic approaches, such as argument-based theories (as surveyed in [29]) or Makinson and van der Torre's input/output logic [22,23,25]. But that task is not undertaken here.

In Section 2 I define the Hohfeldian logic that is critiqued in the remainder of the paper. In Section 3 I present a number of difficulties for the account that relate to the defeasibility of rights reasoning. Section 4 defines the default logic. Section 5 then demonstrates its application to rights reasoning. I conclude by discussing some further challenges raised by the alternative logical account.

2 The Logic of Rights and the Hohfeldian Framework

Philosophical proponents of the Hohfeldian model of rights sometimes argue for the account on the basis of its logical clarity and precision (see especially [17, p. 22, pp. 24-25]). It is curious that the challenges that arise from developing a suitable semantic account of this logic have been mostly neglected in contemporary discussion, at least amongst ethicists and legal theorists. (I don't mean to suggest deontic logicians have neglected these challenges.) In this section I introduce the logic and offer a basic semantic account in a multi-modal setting.

Hohfeld's logic of rights brings the insights of SDL to bear on various 'fundamental legal relations', which are parsed into right-holder and duty-bearer pairings. Hohfeld's logic has been considerably refined by subsequent work in deontic logic, particularly in work on normative positions by Kanger, Lindahl, and Sergot [13,14,15,16,19,37]. These logics lack any treatment of the role of *counterparty* or rights-bearer. As Sergot notes in [37], work in the Kanger-Lindahl tradition has tended to provide 'a typology of duties', rather than an account of rights (see further [20]). Various attempts have been made to account for the counterparty role in these logics, but none of them have been uncontentious (see [8,20,21]). In any case, lawyers and philosophers have largely neglected these refinements, possibly because they add a degree of technicality

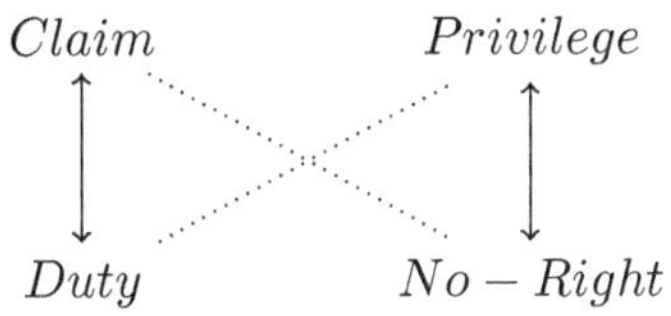

Fig. 1. Hohfeld's system of jural correlates and opposites.

that is deemed unnecessary for their purposes. In this note I will ignore these refinements and focus on the fundamental legal relations as they were initially developed by Hohfeld.

In contemporary ethics and legal theory, the basic Hohfeldian analysis is the one that is still most frequently invoked in the study of rights.[3] Hohfeld's analysis is particularly impressive because, although it drew on the work of other theorists like Bentham, Salmond and Austin, it preceded the formal systems of deontic logic developed by von Wright and others later in the 20th century. According to Hohfeld, our ordinary rights-talk can be systematically disambiguated into eight distinct concepts, distinguished into jural correlates and jural opposites [9]. I will focus here on Hohfeld's four first-order concepts (*claim-right*, *duty*, *privilege*, and *no-right*), since considering his four higher-order concepts (*power*, *liability*, *disability*, *immunity*) presents unnecessary complications.

Adopting a version of the agent-relativised deontic logic originally proposed by Krogh and Herrestad [8], I have chosen to capture Hohfeld's logic in a standard model that relativises the obligation operator to bearer and counterparty pairings (that is, each obligation is relativised to an ordered pair of individuals). My approach differs from Krogh and Herrestad in relativising the deontic logic to ordered pairs of individuals, rather than accounting for Hohfeld's legal relations in terms of ordinary bearer-relativised obligations and some further bridging principle.

I begin by assuming an ordinary propositional language involving the logical connectives $\wedge, \vee, \supset, \neg$. A set of modal operators, $\{O_{i_m,i_n} | \langle i_m, i_n\rangle \in I^2\}$ is added to the language. A set of relativised permission operators $\{P_{i_m,i_n} | \langle i_m, i_n\rangle \in I^2\}$ function as the duals.

Definition 2.1 (Hohfeldian Model). A Hohfeldian model M is a structure $\langle W, F, I, V\rangle$, where W is a set of worlds, I is a set of individuals $\{i_1, i_2, \dots i_n\}$, F is a set of functions $f_{\langle i_m,i_n\rangle} : W \to 2^W$, for each $\langle I_m, I_n\rangle \in I^2$, and V is an ordinary valuation function. We add the requirement that for any function $f_{\langle i_m,i_n\rangle} \in F$, $f_{\langle i_m,i_n\rangle} \neq 0$.

An intuitive reading of the sentence $O_{i_m,i_n}\alpha$ is that the proposition ex-

[3] The body of rights literature in both ethics and legal theory is now so large that it would be futile to hope to give a complete survey, but representative works include [12,17,38,40,42,45,46,47,48,49,50].

pressed by the sentence α, is obligatory between a duty bearer and a counterparty. The permission operator likewise expresses what is permissible between the two.

Definition 2.2 (Evaluation Rules). The evaluation rules for the deontic operators are as follows:

- $M, w \models O_{i_m,i_n}\alpha$ iff for each $\beta \in f_{\langle i_m, i_n\rangle}(w)$, $\beta \models \alpha$
- $M, w \models P_{i_m,i_n}\alpha$ iff there is a $\beta \in f_{\langle i_m,i_n\rangle}(w)$ and $\beta \models \alpha$

The evaluation rules are otherwise defined in the usual fashion over the classical connectives.

Hohfeld's correlative pairings can then be defined in terms of the bearer-counterparty relativised obligation and permission operators. Assume the following additional operators in the language for each of Hohfeld's juridical concepts: *Duty*, *Claim*, *Privilege*, and *No-Right*. (Let the operator $Duty_{i_m,i_n}(\alpha)$ indicate that an agent i_m owes a duty to i_n that α, for example). According to Hohfeld, $Duty_{i_m,i_n}(\alpha)$ is just equivalent to a $Claim_{i_n,i_m}(\alpha)$, while a $Privilege_{i_m,i_n}(\alpha)$ is equivalent to $No - Right_{i_n,i_m}(\alpha)$. (Each is the other's converse relation.)

Definition 2.3 (Correlatives).

- $Claim_{i_n,i_m}(\alpha) =_{def} Duty_{i_m,i_n}(\alpha) =_{def} O_{i_m,i_n}\alpha$; and
- $Privilege_{i_n,i_m}(\alpha) =_{def} NoRight_{i_m,i_n}(\alpha) =_{def} P_{i_m,i_n}\alpha$.

These are Hohfeld's two fundamental correlative pairings—the claim-duty pairing and the no-right-privilege pairing. From these two definitional stipulations, it is easy to show that Hohfeld's jural opposites are a feature of the logic.

Fact 2.4 *The following two schemas, which correspond to Hohfeld's jural opposites, are valid in the class of all Hohfeldian models:*

- $Duty_{i_m,i_n}(\alpha) \longleftrightarrow \neg Privilege_{i_m,i_n}(\neg\alpha)$
- $Claim_{i_m,i_n}(\alpha) \longleftrightarrow \neg NoRight_{i_m,i_n}(\neg\alpha)$

I think this logic captures the informal analysis offered by Hohfeld in model theoretic terms. This version of Hohfeldian logic can be criticised for failing to accommodate agency, since it doesn't use an action operator, like other formal attempts to model rights in the Kanger-Lindahl tradition. In this note I want to focus on the problem that I take to be most pressing—that the Hohfeldian analysis obscures the proper justificatory relationship between rights and our all things considered practical judgements. Rights are defeasible considerations that count for or against certain kinds of behaviour.

A number of authors have noted the defeasible character of rights reasoning. This aspect of rights reasoning is particularly evident in Raz's informal treatment of moral and legal rights [30,31]. Raz charges that the Hohfeldian account of rights obscures their 'dynamic aspect' [31, p. 212]. He notes that rights can be defeated by 'conflicting considerations of greater weight', and that they may support a variety of different duties in different circumstances [31, p. 200]. Judith Thomson likewise argues that while rights act as a certain kind of constraint on behaviour, and in that sense are correlative with a type of constraint or commitment, they are non-absolute [40, pp. 61-104]. There can be rights that it is appropriate not to act upon in the circumstances, either because they are defeated by other considerations, or because they are conditional on the absence of certain disabling factors.

These objections demonstrate the need for a logic that accommodates rights' function as reasons within practical deliberation, rather than as decisive conclusions. In the following section, I introduce three examples of rights reasoning that are not suitably captured by the basic Hohfeldian model. I will then demonstrate that embedding rights within default logic offers a promising solution to these weaknesses.

3 Defeasibility in the Logic of Rights

The various problems for the Hohfeldian account of rights that I want to consider can be characterised as due to the monotonicity of the logic. In Hohfeldian models, if $\alpha \models_{M,w} O_{i_m,i_n}\gamma$, then $\alpha, \beta \models_{M,w} O_{i_m,i_n}\gamma$. Several considerations demonstrate the need for a logic of rights that accounts for the non-monotonic, or defeasible character of our rights reasoning. I will focus on three particular examples here. The first is the prominence of a certain kind of reasoning with conflicting rights, which I will refer to as 'balancing'. The second is the possibility of rights that are conditional, in that they are capable of being disabled or undercut. The third is the need to accommodate the distinction between strong and weak permissive rights.

3.1 Conflicting Rights and Balancing

Rights appear to conflict with one another. One party might have a right that a certain state of affairs obtain at the same time that another party has a right that the contrary state of affairs obtain. It is important to recognise that the Hohfeldian logic does not rule out the possibility of these conflicts. The model trivially allows for the truth of conflicting rights claims. There are Hohfeldian models in which $\models_{M,w} O_{i_m,i_n}\alpha \wedge O_{i_j,i_k}\neg\alpha$, so in this sense conflicting rights are clearly possible. What the logic cannot do is identify the sense in which these rights conflicts can be resolved by appealing to a salient ordering on the rights. Most theorists accept not only that rights may conflict, but that rights in conflict may be balanced against one another in order of priority across different contexts (see especially [12, pp. 262-301], [40, pp. 149-176], [43]).

Example 3.1 For instance, to provide a simplified example adapted from a discussion in [29], a Minister's right to privacy can conflict with the public's right to have information about any activities of the Minister that affect his public responsibilities. Consider the following plausible chain of reasoning concerning whether or not to publish details concerning the Minister's health.

(1) The Minister has a right to privacy.

(2) The public has a right to information.

(3) If the Minister has a right to privacy, then we ought not to publish his health details.

(4) If the public has a right to information, then we ought to publish the Minister's health details.

(5) If the information relates the to the Minister's ability to perform his responsibilities, the public's right to information relating to the Minister's performance of his responsibilities has priority over the Minister's right to privacy.

(6) The information relates to the Minister's ability to perform his responsibilities.

(7) Therefore, the information ought to be published.

In this argument, the right to privacy in (1) defeasibly entails that the Minister's information not be published (as stipulated in premise (3)), while the public's right to information in (2) defeasibly entails the opposite conclusion (as stipulated in (4)). These sort of rights conflicts are resolved by further defeasible inference rules that allow us to balance one right against the other, as the conjunction of premises (5) and (6) suggests. Without amendment, the standard Hohfeldian logic is silent as to the possibility of ordering various rights in terms of their priorities in order to resolve these conflicts.

It is worth noting that the reasoning process represented in this example is still idealised in at least two ways. First, it ignores the question of the grounds and content of rights—the existence and effects of the rights to privacy and to information are treated as self-evident, and the question of their justification is ignored. In addition to capturing the defeasibility of premises like (3) and (4), a complete logic of rights will capture the sense in which these premises are themselves the product of a reasoning process. Legal reasoning sometimes involves reflecting on the underlying interests that ground the right in question in order to determine the scope of the duties that it justifies (see [30], [35, pp. 211-214]). Second, and relatedly, the reasoning process above ignores the various meta-normative interpretive processes that the resolution of these conflicts often involves, including the appeal to the values that underlie the rights (for discussion see [35]). It is plausible, for instance, that the priority rule expressed in (5) is justified by appealing to the weight of the interests that the rights expressed in (1) and (2) reflect. Ultimately a desirable logic of rights will need to accommodate these further aspects of rights reasoning. These issues are discussed further when I present the alternative logic.

3.2 Disabling Conditions on Rights

The Hohfeldian account has difficulty accommodating the way that we reason with certain kinds of rights that are conditional on the presence or absence of certain facts. We want to say that certain propositions operate as cancelling considerations, or disablers. Certain propositions make it the case that a right that otherwise would have obtained fails to obtain.

Example 3.2 Consider an example initially offered by Thomson [40, pp. 313-316]. Ordinarily, we have a right to have promises kept. But suppose that someone promises to me that they will assassinate the President of the United States. It appears that there is no right to have this particular promise kept. Moreover, this is not just a case of a right that is defeated by countervailing considerations. It is a case where, in Thomson's words, the right is 'stillborn, forfeit from conception' [40, p. 315]. Thomson's particular explanation is that where it would be impermissible for someone to accept a promise, the right to have the promise kept is disabled.[4] The example can be represented in something like the following chain of reasoning:

(1) John has a right to have promises kept.

(2) If John has a right to have promises kept, and Kelly has made a promise to John, then Kelly ought to keep her promise.

(3) Kelly has promised John to assassinate the President.

(4) It would be impermissible for John to accept Kelly's promise.

(5) Therefore, Kelly ought not to keep her promise.

The right stated in (1) appears to be undercut or excluded by the nature of the promise in (3). The observation in (4) that it would be impermissible for John to accept the promise seems to undercut or cancel any inference from the existence of the right to the conclusion that the promise ought to be kept.

3.3 Strong and Weak Permissions

The standard Hohfeldian models also neglect the distinction between strong (or 'positive') and weak ('negative') permissions. An agent may hold either of two types of privilege against another agent. They may have a weak privilege, in the sense that there is no norm requiring them to act in a certain way. Alternatively, they may benefit from an explicit norm that purports to either exclude or override another obligatory norm (on the distinction between strong and weak permissions see *inter alia* [1,41]; for recent work on the topic see [3,7,24,39]). Some rights appear to amount to strong rather than weak permissions.

Example 3.3 Various constitutional rights act to exclude the operation of otherwise validly enacted laws that are incompatible with those freedoms. In the Australian case of *Coleman v Power*, for instance, the High Court held that the right to freedom of political communication exempted Coleman from certain laws prohibiting insulting or offensive speech.[5] Coleman's right to freedom of political communication excluded the operation of an express prohibition of offensive or insulting speech. The prohibition of offensive or insulting speech was valid, but did not apply in the context of Coleman's political communication.

[4] I assume that Thomson's explanation is correct for the purposes of discussion.

[5] *Coleman v Power* (2004) 220 CLR 1.

In Hohfeldian models, there is no distinction between weak and strong permissions. Since privileges are definitionally equivalent to the absence of duties not to do an act, all permissions are defined as weak permissions. Hohfeldian models are unable to capture the sense in which someone may have a right that amounts to permission to do something that they are also *pro tanto* obliged not to do.

Recent theoretical work undertaken with respect to input/output logic has focussed attention on at least two distinct types of positive permission [3,24,39]. The first are what are sometimes called 'exemptions'—exceptional rules that act on specific existing prohibitions to permit a certain restricted kind of behaviour. The second are what are known as 'dynamic' or 'antithetic' permissions. Antithetic permissions overrule any incompatible norms, rather than just working on individual norms. The two types of permission are closely related, though constitutional freedoms appear to be better regarded as antithetic provisions [39, pp. 98-99].

4 The Default Logic

Having provided a number of examples of the difficulties that face the traditional logic of rights, I will now define a default logic that is suitable for capturing these aspects of rights reasoning. The default theories in question are suitable for modelling both undercutting defeat and rebutting defeat. I describe these theories, following Horty [11, chapter 5], as *exclusionary variable priority default theories.*

4.1 Exclusionary Variable Priority Default Theories

Assume an ordinary propositional language W containing the connectives $\wedge, \vee, \supset$, and $\neg$. A default δ is a rule connecting sentences in W, written $X \rightarrow Y$ for any two sentences X and Y. Two functions, $Premise$ and $Conclusion$ identify the sentences connected by a default rule.

A theory consists of a set of defaults D and a set of sentences W. Each default in D is given a unique index, so $D = \{\delta_1, \delta_2 ... \delta_n\}$. Indexed defaults in D are also represented in the background language by assuming that each $\delta_m \in D$ is assigned a unique name d_m. Priority relations between named defaults are also expressed in the background language using the predicate $\prec$. We further assume that amongst the sentences in W are all instances of the irreflexivity schema $(\neg(d' \prec d'))$ and transitivity schema $(d' \prec d'' \wedge d'' \prec d''') \rightarrow d' \prec d''')$, where the variables are replaced with a named default for each named default in D. A notion of exclusion (which will be needed to model strong permissive rights and disabling conditions) is captured by assuming that the background language contains a predicate of the form $Out(d_m)$ for each named default.

Definition 4.1 (Exclusionary variable priority default theory). An exclusionary variable priority default theory Δ is a pair $\Delta = \langle W, D \rangle$ where W is a set of sentences and D is a set of defaults, such that: (1) each default δ_m is assigned a unique name d_m; (2) W contains the predicate $\prec$; (3) W contains all instances of the irreflexivity and transitivity schemas for each named default in D; (4) W contains the predicate *Out*.

A scenario S is a subset of the defaults contained in Δ. Against the background of an exclusionary variable priority default theory, a priority ordering is derived for a scenario based on that theory. This allows for the priority relations expressed with respect to named defaults in the object language to be raised into the metalanguage and used in the default reasoning.

Definition 4.2 (Derived priority ordering). Where $\Delta = \langle W, D\rangle$ is an exclusionary variable priority default theory, and S is a scenario based on Δ, a derived priority ordering $<_S$ is defined for S against the background of Δ by taking:

$$\delta' <_S \delta'' \text{ if and only if } W \cup Conclusion(S) \vdash d' \prec d''.$$

A set of defaults excluded at a scenario S is likewise defined against the background of our default theory.

Definition 4.3 (Excluded defaults). Where $\Delta = \langle W, D\rangle$ is an exclusionary variable priority default theory, and S is a scenario based on Δ, a set of excluded defaults is defined for S against the background of Δ by taking:

$$\delta \in Excluded_S \text{ if and only if } W \cup Conclusion(S) \vdash Out(d)$$

Definition 4.4 (Triggered defaults). Where $\Delta = \langle W, D\rangle$ is an exclusionary variable priority theory, and S is a scenario based on Δ, a default in D is *triggered* at S just in case the default belongs to the set of premises entailed by S, and is not excluded. More formally the set of triggered defaults is defined as:

$$\begin{aligned} Triggered_{W,D}(S) = \{\delta \in D : \delta \notin Excluded_S \\ \text{and } W \cup Conclusion(S) \vdash Premise(\delta)\}\end{aligned}$$

Defaults are conflicted in a given scenario just where the conclusion of a default negates a proposition that belongs to that scenario.

Definition 4.5 (Conflicted defaults). Where $\Delta = \langle W, D\rangle$ is an exclusionary variable priority theory, and S is a scenario based on Δ, we define a set of defaults in D conflicted at S as:

$$Conflicted_{W,D}(S) = \{\delta \in D : W \cup Conclusion(S) \vdash \neg Conclusion(\delta)\}$$

In an exclusionary variable priority default theory, a default is defeated in a given scenario in case its conclusion is a negation of another default triggered at that same scenario that is stronger according to the derived priority ordering. (For the sake of simplicity, I ignore the possibility of defeat by a *set* of defaults discussed in [4,10,28].)

Definition 4.6 (Defeated defaults). Where $\Delta = \langle W, D\rangle$ is an exclusionary variable priority theory, and S is a scenario based on Δ, we define a set of defaults in D defeated at S as:

$$\begin{aligned}Defeated_{W,D,<_s}(S) = \{\delta \in D : &\text{ there is a } \delta' \in Triggered_{W,D}(S) \\ &\text{and } \delta <_s \delta' \text{and } W \cup Conclusion(\delta) \vdash \neg Conclusion(\delta')\}\end{aligned}$$

It will be useful to appeal to the notion of a proper scenario, which informally represents the set of defaults that might be accepted at the conclusion of an ideal reasoning process (one that rules out inconsistent and excluded reasons). In order to define a proper scenario, we first define the notion of an approximating sequence, defined against a theory Δ and constrained by a given scenario S.

Definition 4.7 (Approximating sequence). Where $\Delta = \langle W, D\rangle$ is an exclusionary variable priority default theory, S is a scenario based on Δ, and $<_S$ is a priority order derived from S, a sequence $S_0, S_1, S_2, \ldots$ is an approximating sequence based on Δ and constrained by S if and only if:

$$\begin{aligned}S_0 &= \emptyset \\ S_{i+1} &= \{\delta \in D : \delta \in Triggered_{W,D}(S_i), \\ &\qquad \delta \notin Conflicted_{W,D}(S), \\ &\qquad \delta \notin Defeated_{W,D,<_S}(S)\}\end{aligned}$$

Informally, the approximating sequence represents the process by which an ideal reasoner arrives at set of defaults by beginning with an empty set of defaults and then at each successive stage adding defaults that are neither conflicted nor defeated in the previous constraining set. With this definition of an approximating process in place, a definition of a proper scenario can be offered.

Definition 4.8 (Proper scenario). Where $\Delta = \langle W, D\rangle$ is an exclusionary variable priority default theory, and S is a scenario based on Δ, S is a proper scenario if and only if for an approximating sequence constrained by S against Δ, $S = \bigcup_{i\geq 0} S_i$.

The extension of a default theory is just the set of outcomes generated by a proper scenario. Informally, the extension of a theory represents the belief set at the conclusion of a reasoning process undertaken by an ideal reasoner.

Definition 4.9 (Extension of a theory). Let the operator $Cn(X)$ denote the closure of some set of formulae X under logical consequence. For an exclusionary variable priority default theory $\Delta = \langle W, D\rangle$, the extension $\mathcal{E}$ of Δ is defined as as $\mathcal{E} = Cn(W \cup Conclusion(S))$ where S is a proper scenario.

With these definitions in place, I will now turn to examination of the application of exclusionary variable priority default theories to rights reasoning.

5 Rights in Default Logic

It is helpful to see rights in default-theoretic terms. Rights are ordinarily understood as providing reasons that count for or against an action, rather than having a conclusory or verdictive character [49]. Since default logic provides an attractive framework for modelling the logic of practical reasoning (see especially [11]), it offers a promising method of formalising the role of rights in defeasible reasoning.

For the most part, embedding rights within default logic is a relatively straightforward matter, since no extension of the familiar default theoretic apparatus is needed. The choice I have made here is to encode the existence of rights in the background set of sentences in the default theory. I assume that in our language is a predicate of the form $Right(i, \Gamma)$ expressing propositions that state what rights individuals have (where i is an individual and Γ is the content of the right).

Note that, unlike in Hohfeld's original schema, individuals' rights need not be relativised to any individual bearer of a correlative duty. Rights can constitute abstract entitlements to a certain object or good that defeasibly entail certain outcomes. Rights are the premises of defaults. The same right can be the premise for a variety of different defaults. So, for instance, someone's right to privacy can be a premise of a default to the effect that we not publish information about them, and to the effect that we not monitor their phone calls. In other words, the same right might defeasibly entail a variety of possible conclusions [31, pp. 199-200]. This model of rights is meant to capture Raz's characterisation of rights as 'intermediate conclusions' in the process of making all things considered practical judgments—when their existence is contained within the reasoning subject's background information, they defeasibly entail a certain action or set of actions [31, p. 195]. Rights are, on this model, reasons to conclude that someone is under an obligation to act in a certain way. I believe that the logic of rights I propose at least partly captures the informal characterisation of rights that Raz had in mind.

5.1 Balancing of Rights

In an exclusionary variable priority default theory, we model conflicts of rights in terms of scenarios in which the triggered defaults with which they are associated contain conflicting conclusions. We can further model the way in which rights conflicts are usually resolved in terms of derived priority orderings on the relevant defaults. A rights conflict will be resolved if a further proposition is triggered that contains in its conclusion an ordering of the defaults associated with the conflict.

Return to the example of a rights conflict introduced above (example 3.1). Let $Right(Pub, I)$ be the sentence expressing the proposition that the public has a right to information, and $Right(M, Priv)$ be the sentence expressing the proposition that the Minister has a right to privacy. Further, let P be the sentence expressing the proposition that the information is published, and let Q be the sentence expressing the proposition that the information relates to the Minister's ability to perform their responsibilities. A default $\delta_1 : Right(M, Priv) \rightarrow \neg P$ encodes the inference from the Minister's right to privacy through to the conclusion that the information should not be published.

Another default $\delta_2 : Right(Pub, I) \rightarrow P$ encodes the inference from the public's right to information through to the conclusion that the information should be published. A third and final default $\delta_3 : Q \rightarrow d_1 \prec d_2$ represents the rule that if the information relates to the Minister's ability to perform their role, the right to information will prevail over the right to privacy. It is a default that encodes a priority on the other two rights defaults in our theory.

The reasoning process outlined informally in example 3.1 can be illustrated in a exclusionary variable priority default theory $\Delta_1 = \langle W, D \rangle$, where $W = \{Q, Right(Pub, I), Right(M, Priv)\}$ and $D = \{\delta_1, \delta_2, \delta_3\}$. Note that on the derived priority ordering in our proper scenario, $<_{S_1}$, $\delta_1 <_{S_1} \delta_2$. The public's right to information takes priority over the Minister's right to privacy. The unique proper scenario yielded by the theory Δ_1 is the scenario $S_1 = \{\delta_2, \delta_3\}$, supporting the conclusion P, expressing the proposition that the information is published. Embedding rights within an exclusionary variable priority default theory thus allows us to demonstrate the way in which rights might be balanced against one another in order of priority to resolve a conflict.

Although it copes reasonably well with the process of informal reasoning outlined in example 3.1, the theory does not yet capture the initial reasoning processes allowing the deduction of rights from a set of other interests or values. Nor does it capture any process of reasoning involving more general interpretive norms concerning the priority of the rights values at stake. There is no reason to think that the language of default logic could not be enriched to enable the use of both of these sorts of meta-norms (for treatments of interpretive norms in a variety of different settings see [2,26,27,34,44]). I leave this as a subject for future research.

5.2 Disabling Conditions on Rights

The presence of disabling conditions on certain rights can be modelled as a kind of exclusion. This can be illustrated if we return to the example of a promissory right that is undercut (example 3.2). The sentence F, expressing the proposition that it would be impermissible to accept the promise, can be thought of as the premise in a default rule the conclusion of which recommends the exclusion of the default for which the promissory right is a premise. Let $R(I, P)$ express an individual's right that the promise be kept, and let K express the proposition that the promise is kept.

The situation can be illustrated in an exclusionary variable priority default theory $\Delta_2 = \langle W, D \rangle$ in which $W = \{F, R(I, P)\}$ and $D = \{\delta_1, \delta_2\}$, where $\delta_1 : R(I, P) \rightarrow K$ and $\delta_2 : F \rightarrow Out(d_1)$. It is easy to verify that the unique proper scenario in Δ_2 is the scenario $S = \{\delta_2\}$, such that the promissory right is excluded and no additional action is recommended. Informally, the nature of the promise acts as a reason for excluding a right that ordinarily would be a reason to keep the promise (the general right to have promises kept). The general right is conditional in the sense that its particular weight or force is dependent on the absence of these considerations.

5.3 Rights as Strong Permissions

To begin with, strong permissions can be modelled in our default logic as the premises of defaults that exclude other defaults—what Raz calls 'exclusionary permissions' [32, pp. 89-91]. The invocation of a right as a strong permission excluding the operation of another explicit norm can be illustrated within a default theory.

Return to the example of a strong permissive right offered above (example 3.3). Let I be the sentence expressing the proposition that an individual uses insulting words, and $Right(C, S)$ be the sentence expressing the proposition that Coleman has a right to speak freely on political matters. The prohibition of insulting speech can be thought of as a default of the form $\delta_1 : \top \rightarrow \neg I$. Coleman's right to speak freely is the premise in a default $\delta_2 : Right(C, S) \rightarrow Out(d_1)$.

This scenario can be depicted in an exclusionary variable priority default theory $\Delta_3 = \langle W, D \rangle$, where $W = \{Right(C, S)\}$ and $D = \{\delta_1, \delta_2\}$. Our default theory Δ_3 yields as its only proper scenario $S_1 = \{\delta_2\}$, meaning the conclusion $Out(d_1)$ is in the extension of the theory. Note that in our default theory, the reasoning process supports neither the conclusion that I or $\neg I$. The theory thus appropriately captures the sense in which a strong permissive right is not in and of itself something that recommends a certain action, it merely excludes the operation of norms that would ordinarily count against the action in question.

There appears to be some connection between this notion of rights as premises in exclusionary defaults and the permissions that others have identified in the context of input/output logic as 'exemptions' [24,39]. The right acts as the premise in a a default that explicitly excludes an already existing default. The default theory Δ_3 does not yet capture the sense in which the right to freedom of speech can operate as a 'dynamic positive permission' or 'antithetic permission' [3,24,39].

In order to model rights as the grounds of dynamic positive permissions, it is necessary to introduce an appropriate meta-norm which captures the (defeasible) inference from the existence of a conflict between a default and the right in question to the conclusion that the default is excluded. This might be done by assuming that in the exclusionary variable priority default theory is a predicate of the form $Incompatible(d_m, R(I, \Gamma))$ which identifies named defaults in W that are incompatible with rights in W, and a set of instances of the general rule $Incompatible(d_m, R(I, \Gamma)) \rightarrow Out(d_m)$, which exclude incompatible defaults.

For instance, suppose there are two defaults $\delta_1 : Incompatible(d_2, Right(C, S)) \rightarrow Out(d_2)$ and $\delta_2 : \top \rightarrow \neg I$. Now suppose we have a default theory $\Delta_4 = \langle W, D \rangle$ where $W = \{Incompatible(d_2, Right(C, S)), Right(C, S)\}$ and $D = \{\delta_1, \delta_2\}$. Δ_4 yields as its only proper scenario $S_1 = \{\delta_1\}$ meaning $Out(d_2)$ is in the extension of the theory. Informally, the theory Δ_4 involves the use of an exclusionary meta-norm δ_1 that requires the exclusion of prohibitions that are incompatible with fundamental rights.

6 Conclusion

This paper has introduced the basic formalisms necessary for a logical account of rights within the framework of a default logic. The previous section demonstrated the logic's application to rights reasoning. If the approach is plausible, it shows that there are suitable logical alternatives to the relativised deontic logic that was codified by Hohfeld. Rights are better understood from within a theory of practical reasoning. Reasons themselves can be modelled in terms of default logic, so it is tempting to think that a logical account of rights should embed them within a default logic. I have developed one possible logic here.

It is possible to identify several areas for further inquiry. One issue already raised in discussion concerns how rights and their contents are justified by appealing to certain underlying interests or priorities. These teleological factors also seem to influence the interpretation of rights. Questions relating to the priority of rights, as well as their expansion into new areas of application, often appear to be resolved by appealing to the nature and weight of the underlying values in question. It remains to be seen how far the logic could be enriched to accommodate these aspects of rights reasoning.

One weakness of the proposed logic is that it does not make any allowance for agency. A great strength of the Kanger-Lindahl tradition has been its ability to incorporate concepts of agency into the articulation of the normative positions. Extending default logic to include expressions of agency would present obvious difficulties, but the prospects of such an approach seem worth exploring. Relatedly, the logic makes no distinction between rights that act as deontic constraints on actions and rights that express goals. A number of authors have noted that the language of rights can be used to state goals or values, rather than stating reasons for action of any sort (see [6, chapter 11], [35,36]). Default logic seems to offer a promising framework for considering the distinction between both types of rights, but the issue merits further attention.

Finally, it is worth considering the relationship between the Hohfeldian account of fundamental legal relations and the default logic introduced here. It seems plausible to suggest that the Hohfeldian logic is an account of the conclusions of our practical deliberation, not our practical reasoning involving intermediate entitlements (see [5, pp. 218-221]). If this is the case, then the Hohfeldian logic and the default logic introduced here may well be complementary, since Hohfeld's logic can be taken to be a particular deontic interpretation of the default logic.

References

[1] Alchourron, C. and E. Bulygin, *Permission and permissive norms*, in: *Theorie der Normen*, Duncker and Humblot, Berlin, 1984 pp. 349–371.

[2] Bench-Capon, T. and G. Sartor, *A model of legal reasoning with cases incorporating theories and values*, Artificial Intelligence **150** (2003), pp. 97 – 143.

[3] Boella, G. and L. van der Torre, *Institutions with a hierarchy of authorities in distributed dynamic environments*, Artificial Intelligence and Law **16** (2008), pp. 53–71.

[4] Delgrande, J. P. and T. Schaub, *Reasoning with sets of defaults in default logic*, Computational Intelligence **20** (2004), pp. 56–88.
[5] Finnis, J., "Natural Law and Natural Rights," Oxford University Press, 2011, 2nd edition.
[6] Forrester, J. W., "Being Logical and Being Good," M.E Sharpe, 1996.
[7] Governatori, G., F. Olivieri, A. Rotolo and S. Scannapieco, *Computing strong and weak permissions in defeasible logic*, Journal of Philosophical Logic **42** (2013), pp. 799–829.
[8] Herrestad, H. and C. Krogh, *Deontic logic relativised to bearers and counterparties*, in: *25 Years Anthology-Norwegian Research Center for Computers and Law*, Tano forlag, 1995 pp. 453–522.
[9] Hohfeld, W. N., "Some Fundamental Legal Conceptions as Applied in Judicial Reasoning and other Legal Essays," Yale, 1919.
[10] Horty, J., *Defaults with priorities*, Journal of Philosophical Logic **36** (2007), pp. 367–413.
[11] Horty, J., "Reasons as Defaults," Oxford University Press, 2012.
[12] Kamm, F., "Intricate Ethics," Oxford University Press, 2007.
[13] Kanger, S., "New Foundations for Ethical Theory," Stockholm University, 1957.
[14] Kanger, S., *Law and logic*, Theoria **38** (1972), pp. 105–132.
[15] Kanger, S., *On realisation of human rights*, in: S. L. G. Hollmstrom-Hintikka and R. Sliwinski, editors, *Collected Papers of Stig Kanger with Essays on his Life and Work*, Studies in Epistemology, Logic, Methodology and the Philosophy of Science **1**, Kluwer, Dordrecht, 2001 pp. 179–185.
[16] Kanger, S. and H. Kanger, *Rights and parliamentarianism*, in: S. L. G. Hollmstrom-Hintikka and R. Sliwinski, editors, *Collected Papers of Stig Kanger with Essays on his Life and Work*, Studies in Epistemology, Logic, Methodology and the Philosophy of Science **1**, Kluwer, Dordrecht, 2001 pp. 120–145.
[17] Kramer, M., *Rights without trimmings*, in: H. S. M. Kramer, N. E. Simmonds, editor, *A Debate Over Rights*, Oxford University Press, 1998 pp. 7–111.
[18] Kramer, M. H., "Where Law and Morality Meet," Oxford University Press, 2008.
[19] Lindahl, L., "Position and Change: A Study in Law and Logic," Synthese, 1977.
[20] Lindahl, L., *Stig Kanger's theory of rights*, in: G. Hollmstrom-Hintikka, S. Lindstrom and R. Sliwinski, editors, *Collected Papers of Stig Kanger with Essays on his Life and Work*, Studies in Epistemology, Logic, Methodology and the Philosophy of Science **2**, Synthese, 2002 pp. 151–171.
[21] Makinson, D., *On the formal representation of rights relations*, Journal of Philosophical Logic **15** (1986), pp. 403–425.
[22] Makinson, D. and L. van der Torre, *Input/output logics*, Journal of Philosophical Logic **29** (2000), pp. 383–408.
[23] Makinson, D. and L. van der Torre, *Constraints for input/output logics*, Journal of Philosophical Logic **30** (2001), pp. 155–185.
[24] Makinson, D. and L. van der Torre, *Permission from an input/output perspective*, Journal of Philosophical Logic **32** (2003), pp. 391–416.
[25] Makinson, D. and L. van der Torre, *What is input/output logic?*, in: B. Löwe, W. Malzkom and T. Räsch, editors, *Foundations of the Formal Sciences II: Applications of Mathematical Logic in Philosophy and Linguistics, Papers of a Conference held in Bonn, November 10–13, 2000* (2003), pp. 163–174.
[26] Prakken, H., "Logical Tools for Modelling Legal Argument," Kluwer, 1997.
[27] Prakken, H., *An exercise in formalising teleological case-based reasoning*, Artificial Intelligence and Law **10** (2002), pp. 113–133.
[28] Prakken, H., *A study of accrual of arguments, with applications to evidential reasoning*, in: *Proceedings of the 10th International Conference on Artificial Intelligence and Law (ICAIL05)* (2005), pp. 85–94.
[29] Prakken, H. and G. Vreeswijk, *Logics for defeasible argumentation*, in: D. M. Gabbay and F. Guenthner, editors, *Handbook of Philosophical Logic*, Springer Netherlands, Dordrecht, 2002 pp. 219–318.
[30] Raz, J., *Legal rights*, Oxford Journal of Legal Studies **4** (1984), pp. 1–21.
[31] Raz, J., *On the nature of rights*, Mind **93** (1984), pp. 194–214.
[32] Raz, J., "Practical Reason and Norms," Oxford University Press, 1999, 2nd edition.

[33] Reiter, R., *A logic for default reasoning*, Artificial Intelligence **13** (1980), pp. 81–137.
[34] Rotolo, A., G. Governatori and G. Sartor, *Deontic defeasible reasoning in legal interpretation: Two options for modelling interpretive arguments*, in: *Proceedings of the 15th International Conference on Artificial Intelligence and Law*, ICAIL '15 (2015), pp. 99–108.
[35] Sartor, G., *Doing justice to rights and values: Teleological reasoning and proportionality*, Artificial Intelligence and Law **18** (2010), pp. 175–215.
[36] Sen, A., *Elements of a theory of human rights*, Philosophy and Public Affairs **32** (2004), pp. 315–356.
[37] Sergot, M., *Normative positions*, in: D. M. Gabbay, J. Horty, R. van der Meyden, X. Parent and L. van der Torre, editors, *Handbook of Deontic Logic and Normative Systems*, College Publications, 2013 pp. 353–406.
[38] Steiner, H., "An Essay on Rights," Blackwell Oxford, 1994.
[39] Stolpe, A., *A theory of permission based on the notion of derogation*, Journal of Applied Logic **8** (2010), pp. 97 – 113.
[40] Thomson, J. J., "The Realm of Rights," Harvard University Press, 1990.
[41] von Wright, G. H., "Norm and Action: A Logical Enquiry," Routledge and Kegan Paul, 1963.
[42] Waldron, J., "Theories of Rights," Oxford University Press, 1984.
[43] Waldron, J., *Rights in conflict*, Ethics **99** (1989), pp. 503–519.
[44] Walton, D., G. Sartor and F. Macagno, *An argumentation framework for contested cases of statutory interpretation*, Artificial Intelligence and Law **24** (2016), pp. 51–91.
[45] Wellman, C., "A Theory of Rights: Persons Under Laws, Institutions, and Morals," Rowman & Allanheld, 1985.
[46] Wellman, C., "Real Rights," Oxford University Press New York, 1995.
[47] Wellman, C., "An Approach to Rights: Studies in the Philosophy of Law and Morals," Springer Science & Business Media, 1997.
[48] Wenar, L., *The nature of rights*, Philosophy and Public Affairs **33** (2005), pp. 223–253.
[49] Wenar, L., *Rights*, in: E. N. Zalta, editor, *The Stanford Encyclopedia of Philosophy*, 2011, fall 2011 edition .
[50] Wenar, L., *The nature of claim rights*, Ethics **123** (2013), pp. 202–229.

Logical Inference and Its Dynamics

Carlotta Pavese [1]

Duke University
Philosophy Department

Abstract

This essay advances and develops a dynamic conception of inference rules and uses it to reexamine a long-standing problem about logical inference raised by Lewis Carroll's regress.

Keywords: Inference, inference rules, dynamic semantics.

1 Introduction

Inferences are linguistic acts with a certain dynamics. In the process of making an inference, we add premises incrementally, and revise contextual assumptions, often even just provisionally, to make them compatible with the premises. Making an inference is, in this sense, moving from one set of assumptions to another. The goal of an inference is to reach a set of assumptions that supports the conclusion of the inference.

This essay argues from such a dynamic conception of inference to a dynamic conception of *inference rules* (section §2). According to such a dynamic conception, inference rules are special sorts of *dynamic semantic values.* Section §3 develops this general idea into a detailed proposal and section §4 defends it against an outstanding objection. Some of the virtues of the dynamic conception of inference rules developed here are then illustrated by showing how it helps us re-think a long-standing puzzle about logical inference, raised by Lewis Carroll [3]'s regress (section §5).

2 From The Dynamics of Inference to A Dynamic Conception of Inference Rules

Following a long tradition in philosophy, I will take inferences to be *linguistic acts.* [2] Inferences are *acts* in that they are conscious, at person-level, and

[1] I'd like to thank Guillermo Del Pinal, Simon Goldstein, Diego Marconi, Ram Neta, Jim Pryor, Alex Rosenberg, Daniel Rothschild, David Sanford, Philippe Schlenker, Walter Sinnott-Armstrong, Seth Yalcin, Jack Woods, and three anonymous referees for helpful suggestions on earlier drafts. Special thanks go to Malte Willer and to all the organizers of DEON 2016.

[2] For example, see [6], [12], [20], [2], [21], [26], and [27], among others.

intentional. They are *linguistic*, in that they consist in the utterance of a list of sentences. These linguistic acts may be private, as when we argue to ourselves, or public, as when we try to convince others that they should endorse a certain conclusion through an argument. Inferences divide into *inductive* and *deductive* inferences, but only deductive inferences are the focus here. In this case, inferences consist in the utterance (mental or public) of a list of premises $\phi_1, \ldots, \phi_n$ and a conclusion ψ of the form $\ulcorner \phi_1, \ldots, \phi_n$; *therefore*, $\psi \urcorner$.

Now, as linguistic acts, inferences have a *dynamic* aspect. In the process of making an inference, we add premises incrementally, and revise contextual assumptions, often even just provisionally, to make them compatible with the premises. For example, suppose I argue as follows:

(i) If Marco were in Italy, he would inform me;

(ii) He has not informed me;

(iii) Hence, he must not be in Italy.

In making this argument, I add premises to the set of assumptions that I and my listeners already accept, or provisionally revise that set of assumptions in order to make it compatible with the newly introduced premises. If my listeners and I were previously assuming that Marco was in Italy, by uttering the premises (i) and (ii), I am in effect asking to provisionally suspend those assumptions from the initial set and to consider revising them in light of my argument.

Making an inference is, in this sense, moving from one set of assumptions to another — in this case, from a set of assumptions that may or may not be opinionated about Marco's whereabouts — to a set of assumptions that includes both (i) and (ii) and is adjusted for coherence. The *goal* of an inference is to reach a set of assumptions that *supports* the conclusion of the inference — in this case, to reach a set of assumptions that entails that Marco is not in Italy.

Nothing thus far is particularly surprising. Several have highlighted the dynamic aspect of inferences.[3] What is less commonly explored is what this dynamic conception of inference tells us about *inference rules* — such as the rule of *modus ponens* or conjunction introduction.

There seems to be a natural argument from a dynamic conception of inference to a *dynamic conception of inference rules*. The first premise is the dynamic conception of inference:

Premise 1 *An inference is a matter of moving from a set of assumptions to another set of assumptions which is meant to license the conclusion.*

Now, what is the relation between inferences dynamically conceived and inference rules? Inferring is, just like asserting, a linguistic act. And just like assertion, inferring is subject to *rules or norms*. Inference rules codify our inferential practices along certain structural dimensions. The rules of the propositional calculus codify our inferential practices along their truth-functional

---

[3] For example, see [29]'s notion of a reasonable inference and [39]'s notion of informational consequence.

dimensions, whereas the rules for the quantifiers do so along the predicative dimension. That gives us the second premise:

Premise 2 *Inference rules codify our inferences along certain structural dimensions.*

The first and the second premise together lead us to think of the rules that govern inferences *as telling us how to update* a set of assumptions in such a way as to reach another set of assumptions that supports the conclusion:

Premise 3 *Inference rules are rules to move from a set of assumptions to another set of assumptions.*

But note that, according to dynamic semantics, that is exactly what *dynamic semantic values* are supposed to be ([13], [16], [17], [35], [36], and [11]).[4] Dynamic semantic values are precisely rules to update sets of assumptions and are modeled as *functions* from sets of assumptions to sets of assumptions. This modeling claim is the fourth premise:

Premise 4 *Rules to update sets of assumptions can be modeled as functions from sets of assumptions to sets of assumptions — i.e., as* dynamic semantic values.

Premise 1-Premise 4 yield the dynamic conception of inference rules:

Conclusion *Inference rules can be modeled as functions from sets of assumptions to sets of assumptions — i.e., as dynamic semantic values.* [**Premise 1-Premise 4**]

So, the dynamic conception of inference motivates a dynamic conception of inference rules as *dynamic semantic values.* The next question is: what kinds of dynamic semantic values? Section §3 develops the proposal in some detail.

3 Towards the proposal

3.1 Dynamic Semantics

Sets of assumptions are often referred to as *contexts.* A *context* is a set of assumptions that are mutually shared by the participants of a conversation or that characterize a subject's mental state.[5] Contexts could be modeled linguistically, as a set of sentences or as a set of linguistically structured propositions. In this section, in order to flesh out my proposal in some detail, it is very convenient to follow Stalnaker and most dynamic semanticists in taking a context to be *a set of possible worlds* — those worlds where *every* proposition in some

[4] Other advocates of dynamic semantics are, among others, [8], [9], [38], [33], and [34].

[5] [30], [31], and [32]. On modeling (private) mental states as sets of assumptions, see [40]. This notion of context is to be distinguished from another familiar notion of context, the Kaplanian notion, on which context is whatever is relevant to fix the meaning of context-sensitive expressions of a language. See [25], pp. 3-4 for a brief comparison of these two notions of context.

given set of assumptions is true. [6] In the last section, I will explain how such a coarse-grained conception of context is not at all required by my proposal, that can be developed also within a more fine-grained notion of context.

The dynamic semantic value of a sentence ϕ is a function from contexts to contexts. More precisely, let $\langle p \rangle$ be the set of possible worlds where p — i.e., *the set of p-worlds*. The dynamic semantic value of a sentence σ — which I will indicate by '$[\sigma]$' — is a function from takes a context as argument and outputs a context as value — i.e., '$c[\sigma]$'. The inductive definition is as follows:

Definition 3.1 (**Dynamic Semantics**)

(i) If σ has the form p, $c[\sigma] = \{w \in c\colon\ w \in \langle p \rangle\}$;

(ii) If σ has the form $\neg\phi$, $c[\sigma] = c - c[\phi]$;

(iii) If σ has the form $\phi \mathbin{\&} \psi$, $c[\sigma] = c[\phi][\psi]$;

(iv) If σ has the form $\phi \vee \psi$, $c[\sigma] = c[\phi] \cup c[\neg\phi][\psi]$.

The dynamic meaning of an atom p is the function that takes a context c into another context c' that includes all and only the p-worlds from c. The dynamic meaning of a negation $\neg\phi$ is a function that takes a context c into another c' that results from eliminating from c all the ϕ-worlds ($= c - c[\phi]$). The dynamic meaning of a conjunction $\phi \mathbin{\&} \psi$ is a function that takes a context c into the result of updating c first with ϕ and then with ψ ($= c[\phi][\psi]$). [7] The dynamic meaning of a disjunction $\phi \vee \psi$ is the dual of the conjunction's dynamic meaning ($= c[\phi] \cup c[\neg\phi][\psi]$). [8]

Standard presentations of dynamic semantics define a relation of *support* between contexts and sentences of our language. [9] A context c *supports* ϕ just in case the result of updating c with ϕ is c itself — just in case every world in c is a ϕ-world:

Definition 3.2 (**Support**) *c supports* ψ ($c \vDash \phi$) iff $c[\phi] = c$.

Dynamic Entailment ($\vDash_{DE}$) is instead a semantic relation holding between sentences:

---

[6] This idealization risks narrowing down the scope of my proposal to inferential relations between sentences that are contingently true — and so ruling out mathematical inferences, the sort of inferences that hold between necessarily true (or necessarily false) sentences. The current literature discusses several ways in which one could tweak the current apparatus to make it encompass mathematical inferences. One is to appeal to linguistically structured contexts (see [25] for some discussion). Another is to appeal to metalinguistic propositions, following [30]'s solution to the problem of logical omniscience. Finally, another approach, recently explored by [24] for the specific case of mathematics, consists in adding non-linguistic structure (in particular the structure of *subject matters*), to logical space and to contexts. I will return to this issue in the last section when discussing Lewis Carroll's regress. A more fine-grained notion of context will turn out to be better suited to apply my proposed conception of inference rules and inferences to Carroll's paradox.

[7] [35], p. 18.

[8] See [18] and [37], p. 10 for discussion of this entry for disjunction.

[9] for example, [36], p. 221-222

Definition 3.3 (Dynamic Entailment) $\phi_1, \ldots, \phi_n \vDash_{DE} \psi$ iff $\forall c$: $c[\phi_1] \ldots [\phi_n] \vDash \psi$.

In other words, a set of sentences $\phi_1, \ldots, \phi_n$ dynamically entails ($\vDash_{DE}$) ψ just in case, for every context c, the result of updating c successively with all the premises is a context that supports ψ.

The last dynamic notion that we need is that of *a test*. An expression is a test just in case its dynamic role is *that of checking whether that context satisfies certain conditions*. If the context does satisfy those constraints, the test will return the context itself; otherwise, the test will return the absurd context — i.e., the empty set. So, for example, following Veltman ([36], p. 228), one can think of a sentence containing an epistemic modal such as *must*-ϕ as a test that checks whether ϕ is supported by the current context, in which case it returns the context itself as value; else, it returns the absurd context (— i.e., the empty set):

Definition 3.4 (Example of a Test)

$$\text{If } \sigma \text{ has the form } must\text{-}\phi,\ c[\sigma] = \begin{cases} c & \text{if } c \vDash \phi \\ \emptyset & \text{if } c \nvDash \phi \end{cases}$$

3.2 A Dynamic Conception of Inference Rules

We observed that inferences have a *dynamic aspect*. Here is van Benthem ([35], p. 11) explicitly highlighting the dynamic nature of ordinary inferences:

> The premises of an argument invite us to update our initial information state, and then, the resulting transition has to be checked to see whether it 'warrants' the conclusion (in some suitable sense).

Following van Benthem, we can distinguish between *two* aspects of an inference:

(i) **update**: updating the initial set of assumptions;

(ii) **test**: checking whether the update has resulted in a set of assumptions that 'warrants', or supports, the conclusion.

For example, consider the inference: ⌜ϕ_1, ϕ_2, ϕ_3; *therefore*, ψ⌝. The first three premises correspond to **update**: uttering them is an invitation to update the current context sequentially with ϕ_1, ϕ_2, and ϕ_3. It is, moreover, plausible that the phrase ⌜*therefore*, ψ⌝ plays the role of the **test** part. As argued by Neta ([21], p. 399), '*therefore*' is a deictic expression in that it refers back to the utterance of certain premises. Because of that, a dynamic interpretation of '*therefore*' — i.e., an interpretation that highlights the role played by the expression within a discourse — seems to be particularly fitting. Moreover, as famously argued by Grice [10], in a sentence such as "John is English and therefore brave," '*therefore*' does not seem to add anything to the core content of a sentence such as "John is English and brave." The two sentences may well have the same truth conditions, even though the former also signals (or *conventionally implies*) that John's being brave follows from his being English. Thinking of '*therefore*' as a test captures the Gricean insight that in some sense, '*therefore*' is informationally empty. A test is in a similar sense informationally empty: its utterance does not alter the context by eliminating assumptions. Rather, it checks that the context satisfies certain constraints. If so, the overall meaning of a one-premise argument of the form ⌜ϕ; *therefore* ψ⌝ can be thought of as a function that checks whether the context created by the utterance of the premise ϕ supports the conclusion ψ.

So, an inference of the form ⌜ϕ_1, ϕ_2, ϕ_3; *therefore,* ψ⌝ has an **update** part and a **test** part. My proposal is that we use these two parts of an inference to characterize both inferences and inference rules. As pointed out by Rumfitt ([27], p. 53), the horizontal line in an inference can be thought along the same lines as the English '*therefore*' or '*thus*' — as a function that checks whether the context created by the premises supports the conclusion. This leads us to think of an inference as the *composition* of **update** and **test**:

$$\textbf{test}\left\{\frac{\left.\phi_1\,,\ldots,\,\phi_n\right\}\ \textbf{update}}{\psi}\right.$$

This composite function will return the result of updating c successively with the premises $\phi_1, \ldots, \phi_n$ if the resulting context c' supports the conclusion; else it returns the empty set (or absurd context).

This is a very natural semantic interpretation of an inference (or an argument). Along the same lines, an inference (or an argument) schema corresponding to a certain inference rule can be thought of as a composite function that takes as arguments a context and one or more sentences and returns another context. Let us consider, as a first example, the rule of &-elimination, represented here as the inference schema that takes a conjunction into either conjunct:

$$\frac{\phi_1 \,\&\, \phi_2}{\phi_1, \phi_2}\ \textbf{\&-E}$$

In this case, the **update** part is a function that takes the sentence $\phi_1 \,\&\, \phi_2$, and a context c, into the result of updating c with that sentence (= $c[\phi_1 \,\&\, \phi_2]$). Given **Definition 3.1**(iii), the result of so updating will be a context c' that results from sequentially updating c with ϕ_1 and then ϕ_2. The second part of the rule is the **test** corresponding to the horizontal line (/) and the conclusions ϕ_1 and ϕ_2 — a function that takes the sentences ϕ_1, ϕ_2 and a context c into c having checked that c supports both ϕ_1 and ϕ_2:

$$\textbf{Test (for \&-E)}\ c[/\phi_1, \phi_2] = \begin{cases} c & \text{if } c \models \phi_1 \quad \& \quad c \models \phi_2 \\ \emptyset & \text{if either} \quad c \not\models \phi_1 \quad \text{or} \quad c \not\models \phi_2 \end{cases}$$

It then becomes very natural to think of the &-elimination rule as the *composition* of these two different functions:

Definition 3.5 (**&-Elimination**)

$$c[\phi_1 \,\&\, \phi_2] \circ [/\phi_1, \phi_2]) = \begin{cases} c[\phi_1 \,\&\, \phi_2] & \text{if } c[\phi_1 \,\&\, \phi_2] \vDash \phi_1,\, \phi_2 \\ \emptyset & \text{if } c[\phi_1 \,\&\, \phi_2] \nvDash \phi_1,\, \phi_2 \end{cases}$$

The resulting composite dynamic value will take two sentences ϕ_1 and ϕ_2 and a context c and will return the result of updating c successively with ϕ_1 and ϕ_2 if the new context supports both ϕ_1 and ϕ_2. Otherwise, it will return the absurd context.

Let us consider a second example. Here is &-introduction:

$$\frac{\phi_1, \phi_2}{\phi_1 \,\&\, \phi_2} \;\textbf{\&-I}$$

Taking the comma above the horizontal line to mean conjunction, we can think of the **update** part of this rule as the function that, for any two sentences ϕ_1 and ϕ_2, takes a context c into a new context c' that results from updating c successively with ϕ_1 and then ϕ_2:

Update (for &-I) $c[\phi_1,\, \phi_2] = c[\phi_1][\phi_2]$

The second part of the rule checks whether the resulting context supports ϕ_1 & ϕ_2:

Test (for &-I) $c[/\phi_1 \,\&\, \phi_2] = \begin{cases} c & \text{if } c \vDash \phi_1 \,\&\, \phi_2 \\ \emptyset & \text{if } c \nvDash \phi_1 \,\&\, \phi_2 \end{cases}$

&-introduction corresponds to the composition of these two functions: [10]

Definition 3.6 (**&-Introduction**)

$$c([\phi_1,\, \phi_2] \circ [/\phi_1 \,\&\, \phi_2]) = \begin{cases} c[\phi_1][\phi_2] & \text{if } c[\phi_1][\phi_2] \vDash \phi_1 \,\&\, \phi_2 \\ \emptyset & \text{if } c[\phi_1][\phi_2] \nvDash \phi_1 \,\&\, \phi_2 \end{cases}$$

In order to specify the dynamic semantic value corresponding to *modus ponens*, I would have to settle the thorny issue of what the dynamic semantic value of the English conditional is. Here, I will set for myself a much less ambitious task, and will only consider the elimination rule for the *material conditional* '$\to$': [11]

[10] Note that although the function $[\phi_1,\, \phi_2]$ will return the same value as the function $[\phi_1 \,\&\, \phi_2]$ for any context c, the two functions are *intensionally* different, just like the function 'x +1' is different from the function '$x + 1 + 0$', even though they will return the same value for any argument x. By characterizing functions intensionally, rather than extensionally, on my proposal, &-**introduction** is a different function from &-**elimination**.

[11] That does not mean I endorse a material conditional analysis of the indicative conditional. I do not. Proponents of the material conditional analysis of the English indicative conditional are, notoriously, [10], [14], [15], and [19]. Among its many detractors, see [29], [5], [1], [7], [9], [33], and [34].

$$\frac{\phi \quad \phi \to \psi}{\psi}\ \mathbf{MP}$$

The **update** part of this rule is a function that, for any two sentences ϕ and ψ, takes a context c into a context c' that results from successively updating c with ϕ and $\neg\phi \vee \psi$:

Update (for Modus Ponens for the Material Conditional)
$c[\phi, \phi \to \psi] = c[\phi][\neg\phi \vee \psi]$

The composition of this **update** part with the **test** part gives us the following composite dynamic semantic value:

Definition 3.7 (**Modus Ponens for the Material Conditional**)

$$c([\phi, \phi \to \psi] \circ [/\psi]) = \begin{cases} c[\phi][\neg\phi \vee \psi] & \text{if } c[\phi][\neg\phi \vee \psi] \vDash \psi \\ \emptyset & \text{if } c[\phi][\neg\phi \vee \psi] \nvDash \psi \end{cases}$$

Generalizing, a rule of the following form:

$$\frac{\phi_1, ..., \phi_n}{\psi}$$

is a function that takes sentences $\phi_1, \ldots, \phi_2$ and ψ into the following dynamic semantic value:

Definition 3.8 (**Schema**)

$$c([\phi_1, \ldots, \phi_n] \circ [/\psi]) = \begin{cases} c[\phi_1][\phi_2]\ldots[\phi_n] & \text{if } c[\phi_1, \ldots, \phi_n] \vDash \psi \\ \emptyset & \text{if } c[\phi_1, \ldots, \phi_n] \nvDash \psi \end{cases}$$

So we get that each instance of a rule is a dynamic semantic value. **Schema** works well to characterize simple inference rules, such as the ones just considered, and can be used to capture both the rules of the propositional calculus and those of the predicative calculus. What about meta-rules, such as conditional proof, *reductio ad absurdum*, and argument by cases?

It turns out that we can model meta-rules simply by generalizing **Schema** to cover *a new kind of update*. To see this, start by considering conditional proof:

$$\begin{array}{c} [\phi] \\ \vdots \\ \psi \\ \hline (\phi \to \psi) \end{array} \to I$$

In this schema, a *sub-proof* seems to play the role of the premise of the argument schema. So the **update** part of this rule must consist in updating a context with *a sub-proof*. But what could updating a context with a *sub-proof* amount to?

Let me make two different suggestions and let me point out one important reason to prefer one of the two. On the first suggestion, updating a context c with a sub-proof, say with a premise ϕ and conclusion ψ, updates the context with the conditional $\phi \rightarrow \psi$:

Definition 3.9 (**Update for sub-proofs I**)
If σ has the form $\phi_1, \ldots, \phi_n/\psi$, $c[\sigma] = c[(\phi_1 \,\&\, \ldots \,\&\, \phi_n) \rightarrow \psi]$.

According to **Update for sub-proofs I**, the dynamic meaning of a sub-proof like ϕ/ψ is the same as the dynamic meaning of the conditional $\phi \rightarrow \psi$ — their difference is merely syntactic. Observe that **Update for sub-proofs I** treats the sub-proof as a *real* premise. For in the dynamic settings explored here, using a sub-proof as a premise means *updating* the current context with it. And note that that is exactly what **Update for sub-proofs I** instructs us to do with subproofs. The main problem with this is that unless we take the dynamic meaning of the conditional $\phi \rightarrow \psi$ itself to be that given by **Schema**, **Update for sub-proofs I** does not make clear how the dynamic meaning of a subproof such as ϕ/ψ arises from the dynamic meaning of an argument from ϕ to ψ, that (as we have just seen) consists in a composite function first updating the context with the premise ϕ and then checking whether the conclusion ψ follows. If we want the meaning of the subproof ϕ/ψ to be compositional on the meaning of the argument from ϕ to ψ, a better solution is to take the dynamic meaning of the subproof ϕ/ψ to result from combining an instruction to first update the context with ϕ and to then check whether ψ follows (just as **Schema** would instruct) with an instruction to *discharge the assumption* ϕ. The upshot of this composite instruction is the same as that of a test — an instruction to return to the context c having checked that $c[\phi]$ supports ψ:[12]

Definition 3.10 (**Update for sub-proofs II**)
If σ has the form $\phi_1, \ldots, \phi_n/\psi$, $c[\sigma] = \{w \in c\colon c[\phi_1 \,\&\, \ldots \,\&\, \phi_n] \vDash \psi\}$.

We will see in the last section another conceptual advantage of this second option. For now simply note that both **Update for sub-proofs I** and **Update for sub-proofs II** allow us to think of Conditional Proof along the exact same lines as our previous rules —i.e., as the following composite function:

Definition 3.11 (**Conditional Proof**)

$$c([\phi/\psi] \circ [/\phi \rightarrow \psi]) = \begin{cases} c[\phi/\psi] & \text{if } c[\phi/\psi] \vDash \phi \rightarrow \psi \\ \emptyset & \text{if } c[\phi/\psi] \nvDash \phi \rightarrow \psi \end{cases}$$

The discussion of **Conditional Proof** also covers also *Reductio ad Absurdum*, for such a rule can be thought as a special instance of **Conditional**

[12] Accordingly, the subproof $\ulcorner\phi/\psi\urcorner$ represents the combination of the argument from ϕ to ψ with the further instruction to discharge the premise ϕ. Plausibly, the latter instruction to discharge premise ϕ is expressed by the ending of the sub-proof line.

Proof, where $\neg A = A \rightarrow \bot$. Finally, consider Argument by Cases:

$$\dfrac{\phi_1 \vee \phi_2 \quad \begin{array}{c}[\phi_1]\\ \vdots \\ \psi\end{array} \quad \begin{array}{c}[\phi_2]\\ \vdots \\ \psi\end{array}}{\psi}$$

Here, we have three different premises — a sentence $\phi_1 \vee \phi_2$ as well as the sub-proofs ϕ_1/ψ and ϕ_2/ψ. Treating each as an individual premise, the update part of this rule will update the context sequentially with $\phi_1 \vee \phi_2$, ϕ_1/ψ, and ϕ_2/ψ. Just like in the previous case of **Conditional Proof**, this update allows us to define the rule of Argument by Cases as one would expect, given our previous discussion (where $c* = c[\phi_1 \vee \phi_2][\phi_1/\psi][\phi_2/\psi]$):

Definition 3.12 (**Argument by Cases**)

$$c([\phi_1 \vee \phi_2, \phi_1/\psi, \phi_2/\psi] \circ [/\psi]) = \begin{cases} c* & \text{if } c* \vDash \psi \\ \emptyset & \text{if } c* \nvDash \psi \end{cases}$$

We are now in position to generalize the original **Schema** to encompass the case of meta-rules. The only difference with respect to **Schema** is that the first function can now take as arguments not just sentences $\phi_1, \ldots, \phi_n$ but any environment above the horizontal line, including sub-proofs. Hence, by letting P (for premise) vary over both sentences and sub-proofs and by letting C vary over conclusions, we arrive at:

Definition 3.13 (**Schema-General**)

$$c([P_1, \ldots, P_n] \circ [/C]) = \begin{cases} c[P_1][P_2]\ldots[P_n] & \text{if } c[P_1, \ldots, P_n] \vDash C \\ \emptyset & \text{if } c[P_1, \ldots, P_n] \nvDash C \end{cases}$$

The structure of the premises together with **Schema-General** gives us a general recipe to map any schema into the corresponding context-change potential. Note also that the assignment of dynamic semantic values to schemas is compositional, as it is entirely determined by the meanings of their parts, together with the syntax of the schemas (the order and the syntax of their premises, the horizontal line, and the syntax of their conclusion).

4 Classical versus Dynamic Validity

On the current approach, an inferential rule is valid ($\vDash_0$) just in case *the relevant composite function never returns the empty set for any context:*

Definition 4.1 $P_1, \ldots, P_n \vDash_0 C$ iff for every c: $c([P_1, \ldots, P_n] \circ [/C]) \neq \emptyset$.

It is straightforward to prove that an inference rule is valid ($\vDash_0$) just in case it is DE-valid ($\vDash_{DE}$):

Theorem 4.2 $P_1, \ldots, P_n \vDash_0 C$ *iff* $P_1, \ldots, P_n \vDash_{DE} C$.

Proof. Suppose $P_1, \ldots, P_n \vDash_0 C$. Then, by **Definition 4.1**, for every c: $c([P_1, \ldots, P_n] \circ [/C]) \neq \emptyset$. But then, by **Definition 3.13**, for every c: $c[P_1, \ldots, P_n] \vDash C$. Hence, by **Definition 3.3**, $P_1, \ldots, P_n \vDash_{DE} C$. Now, suppose $P_1, \ldots, P_n \vDash_{DE} C$. Then, by **Definition 3.3**, for every c: $c[P_1, \ldots, P_n] \vDash C$. Then, by **Definition 3.13**, for every c: $c([P_1, \ldots, P_n] \circ [/C]) \neq \emptyset$. Then, by **Definition 4.1**, $P_1, \ldots, P_n \vDash_0 C$. □

A possible objection is that this notion of validity ($\vDash_0$) conflates two notions of validity: *classical validity* and *dynamic validity*. As observed by van Benthem ([35], p. 11 and pp. 18-19), these are indeed different notions of validity. For dynamic validity, the order of the premises and the multiplicity of their occurrence matters. That seems to clash with the basic structural rules of standard classical logic.

Can this important distinction between dynamic validity and classical validity be vindicated on the present approach? As Starr ([33], p. 9) has pointed out, classical entailment can be thought of as a special case of dynamic entailment — i.e., as dynamic entailment in *contexts of perfect information.* Contexts of perfect information only include *the world of the context* — no other world is compatible with a set of propositions that completely distinguishes the actual world from any other possible worlds. So, let the context of perfect information relative to w be $\{w\}$. Classical entailment ($\vDash_{CE}$) emerges by focusing on perfect information:

Definition 4.3 $P_1, \ldots, P_n \vDash_{CE} C$ iff $\forall\{w\}$: $\{w\}[P_1] \ldots [P_n] \vDash C$.

Call a function from contexts of perfect information to other contexts of perfect information a *limiting dynamic semantic value.* The limiting dynamic semantic value of a schema is insensitive to the order of its premises. So when we want to highlight the insensitivity of classical structural rules to the order of their premises, we can then think of them as limiting dynamic semantic values. This move preserves van Benthem's distinction, while clinging to the idea according to which classical inference rules are sorts of dynamic semantic values.

5 The Dynamic Conception of Inference Rule and Lewis Carroll's Regress

What does it mean to follow a rule in an argument? According to a very popular diagnosis of Carroll's regress of the premises, following a rule in an argument is not the same as using, or being guided by, a logical truth. The argument to this conclusion goes as follows. Let us start with premises A and *if* A *then* B:

(i) A.
(ii) *If* A *then* B.

How does one get from these premises to the conclusion B? Presumably, by appeal to *modus ponens*. But now, suppose the rule of *modus ponens* were identical to the general principle **LT-mp**:

LT-mp *For every* X, Y, *if* X *and* if X then Y, *then* Y.

or to the conditional schema:

LT-mp-schema *If* X *and* if X then Y, *then* Y.

Then, presumably, following that rule would be a matter of instantiating **LT-mp** or **LT-mp-schema** for the particular case of A and B — adding an instance of theirs as premise. But by instantiating such a logical truth, one only gets to (iii), still short of the conclusion B:

(i) A.
(ii) *If* A *then* B.
(iii) *If* A *and* if A then B *then* B.

How can one get from (i-iii) to the conclusion? Again, by appeal to *modus ponens*, one would guess. The problem is that if following *modus ponens* is the same thing as using **LT-mp** or **LT-mp-schema**, then arguing by *modus ponens* must amount to instantiating either for the particular cases of the premises. But by so doing, one will only get to the following four-premise argument (by taking the conjunction of A and *If* A *then* B as the first premise, and (iii) as the second premise), and still short of the conclusion B. And so on. Therefore, if following *modus ponens* were the same as using a general principle such as **LT-mp** or a conditional schema such as **LT-mp-schema** in the course of an argument, following a rule would trigger a regress, making it impossible to reach a conclusion. But we do routinely succeed at reasoning by *modus ponens.* So, the argument concludes that following *modus ponens* cannot be the same as using **LT-mp** or **LT-mp-schema**:

Diagnosis Following an inference rule in the course of an argument is not the same as using a logical truth.

And from **Diagnosis**, it is a shot step to conclude to:

Rules versus logical truths An inference rule is not the same as a logical truth.

This step is motivated by the thought that *the only ways* a truth can be used in an argument are by instantiation (if the logical truth is general, such as **LT-mp**) or by iteration (if the logical truth is singular such as the conditional '*If A, and if A then B, then B*' or an instance of the conditional schema **LT-mp-schema**). This is plausibly a general claim about truths. This claim is not irresistible but to resist it is not an easy task: defending it would require providing a different model of how truths can be used in the course of an argument, one that to my knowledge nobody has ever provided.

It should not come as a surprise, then, that many have underwritten **Rules versus logical truths** as the correct diagnosis of the regress. For example, Dummett [4], p. 303, observes that the main moral of the regress is that an argument of the form:

(A) *Pietro is Italian; if Pietro is Italian then Andrea is Italian; therefore*

Andrea is Italian.

cannot be identified with the conditional statement:

(C) *If both Pietro is Italian and if Pietro is Italian, then Andrea is Italian, then Andrea is Italian.*

Along similar lines, Ryle ([28], p. 7) argues that knowing a rule is not the same as knowing a truth. Finally, Rumfitt ([26], p. 358) argues that knowledge of a logical truth cannot explain our ability to make deductions and takes that to be the moral of Lewis Carroll's fable.

Although very popular, it should be emphasized that **Rules versus logical truths** is entirely *negative*: it only tells us that rules are not logical truths but it does not tell us *what rules are*. A solution to the puzzle raised by the regress requires replacing a conception of inference rules as logical truths with something else.

The dynamic notion of inference rules I developed in §3 offers a suitable semantic replacement of the notion of rule as logical truth. To see this point, it is now convenient to shift from the coarse-grained notion of context employed so far to a more fine-grained notion of context, as a set of structured propositions. On such a fine-grained notion of context, it makes sense to think of our using a logical truth in the course of an argument as a matter of *adding it to the context as a further premise.*[13] But now suppose inference rules are not logical truths but dynamic semantic values of the sort that I described in the last section. Following an inference rule in this sense is not a matter of adding a logical truth to the context as a further premise. Rather, it is a matter of *implementing a particular function* and, in particular, one whose implementation does *not* require to go through instantiation of the rule itself as an extra premise. To see this, consider **Modus Ponens for the Material Conditional** (**Definition 3.7**). Such rule is a function that, given the premises and a context as arguments, updates the context with the premises, having checked that the result supports the conclusion. So the use of this rule — i.e., the implementation of this function — does not require adding the rule itself to the context nor does it require instantiating the rule itself as an extra-premise.

Relatedly, my proposal, coupled with a suitably fine-grained notion of context, also helps us appreciate Dummett's distinction between an argument such as (A) and a conditional such as (C). Whereas using a conditional such as (C) in the course of an argument plausibly does require to add to the context (C) as an extra premise, on my proposal, using an argument such as (A) does not require to add to the context (A) itself as an extra premise. Here, one might object that arguments do sometimes appear to be treated as premises, as when

[13] Adopting a more fine-grained notion of context makes it easier to appreciate the difference that my proposal draws between, on one hand, using a logical truth in the course of an argument and, on the other, following a rule in the sense of 'rule' here outlined. By contrast, on a coarse-grained conception of context as set of possible worlds, updating a logical truth (a necessary truth) in the course of an argument is an *ineffectual* update — for it does not eliminate any possible world.

we argue by conditional proof. But recall that, even when arguments are *apparently* used as premises in arguments by conditional proof or by argument by cases, my **Update for Sub-Proofs II** does not take them to *work* as real premises. Rather, according to **Update for Sub-Proofs II**, their role as sub-proofs is equivalent to that of a test — an instruction to first check that the context created by adding their premises supports the conclusion and to then return to the original context after the checking.

Note that my claim is not that the implementation of a context-change potential *never* requires adding an extra premise. To see this, consider the following universal principle:

LT-mp-dynamic *For every* X, Y, *if* X *and* if X then Y, *then must*-Y.

On its dynamic interpretation, **LT-mp-dynamic** *also* expresses a dynamic semantic value. But the relevant context-change potential is different from the one defined in **Definition 3.7**, for **LT-mp-dynamic** is a universal claim. So **LT-mp-dynamic**'s dynamic meaning is function also of the dynamic meaning of the universal quantifier. Assuming a certain dynamic treatment of the universal quantifier and a suitably fine-grained notion of context, its dynamic meaning will consist in an update — in adding it (or an instance of it) as a premise to the context. So while implementing the function corresponding **Definition 3.7** does not require instantiating extra premises, implementing the function corresponding to **LT-mp-dynamic** does (plausibly) require going through instantiation for possible assignments to X and Y.

What should we conclude? The conclusion to draw is that *also* **LT-mp-dynamic** is not the right way to think of the inference rule of *modus ponens*. Of course, it does not follow from that my proposal is not correct, for as I just observed, on my proposal *modus ponens* is not the same as the dynamic interpretation of **LT-mp-dynamic** just mentioned. My claim is that *certain* dynamic semantic values (the ones described in section §3) can stop the regress. My claim is *not* that every possible dynamic semantic value will block the regress.

Here is a second potential worry. Could not we imagine a 'dynamic' version of the regress, on which a subject keeps updating the context with more premises, without being able to run a test? If so, how does the dynamic conception of inference rules improve on a conception of rules as logical truths? In response, let me note that the current proposal offers an account of rule-following on which, *if* one can follow a rule at all, one *could not* get stuck in the regress of the premises. It is worth going through this point carefully. On the current proposal, a subject who keeps updating the context with more premises without being able to run a test would prove to be *unable* to follow the relevant rule. That is so because on the current proposal, following an inference rule *requires* being able to run a test. That can be seen from the fact that running a test is a condition for implementing the composite functions with which I have identified inference rules (which are composed both of an **update** part and a **test** part). Hence, if one can follow a rule at all in this sense, one simply cannot get stuck with the regress of the premises. By contrast, if rules are logical truths, following a rule *will* get us stuck in the regress of the premises. Whereas the notion of following a logical truth is regress-triggering and hence paradoxical, the notion of following a rule developed here is not. That is the important respect under which the dynamic conception of inference rules improves on a conception of rules as logical truths.

6 Conclusions

The dynamic conception of inference rules developed in this essay provides a picture of rule-following which blocks Carroll's regress. As argued in §2, such a dynamic conception is independently motivated by a dynamic conception of inference. As shown in §3, it arises from a plausible and compositional assignment of meanings to argument schemas and from an independently motivated dynamic interpretation of the horizontal line and of the English '*therefore*'. Because of that, it captures the distinction Dummett seemed to be after in the passage mentioned: that between an argument of the form ⌜*P; if P then Q; therefore, Q*⌝ and a conditional of the form ⌜*if P, and if P then Q, then Q*⌝. All in all, the dynamic conception of inference rules seems to provide a suitably *semantic* replacement of a conception of inference rules as logical truths.

Some outstanding issues are left open by this essay, that I do not have the space to discuss here (though see [22] for discussion). The first is: how does the dynamic conception of inference rules compare to other conceptions of inferential rules, for example, as syntactic mappings or conditional recommendations of sort? Whether or not these conceptions of inference rules can block Carroll's regress, I argue elsewhere that they all fall short in *other* respects ([22]). Either they do not correspond to a plausible interpretation of argument schemas and/or they are not suitably *semantic* conceptions of inference rules, and because of that, they face a version or another of a problem that I call *the problem of understanding*. If I am right, a stronger case on behalf of my proposal is available than I can make here.

The second outstanding issue is this. On the dynamic conception of inference rules, being competent with an inference rule is a matter of *being competent with a function*. As I argue elsewhere ([23], [22]), being competent with a function plausibly requires the function being representable by a subject in terms of operations that the subject can already perform.[14] So if inference rules are dynamically conceived, our being competent with inference rules requires these functions being representable by us in terms of operations that are performable by us — by individuals of average linguistic competence. How plausible is this claim? Although the issue is of great importance, it is not specific to my dynamic conception of inference rules. It arises for any appeal to dynamic semantics as an explanatory theory of our linguistic competence, for on the dynamic picture, knowledge of meanings *is* in general a matter of being competent with functions. Hence, although a thorough defense of the

[14] By analogy with the notion of *primitive recursive function* in computability theory.

dynamic conception of inference rules does require a detailed defense of the plausibility of this claim, I consider it to be part of a bigger project that will have to be discussed in further work.

References

[1] Bennett, J., "A Philosophical Guide to Conditionals," Oxford University Press, 2003.
[2] Boghossian, P. A., *What is Inference?*, Philosophical Studies **169** (2014), pp. 1–18.
[3] Carroll, L., *What The Tortoise Said to Achilles*, Mind **4** (1895), pp. 278–80.
[4] Dummett, M., "Frege: Philosophy of Language Vol. 2," London: Duckworth, 1981.
[5] Edgington, D., *On Conditionals*, Mind **104** (1995), pp. 235–329.
[6] Frege, G., *Logic. Posthumous Writings* (1979).
[7] Gibbard, A., *Two Recent Theories of Conditionals*, in: *Ifs*, Springer, 1980 pp. 211–247.
[8] Gillies, A. S., *A New Solution to Moore's Paradox*, Philosophical Studies **105** (2001), pp. 237–250.
[9] Gillies, A. S., *Epistemic Conditionals and Conditional Epistemics*, Noûs **38** (2004), pp. 585–616.
[10] Grice, H. P., "Studies in the Way of Words," Harvard University Press, 1991.
[11] Groenendijk, J. A. G. and M. Stokhof, *Dynamic Predicate Logic*, Linguistics and Philosophy (1991), pp. 39–100.
[12] Harman, G., "Change in View: Principles of Reasoning," Cambridge University Press, 2008.
[13] Heim, I., *The Semantics of Definite and Indefinite Noun Phrases* (1982), pHD Dissertation.
[14] Jackson, F., *On Assertion and Indicative Conditionals*, The Philosophical Review **88** (1979), pp. 565–589.
[15] Jackson, F., "Conditionals," Oxford University Press, 1991.
[16] Kamp, H., *A Theory of Truth and Linguistic Representation*, in: *Formal Methods in the Study of Language (Groenendijk, J. A. G and M. Stokhof (eds.), pp. 277-321)*, Mathematish Centrum, Universiteit van Amsterdam, 1981 .
[17] Kamp, H. and U. Reyle, "From Discourse to Logic: Introduction to Model-theoretic Semantics of Natural Language, Formal Logic and Discourse Representation Theory," 42, Springer Science & Business Media, 1993.
[18] Klinedinst, N. and D. Rothschild, *Connectives without truth tables*, Natural Language Semantics **20** (2012), pp. 137–175.
[19] Lewis, D., *Probabilities of Conditional and Conditional Probabilities*, The Philosophical Review **8** (1976), pp. 297–315.
[20] Longino, H. E., *Inferring*, Philosophy Research Archives **4** (1978), pp. 17–26.
[21] Neta, R., *What is an Inference?*, Philosophical Issues **23** (2013), pp. 388–407.
[22] Pavese, C., *Inference rules as dynamic semantic values*, Manuscript `https://www.academia.edu/23462009/Inference_Rules_as_Dynamic_Semantic_Values`.
[23] Pavese, C., *Practical Senses*, Philosophers' Imprint **15**, pp. 1–25.
[24] Pérez Carballo, A., *Structuring Logical Space*, Philosophy and Phenomenological Research (2014).
[25] Rothschild, D. and S. Yalcin, *On the Dynamics of Conversation*, Noûs (2015).
[26] Rumfitt, I., *Inference, Deduction, and Logic*, in: J. Bengson and M. Moffett, editors, *Knowing How: Essays on Knowledge, Mind, and Action*, Oxford University Press, Oxford, 2011 pp. 334–350.
[27] Rumfitt, I., "The Boundary Stones of Thought: An Essay in the Philosophy of Logic," Oxford University Press, USA, 2015.
[28] Ryle, G., *Knowing How and Knowing That: the Presidential Address*, Proceedings of the Aristotelian Society, New Series **46** (1945), pp. 1–16.

[29] Stalnaker, R., *Indicative Conditionals*, in: *Language in Focus: Foundations, Methods and Systems Volume 43 of the series Boston Studies in the Philosophy of Science*, Springer, 1975 pp. 179–196.
[30] Stalnaker, R., "Inquiry," MIT Press, Cambridge, Mass, 1987.
[31] Stalnaker, R., *Context and Content: Essays on Intentionality in Speech and Thought* (1999).
[32] Stalnaker, R., "Context," OUP Oxford, 2014.
[33] Starr, W. B., *A Uniform Theory of Conditionals*, Journal of Philosophical Logic **43** (2014a), pp. 1019–1064.
[34] Starr, W. B., *What 'If'?*, Philosophers' Imprint **14** (2014b).
[35] van Benthem, J., "Exploring Logical Dynamics," CSLI Publications and Stanford/Cambridge University Press, 1996.
[36] Veltman, F., *Defaults in Update Semantics*, Journal of philosophical logic **25** (1996), pp. 221–261.
[37] Willer, M., *An update on epistemic modals*, Journal of Philosophical Logic , pp. 1–15.
[38] Willer, M., *Dynamics of Epistemic Modality*, Philosophical Review **122** (2013), pp. 45–92.
[39] Yalcin, S., *Epistemic modals*, Mind **116** (1995), pp. 983–1026.
[40] Yalcin, S., *Nonfactualism about Epistemic Modality*, in: A. Egan and B. Weatherson, editors, *Epistemic Modality. Oxford University Press*, Oxford University Press, Oxford, 2011 pp. 295–332.

Conditional Normative Reasoning with Substructural Logics

New Paradoxes and De Morgan's Dualities

Clayton Peterson [1]

Munich Center for Mathematical Philosophy
Ludwig-Maximilians Universität München

Piotr Kulicki

Faculty of Philosophy
John Paul II Catholic University of Lublin

Abstract

This paper extends the results presented in [22,20] and explores how new paradoxes arise in various substructural logics used to model conditional obligations. Our investigation starts from the comparison that can be made between monoidal logics and Lambek's [17] analysis of substructural logics, who distinguished between four different ways to introduce a (multiplicative) disjunction. While Lambek's analysis resulted in four variants of substructural logics, namely BL1, BL1(a), BL1(b) and BL2, we show that these systems are insufficient to model conditional obligations insofar as either they lack relevant desirable properties, such as some of De Morgan's dualities or the law of excluded middle, or they satisfy logical principles that yield new paradoxes. To answer these concerns, we propose an intermediate system that is stronger than BL1 but weaker than BL1(a), BL1(b) and BL2.

Keywords: Conditional obligation, Monadic deontic operator, Multiplicative disjunction, Linear logic, Bilinear logic, Lambek calculus, Relevant logic

1 Deontic logic without paradoxes

Recently, Peterson [22,20] introduced a logical framework able to deal with most of the paradoxes of deontic logic (see also [21]). The strategy used to avoid the paradoxes and problems of deontic logic is to modify the underlying logic and use logical systems comparable to various substructural logics. The idea is to model an obligation $O\alpha$ that is conditional to a context φ through

[1] We would like to thank Olivier Roy for feedback on a previous version of this article as well as two anonymous reviewers for comments and suggestions. This research was financially supported by *Deutsche Forschungsgemeinschaft* (RO 4548/4-1) and *Narodowe Centrum Nauki* (UMO-2014/15/G/HS1/04514). The first author also acknowledge financial support from the Social Sciences and Humanities Research Council of Canada.

a conditional connective $\varphi \multimap O\alpha$. Depending on the relationships that can be found between $\multimap$ and the other connectives, one will get different logical systems with different properties. From this standpoint, the paradoxes can be solved by modifying the meaning of the usual propositional connectives together with the rules governing their behavior.

From a foundational point of view, the main idea underlying Peterson's approach is to use category theory (cf. [18]) to analyze the proof-theory of the various deontic logics we find within the literature. Category theory is relevant within this context given that it provides a general standpoint that allows to classify logical systems that *prima facie* appear to be different but that, from a proof-theoretical perspective, share the same categorical structure. Using this approach, most paradoxes in deontic logic can be correlated to specific types of logical systems. Accordingly, building on this framework, the logic $\mathcal{CNR}$ was introduced to model conditional normative reasoning and conditional obligations while avoiding paradoxes.

Despite the fact that $\mathcal{CNR}$ avoids most of the paradoxes of deontic logic, it remains, however, that it also lacks the satisfaction of some logical principles that might seem desirable, such as De Morgan's dualities or the law of excluded middle. In this respect, the aim of the present paper is to analyze how $\mathcal{CNR}$ can be extended with well-known logical principles. To accomplish this, we will first expose the technical background required for our analysis and we will discuss the relationship between this framework and substructural logics. Specifically, we build extensions that are comparable to Lambek's [17] bilinear logics and we argue that these systems are insufficient to model conditional obligations given that they enable the derivation of new paradoxes. As such, we propose an intermediate logic stronger than $\mathcal{CNR}$ that avoids the undesirable paradoxes but that satisfies some intuitively appealing logical principles.

2 Technical background

2.1 Logic for conditional normative reasoning

The deontic logic presented in [22,20] is defined within the framework of monoidal logics, which are comparable to various substructural logics (see [24] for details and a thorough presentation). Using this framework as well as van der Torre's and Tan's [29] analysis of normative conflicts, Peterson [22,20] showed that not only Chisholm's [5] paradox but also other problems, including the problems of augmentation (cf. [15,1]), factual detachment (cf. [30]) and deontic explosion (cf. [8,9]), can be modeled and resolved by using the deductive system $\mathcal{CNR}$ (the acronym stands for *Conditional Normative Reasoning*).

$\mathcal{CNR}$ is an *ought-to-do* deontic logic (cf. [31]) where formulas within the scope of the deontic operators need to be proper action formulas (see [20,23]). As such, it is defined on the grounds of an action logic $\mathcal{AL}$. The action logic that governs the behavior of actions within the scope of deontic operators is defined from a language composed of a collection *Act* of atomic actions a_i, the symbols (,), the connectives $\bullet$ (joint action, with), $\curvearrowright$ (action sequence) and $\ominus$ (without) as well as an action $*$, read 'no change', that is both trivial and impossible to perform.[2] Joint action is represented by $\alpha \bullet \beta$ and means 'α together with β'. It is assumed to be a connective that is commutative and associative but not idempotent. Action conjunction must not be idempotent insofar as performing an action once is not equivalent to performing an action twice (cf. [7]). Sequence of actions is represented by $\alpha \curvearrowright \beta$. It means '$\alpha$ and then β'. The connective is associative but neither commutative nor idempotent. Finally, as it is possible to perform two actions *together*, it is also possible to perform one action *without* another one. The action $\beta \ominus \alpha$ is introduced to represent this phenomenon and means the action 'β without α'. The negation of an action, which is some form of complement, is defined by $\alpha^* =_{df} * \ominus \alpha$, that is, 'no change without α'. In this context, there are no 'disjunctive' actions (see the aforementioned papers for details). Well-formed action formulas are defined recursively as usual.

$\mathcal{AL}$ is defined on the grounds of the following rules and axiom schemas.[3]

$$\frac{}{\alpha \longrightarrow \alpha}\ (1) \qquad \frac{\alpha \longrightarrow \beta \qquad \beta \longrightarrow \gamma}{\alpha \longrightarrow \gamma}\ (\text{cut}) \qquad \frac{\alpha \longrightarrow \beta \qquad \gamma \longrightarrow \delta}{\alpha \bullet \gamma \longrightarrow \beta \bullet \delta}\ (\text{t})$$

$$\begin{array}{c} \delta \longrightarrow (\alpha \bullet \beta) \bullet \gamma \\ \hline\hline \delta \longrightarrow \alpha \bullet (\beta \bullet \gamma) \end{array}\ (\text{a}) \qquad \begin{array}{c} \alpha \bullet \beta \longrightarrow \gamma \\ \hline\hline \beta \longrightarrow \gamma \ominus \alpha \end{array}\ (\text{cl}) \qquad \begin{array}{c} \alpha \longrightarrow \beta \bullet \delta \\ \hline\hline \alpha \longrightarrow \delta \bullet \beta \end{array}\ (\text{b}) \qquad \begin{array}{c} \alpha \longrightarrow \beta \bullet * \\ \hline\hline \alpha \longrightarrow \beta \end{array}\ (\text{r})$$

$$\frac{\alpha \longrightarrow \beta \qquad \gamma \longrightarrow \delta}{\alpha \curvearrowright \gamma \longrightarrow \beta \curvearrowright \delta}\ (\text{t}) \qquad \begin{array}{c} \delta \longrightarrow (\alpha \curvearrowright \beta) \curvearrowright \gamma \\ \hline\hline \delta \longrightarrow \alpha \curvearrowright (\beta \curvearrowright \gamma) \end{array}\ (\text{a}) \qquad \begin{array}{c} \alpha \longrightarrow \beta \curvearrowright * \\ \hline\hline \alpha \longrightarrow \beta \end{array}\ (\text{r})$$

$$\begin{array}{c} \alpha \longrightarrow * \curvearrowright \beta \\ \hline\hline \alpha \longrightarrow \beta \end{array}\ (\text{l}) \qquad \frac{}{\alpha^* \bullet \beta \longrightarrow \beta \ominus \alpha}\ (\text{c1}) \qquad \frac{}{\beta \ominus \alpha \longrightarrow \alpha^* \bullet \beta}\ (\text{c2})$$

The notation $\alpha \longrightarrow \beta$ in $\mathcal{AL}$ can be interpreted in terms of necessary and sufficient conditions for actions. That is, $\alpha \longrightarrow \beta$ means that β is necessary for the realization of α. In this respect, the notation $\alpha \longrightarrow \beta$ can also be interpreted in terms of counts-as conditionals (cf. [12]). For example, consider the actions 'ending someone's life (α)' and 'murder (β)'. Assuming that α counts-as β in a normative system, this can be translated by an axiom $\alpha \longrightarrow \beta$ within the language of $\mathcal{AL}$. Given the purpose of the present paper, we will not consider further the action logic used to model the behavior of the action formulas within the scope of the deontic operators. We refer the reader to [20] for a formal presentation and to [23] for a philosophical analysis of action connectives.

Now, consider a language $\mathcal{L}$ constructed from a collection *Prop* of atomic formulas p_i, the symbols (,), the connectives $\otimes$ (tensor, multiplicative conjunction), $\multimap$ (adjoint, linear implication) and $\oplus$ (co-tensor, multiplicative disjunction), the operators O (obligation) and P_s (strong, explicit permission) as well as the formulas 0, 1 and $\perp$. Weak (implicit) permission and interdiction are

---

[2] This double character of $*$ might seem puzzling at first glance. See [20,23] for details and explanations.

[3] The double line abbreviates two rules, top-down and bottom-up.

defined as follows.

$$P_w\alpha =_{df} \neg F\alpha \qquad\qquad F\alpha =_{df} O\alpha^*$$

Negation is defined by $\neg\varphi =_{df} \varphi \multimap \bot$. While $\bot$ is a constant representing falsehood, 0 is the neutral element of multiplicative disjunction (i.e., it can be absorbed, see the rule (co-r) below). The usual approaches take 0 to be logically equivalent to $\bot$. Nonetheless, defining negation through $\bot$ instead of 0 keeps many of the usual relationships between the logical connectives (see [24]). For instance, we still have contraposition for $\multimap$ as well as $\varphi\otimes\neg\varphi \longrightarrow \bot$. However, the axiom schema $\bot \longrightarrow \varphi$ is not satisfied and, as a result, $\varphi \otimes \neg\varphi \longrightarrow \psi$ is not satisfied either (see [22]). Well-formed formulas are inductively defined as usual, although only action formulas can be within the scope of O and P_s. A consequence relation is represented by an arrow $\varphi \longrightarrow \psi$ between two formulas.

$\mathcal{CNR}$ is defined as a collection of formulas and a collection of equivalence classes of proofs (deductions). It has to satisfy the following rules and axiom schemas.

$$\frac{}{\varphi \longrightarrow \varphi}\,(1) \qquad \frac{\varphi \longrightarrow \psi \qquad \rho \longrightarrow \tau}{\varphi\otimes\rho \longrightarrow \psi\otimes\tau}\,(\mathrm{t}) \qquad \frac{\varphi \longrightarrow \psi \qquad \psi \longrightarrow \rho}{\varphi \longrightarrow \rho}\,(\mathrm{cut})$$

$$\frac{\tau \longrightarrow (\varphi\otimes\psi)\otimes\rho}{\tau \longrightarrow \varphi\otimes(\psi\otimes\rho)}\,(\mathrm{a}) \qquad \frac{\varphi\otimes\psi \longrightarrow \rho}{\varphi \longrightarrow \psi\multimap\rho}\,(\mathrm{cl}) \qquad \frac{\varphi \longrightarrow \psi\otimes\tau}{\varphi \longrightarrow \tau\otimes\psi}\,(\mathrm{b})$$

$$\frac{\varphi \longrightarrow \psi\otimes 1}{\varphi \longrightarrow \psi}\,(\mathrm{r}) \qquad \frac{\varphi \longrightarrow \psi \qquad \rho \longrightarrow \tau}{\varphi\oplus\rho \longrightarrow \psi\oplus\tau}\,(\text{co-t}) \qquad \frac{\tau \longrightarrow (\varphi\oplus\psi)\oplus\rho}{\tau \longrightarrow \varphi\oplus(\psi\oplus\rho)}\,(\text{co-a})$$

$$\frac{\varphi \longrightarrow \psi\oplus\tau}{\varphi \longrightarrow \tau\oplus\psi}\,(\text{co-b}) \qquad \frac{\varphi \longrightarrow \psi\oplus 0}{\varphi \longrightarrow \psi}\,(\text{co-r}) \qquad \frac{}{\neg\neg\varphi \longrightarrow \varphi}\,(\neg\neg)$$

Given these rules, many logical principles well-known in classical propositional logic are not derivable. For instance, strengthening of the antecedent fails for $\multimap$, as copying ($\varphi \longrightarrow \varphi\otimes\varphi$, $\varphi \longrightarrow \varphi\oplus\varphi$) and filtering ($\varphi\otimes\varphi \longrightarrow \varphi$, $\varphi\oplus\varphi \longrightarrow \varphi$) fail for $\otimes$ and $\oplus$. Further, the multiplicative disjunction cannot be defined via $\multimap$, even though the elimination of double negation is satisfied. Finally, the usual rules governing the introduction and the elimination of $\wedge$ and $\vee$ in classical logic fail for $\otimes$ and $\oplus$, including $\varphi\otimes\psi \longrightarrow \varphi$ and $\varphi \longrightarrow \varphi\oplus\psi$. Similarly, *ex falso sequitur quodlibet* and *verum ex quodlibet* are not derivable.

In addition to the aforementioned rules, there are two substitution rules (Σ_O) and (Σ_P), analogous to the rule (RM) in non-normal (monotonic) modal logics (cf. [4]), representing deontic consequence.

$$\frac{\alpha \longrightarrow_{\mathcal{AL}} \beta}{O\alpha \longrightarrow_{\mathcal{CNR}} O\beta}\,(\Sigma_O) \qquad \frac{\alpha \longrightarrow_{\mathcal{AL}} \beta}{P_s\alpha \longrightarrow_{\mathcal{CNR}} P_s\beta}\,(\Sigma_P)$$

From the perspective of category theory, the rules (Σ_O) and (Σ_P) are functors that map arrows in $\mathcal{AL}$ to arrows in $\mathcal{CNR}$. Formally, we have $\Sigma_O(\alpha \longrightarrow \beta) = \Sigma_O(\alpha) \longrightarrow \Sigma_O(\beta)$ and $\Sigma_P(\alpha \longrightarrow \beta) = \Sigma_P(\alpha) \longrightarrow \Sigma_P(\beta)$, with $\Sigma_O(\alpha) = O\alpha$, $\Sigma_P(\alpha) = P_s\alpha$, $\Sigma_O(*) = \bot$ and $\Sigma_P(*) = \bot$ (see [20]). Here are two examples of application of the rule (Σ_O).

$$\frac{\beta \ominus \alpha \longrightarrow_{\mathcal{AL}} \alpha^* \bullet \beta}{O(\beta \ominus \alpha) \longrightarrow_{\mathcal{CNR}} O(\alpha^* \bullet \beta)} (\Sigma_O) \qquad \frac{\alpha \bullet \alpha^* \longrightarrow_{\mathcal{AL}} *}{O(\alpha \bullet \alpha^*) \longrightarrow_{\mathcal{CNR}} \bot} (\Sigma_O)$$

The second example is rather interesting given that it implies that it is necessarily false that the conjunction of an action taken together with its complement is obligatory.

Given these rules, strong permission should be distinguished from other interpretations in the literature. For instance, it should not be interpreted with respect to free-choice permission (e.g. [2,28]). Rather, strong permissions are permissions explicitly stated by authorities. The rule (Σ_P) means that if β is a necessary condition for the realization of α, then, if α is explicitly allowed, so is β.

Finally, the deontic operators in $\mathcal{CNR}$ are governed by the following axiom schemas (from the perspective of category theory, they are natural transformations).

$$\frac{}{O\alpha \longrightarrow P_s\alpha} (\mathsf{D}) \qquad \frac{}{P_s\alpha \longrightarrow \neg O\alpha^*} (\mathsf{P})$$

Taken together, these principles represent normative consistency. By composition of (**D**) and (**P**) through (cut) and from the rule (cl), we can show that it is false that α is both obligatory and forbidden, i.e. $1 \longrightarrow \neg(O\alpha \otimes O\alpha^*)$. An obligation $O\alpha$ that is conditional to a context φ is modeled by $\varphi \multimap O\alpha$.

2.2 Bilinear logic

There are close relationships between monoidal logics and substructural (cf. [26]) as well as display logics (cf. [10]). In [22], Peterson mentioned that $\mathcal{CNR}$ was comparable (but not equivalent) to the multiplicative fragment of Girard's [7] linear logic, where the co-tensor is defined on the grounds of $\multimap$. In $\mathcal{CNR}$, this definition would use the following axiom schemas.

$$\varphi \oplus \psi \longrightarrow \neg\psi \multimap \varphi \qquad (\oplus 1)$$
$$\neg\psi \multimap \varphi \longrightarrow \varphi \oplus \psi \qquad (\oplus 2)$$

As it happens, $\mathcal{CNR}$ is strictly weaker than multiplicative linear logic, and this can be seen in light of the comparison that can be made between monoidal logics and Lambek's [17] analysis of bilinear logic. In [17], Lambek defines four variants of substructural logics, namely BL1, BL1(a), BL1(b) and BL2, each presenting a new way to introduce a multiplicative disjunction. Basically, BL1 corresponds to Lambek's [16] syntactic calculus augmented with a co-tensor and the following rule.

$$\frac{\Gamma \vdash \varphi \qquad \Sigma; \varphi; \Pi \vdash \Delta}{\Sigma; \Gamma; \Pi \vdash \Delta} (\mathsf{CUT}_{BL1})$$

In contrast, BL1(a) is BL1 augmented with Grishin's [11] rule (Ga), BL1(b) is BL1 together with Grishin's rule (Gb) and BL2 is BL1(ab). The commutative and non-bilinear version of BL1(b) corresponds to the full intuitionistic linear logic of Hyland and De Paiva [14], whereas BL2 corresponds to multiplicative linear logic (cf. [7]).

$$\frac{\Pi \vdash (\Gamma;\Sigma) < \Delta}{\Pi \vdash \Gamma;(\Sigma < \Delta)}\ (\mathsf{Ga}) \qquad \frac{\Pi \vdash \Gamma;(\Sigma < \Delta)}{\Pi \vdash (\Gamma;\Sigma) < \Delta}\ (\mathsf{Gb})$$

When translating the language of display logics in the language of monoidal logics (the translation is standard, see [24] for details), we obtain that $\mathcal{CNR}$ is actually comparable to Lambek's BL1 (the translation of CUT_{BL1} can easily be derived) while $\mathcal{CNR}$ satisfying ($\oplus$1) and ($\oplus$2) is comparable to BL2. Grishin's rules can be translated by (Ga_f) and (Gb_f).

$$\frac{\tau \longrightarrow \rho \multimap (\varphi \oplus \psi)}{\tau \longrightarrow \varphi \oplus (\rho \multimap \psi)}\ (\mathsf{Ga}_f) \qquad \frac{\tau \longrightarrow \varphi \oplus (\rho \multimap \psi)}{\tau \longrightarrow \rho \multimap (\varphi \oplus \psi)}\ (\mathsf{Gb}_f)$$

Despite the comparison that can be made between $\mathcal{CNR}$ and BL1, there is an important difference between the two systems insofar as $\mathcal{CNR}$ satisfies the elimination of double negation. In the substructural logic literature, the elimination of double negation is usually taken as a sufficient condition to go from intuitionistic (i.e. BL1) to classical (i.e. BL2) substructural logic (see for example [19], [17] or [10]). Classical substructural logics are known to satisfy the elimination of double negation as well as the law of excluded middle and De Morgan's dualities. Put differently, classical substructural logics have a De Morgan negation (cf. [26,6]). De Morgan's dualities and the law of excluded middle can be translated within the language of monoidal logics as follows.

$$\frac{}{1 \longrightarrow \neg\varphi \oplus \varphi}\ (\mathbf{lem})$$

$$\frac{}{\neg\psi \otimes \neg\varphi \longrightarrow \neg(\varphi \oplus \psi)}\ (\mathbf{dm1}) \qquad \frac{}{\neg(\varphi \oplus \psi) \longrightarrow \neg\psi \otimes \neg\varphi}\ (\mathbf{dm2})$$

$$\frac{}{\neg\varphi \oplus \neg\psi \longrightarrow \neg(\psi \otimes \varphi)}\ (\mathbf{dm3}) \qquad \frac{}{\neg(\psi \otimes \varphi) \longrightarrow \neg\varphi \oplus \neg\psi}\ (\mathbf{dm4})$$

An important aspect of monoidal logics is that it can be proven that the elimination of double negation, even when the law of excluded middle is assumed, does not necessarily imply the satisfaction of De Morgan's dualities (see [24]). All these considerations are relevant to our analysis given that, in analogy with Lambek's bilinear logics, this implies four different possible formulations of $\mathcal{CNR}$, each having specific logical properties. In addition to $\mathcal{CNR}$ as defined above, let us consider $\mathcal{CNR}(a)$, $\mathcal{CNR}(b)$ and $\mathcal{CNR}(ab)$ defined as follows.

Definition 2.1 $\mathcal{CNR}(a)$ is $\mathcal{CNR}$ with (Ga_f).

Definition 2.2 $\mathcal{CNR}(b)$ is $\mathcal{CNR}$ with (Gb_f) (with $\neg\varphi =_{df} \varphi \multimap 0$).

Definition 2.3 $\mathcal{CNR}(ab)$ is $\mathcal{CNR}$ with (Gb_f) and (Ga_f).[4]

In definition 2.2, negation is defined on the grounds of 0 rather than $\perp$. This requirement, which is derivable within $\mathcal{CNR}(ab)$, yields that (Gb_f) implies ($\oplus$1) in $\mathcal{CNR}(b)$.

These definitions should be understood in analogy with Lambek's [17] definition of (non-bilinear and commutative) BL1, BL1(a), BL1(b) and BL2. In our case, however, we consider extensions built using deontic operators and,

[4] $\mathcal{CNR}$ with ($\oplus$1) and ($\oplus$2) provides an alternative definition for $\mathcal{CNR}$(ab).

further, we assume the elimination of double negation in each case (this latter difference should be emphasized).

As shown in [24], some notable relationships between the principles involved within these definitions are that, in $\mathcal{CNR}$, i) ($\oplus$1) is logically equivalent to (**dm1**), which in turn is logically equivalent to (**dm3**), ii) ($\oplus$2) is logically equivalent to (**dm2**), which in turn is logically equivalent to (**dm4**) and iii) when 0 is logically equivalent to $\bot$, (Gb_f) implies ($\oplus$1).

Further, Grishin's rule (Gb) is related to an important characteristic of substructural logics. Indeed, it is related to the property of weak distributivity (a.k.a. mixed associativity or linear distributivity) as well as the usual (CUT) rule in substructural logics (see [17]).

$$\frac{\Gamma; \varphi \vdash \Delta \qquad \Sigma \vdash \varphi; \Pi}{\Gamma; \Sigma \vdash \Delta; \Pi}\ (\mathsf{CUT})$$

Weak distributivity and (CUT) can be translated as follows in $\mathcal{CNR}$.

$$\frac{}{\varphi \otimes (\psi \oplus \rho) \longrightarrow (\varphi \otimes \psi) \oplus \rho}\ (\mathbf{wd}) \qquad \frac{\varphi \otimes \pi \longrightarrow \rho \qquad \psi \longrightarrow \pi \oplus \tau}{\varphi \otimes \psi \longrightarrow \rho \oplus \tau}\ (\mathsf{CUT}_f)$$

While (CUT) concerns substructural logics and can be used with structures, (CUT_f) can only be applied to formulas and (cut) simply represents the transitivity of the consequence relation. In the context of monoidal logics, it can be proven that (**wd**), (CUT_f) and (Gb_f) are all logically equivalent (see [24] for the demonstration). The relationships between the principles discussed are summarized in table 1, section 8.

Even though it can be argued that $\mathcal{CNR}$ is a paradox free deontic logic, it can still be objected that it does not have enough deductive power. Indeed, some might object that it lacks some desirable characteristics, as, for instance, the satisfaction of the law of excluded middle or some of De Morgan's dualities. The question we wish to address within the remainder of this paper is which of these systems, if any, should be used to model conditional normative reasoning. To provide an answer to this question, we will examine whether new paradoxes arise from the logical principles involved within these definitions and whether some intuitively appealing consequences require these principles.

3 Weak augmentation and conditionalization

As we previously mentioned, (**wd**), (CUT_f) and (Gb_f) are logically equivalent. It can be argued, however, that there are problems with these principles if one tries to model conditional obligations using $\multimap$. Indeed, these principles induce a specific relationship between $\multimap$ and $\oplus$, which allows to transform any unconditional obligation into a conditional obligation that is part of a disjunction. Consider the following derivation.

$$\frac{1 \longrightarrow O\alpha \oplus (\varphi \multimap O\beta)}{1 \longrightarrow \varphi \multimap (O\alpha \oplus O\beta)}\ (\mathsf{Gb}_f)$$

From this derivation, it follows that when one has either an unconditional obligation or a conditional obligation that holds under context φ, then the unconditional obligation can be transformed into a conditional obligation that is the member of a disjunction that holds under that same context. We propose to label this as the *conditionalization paradox.* Though one might argue that this is not undesirable, for if $O\alpha$ is unconditional, then we might expect that it also holds under all contexts, there are further problems with these logical principles. Consider the following example taken from the Canadian legislation.

Example 3.1 During the absence or incapacity to act of the chair (φ), his or her powers and duties shall be exercised and performed by the vice-chair ($O\alpha$) or, if he or she should also be absent or unable to act (ψ), they should be performed by some other member designated by resolution of the commission ($O\beta$).[5]

Formally, example 3.1 is translated by $1 \longrightarrow \varphi \multimap (O\alpha \oplus (\psi \multimap O\beta))$ in $\mathcal{CNR}$.[6] That is, under context φ, either $O\alpha$ holds or, if in addition we are under context ψ, $O\beta$ holds. There is a problem, however, insofar as adding the rule (Gb_f), used in combination with (cl), would allow us to derive $1 \longrightarrow (\varphi \otimes \psi) \multimap (O\alpha \oplus O\beta)$. In words, under the context where both the chair and the vice-chair are unable to perform their duties, either the vice-chair should perform them or some other member designated by the commission should. This conclusion is undesirable considering that it is not faithful to the intended meaning of the regulation. The regulation specifies that under the context $\varphi \otimes \psi$, where both the chair and the vice-chair are unable to perform their duties, then it is $O\beta$ that holds, not '$O\beta$ or $O\alpha$'.

At this stage, one might argue that the problem only comes from a bad translation from natural language to $\mathcal{CNR}$. For instance, one might argue that the meaning of the regulation is rather $1 \longrightarrow (\varphi \multimap O\alpha) \oplus ((\varphi \otimes \psi) \multimap O\beta)$, that is, either $O\alpha$ holds under context φ or $O\beta$ holds under context $\varphi \otimes \psi$. However, even with such a translation, we would still be able to derive $1 \longrightarrow (\varphi \otimes \psi) \multimap ((\varphi \multimap O\alpha) \oplus O\beta)$ and, again, this would not be faithful to the intended meaning of the regulation. Indeed, the conditional obligation of the vice-chair ($\varphi \multimap O\alpha$) is not supposed to hold under the context $\varphi \otimes \psi$. As such, we should not be able to conditionalize obligations and (Gb_f) should be rejected.

The undesirable character of (**wd**), (CUT_f) and (Gb_f) can be emphasized if we consider the relationship between these principles and ($\oplus 1$). When negation is defined through 0, these principles imply ($\oplus 1$), which in turn enables the derivation of the following inference schema (τ, ρ and σ stand for arbitrary formulas).

$$\frac{\sigma \longrightarrow (\tau \multimap O\alpha) \oplus \rho}{\sigma \longrightarrow (\neg\rho \otimes \tau) \multimap O\alpha}\ (w\text{-}aug)$$

We propose to label this as the paradox of *weak augmentation.* It is a form of augmentation that comes from weak distributivity. The problem

[5] *The Securities Act*, Consolidated Statutes of Manitoba, c S50.

[6] For an analysis of how $\mathcal{CNR}$ deals with factual detachment as well as an analysis of contraposition of deontic conditionals, see [22].

comes from the fact that ($\oplus$1) implies that we can strengthen the antecedent of a deontic conditional when a disjunctive context is assumed. Strengthening the antecedent of a deontic conditional, however, is a problem that needs to be avoided if one aims to model conditional normative reasoning. To see why this yields paradoxical results, consider the following derivation. From $1 \longrightarrow \varphi \multimap (O\alpha \oplus (\psi \multimap O\beta))$ as well as (*w-aug*) and (cl), we can derive $1 \otimes \varphi \longrightarrow (\neg O\alpha \otimes \psi) \multimap O\beta)$, from which we can derive $1 \longrightarrow (\varphi \otimes \neg(\psi \multimap O\alpha)) \multimap O\beta$ using (cl), associativity, commutativity and the equivalence between $\neg(\psi \multimap O\alpha)$ and $\psi \otimes \neg O\alpha$.

Again, this last derivation does not reflect the intended meaning of the regulation in example 3.1. First, note that $\neg(\psi \multimap O\alpha)$ means that it is not the case that the vice-chair has a conditional obligation to perform the chair's duties when the vice-chair is unable to do so. Given the meaning of the regulation, this is always true and it can always be assumed: it is false that, in a context where the vice-chair is unable to perform the chair's duties, the vice-chair has an obligation to do so. Hence, it is false that some other member should perform the chair's duties in a situation where i) the chair is unable to do so and ii) it is not the case that the vice-chair has a conditional obligation to perform the chair's duties when the vice-chair is unable to do so. Formally, it is not $O\beta$ that holds under the context $\varphi \otimes \neg(\psi \multimap O\alpha)$. Rather, $O\alpha$ holds under that context, unless ψ is also the case. Accordingly, (*w-aug*) allows us to derive formal consequences that, from the perspective of natural normative language, should not be derivable.

4 Unconditional conditional obligations

We mentioned at the beginning of section 3 that, given an unconditional obligation that is part of a disjunction with a conditional obligation, it is always possible to conditionalize the unconditional obligation via (Gb_f). Though it might be argued that it is not necessarily undesirable, given that unconditional obligations are assumed to hold across contexts, we can see that the inverse of this phenomenon, which is derivable through (Ga_f), is more problematic.

$$\frac{1 \longrightarrow \varphi \multimap (O\alpha \oplus O\beta)}{1 \longrightarrow O\alpha \oplus (\varphi \multimap O\beta)}\ (\mathsf{Ga}_f)$$

Indeed, given either $\mathcal{CNR}$(a) or $\mathcal{CNR}$(ab), one can take a conditional obligation that is part of a disjunction out of the context and transform it into an *un*conditional obligation. That is, one can unconditionalize conditional obligations when they are part of a disjunction. This yields another new paradox.

Example 4.1 Every passenger shall, when required to do so (φ), either deliver his ticket ($O\alpha$) or pay the fare for his passage ($O\beta$).[7]

Formally, this regulation is translated by $1 \longrightarrow \varphi \multimap (O\alpha \oplus O\beta)$. From ($\mathsf{Ga}_f$), we obtain that either the obligation for any passenger to deliver his ticket

[7] Adapted from the *Grand River Railway Traffic Rules and Regulations*, paragraph 3, C.R.C., c. 1379 (Canada Transportation Act).

holds unconditionally or there is a conditional obligation to pay the fare for his passage when required to do so. This, however, does not reflect the intended meaning of the regulation insofar as $O\alpha$ is not meant to hold unconditionally: One is only expected to deliver one's ticket under context φ.[8]

The rule (Ga_f) is problematic given that $\multimap$ is used to model conditional obligations. This rule allows taking a conditional obligation out of its context and considering it as an unconditional obligation when it is part of a disjunction. Taken literally, this means that (Ga_f) implies that if there is a disjunction of obligations conditional to some context φ, then one can take out of that context one of the two obligations and consider it as an unconditional obligation. We propose to label this as the *unconditionalization paradox.*

5 Disjunctive syllogism

The previous considerations suggest that conditional normative reasoning should not be modeled using $\mathcal{CNR}$(b) or $\mathcal{CNR}$(ab). This, however, leads us to a new dilemma. Consider disjunctive syllogism. Disjunctive syllogism, it has been argued, is a criterion that should be satisfied by logics that aim to model conditional normative reasoning. It amounts to the inference of φ from the assumption that either φ or ψ is the case together with $\neg\psi$. Formally, it is represented by (**ds**).

$$(\varphi \oplus \psi) \otimes \neg\psi \longrightarrow \varphi \qquad \textbf{(ds)}$$

Goble [8] argued that disjunctive syllogism is an intuitive inference pattern. The relevance of disjunctive syllogism with respect to normative reasoning was highlighted by Horty [13] via the Smith argument, the satisfaction of which is taken by Goble [9] to be a criterion that any logic that aims at modeling conditional normative reasoning should satisfy. The Smith argument is formulated as follows:

1.	Smith ought to fight in the army or else perform alternative service to his country.
2.	Smith ought to not fight in the army (e.g. because of his religion).
∴	Smith ought to perform alternative service to his country.

In $\mathcal{CNR}$, the Smith argument is formulated as follows:

1.	$O\alpha \oplus O\beta$
2.	$O\alpha^*$
∴	$O\beta$

[8] One might argue that $\mathcal{CNR}$ (precisely $\mathcal{AL}$) is not faithful to natural language given that it does not allow disjunctive actions. This would be misleading. We argued elsewhere (cf. [23,20]) that there are not such things as disjunctive actions. The rationale behind that conception is that one cannot *perform* a disjunctive action. One can face a choice and one can choose, but one does not *perform* an act 'choice' represented by a disjunctive action. A disjunctive action or, to be precise, a disjunction of actions, only makes sense if actions are understood as declarative sentences. $\mathcal{CNR}$, however, deals with *ought-to-do* rather than *ought-to-be* operators.

Given the logical rules used to model the behavior of the action formulas within the scope of the deontic operators, we have $O\alpha \longrightarrow O\alpha^{**}$. Hence, given that an analogue to contraposition is also satisfied, we also have $\neg O\alpha^{**} \longrightarrow \neg O\alpha$, which by (**D**), (**P**) and (cut) yield $O\alpha^{*} \longrightarrow \neg O\alpha$. Consequently, the Smith argument can be established via disjunctive syllogism [9]:

$$\dfrac{\dfrac{\dfrac{}{O\alpha \oplus O\beta \longrightarrow O\alpha \oplus O\beta}(1) \quad \dfrac{\vdots}{O\alpha^{*} \longrightarrow \neg O\alpha}}{(O\alpha \oplus O\beta) \otimes O\alpha^{*} \longrightarrow (O\alpha \oplus O\beta) \otimes \neg O\alpha}(\mathrm{t})}{(O\alpha \oplus O\beta) \otimes O\alpha^{*} \longrightarrow O\beta}(\mathbf{ds}, \mathrm{cut})$$

Accordingly, if disjunctive syllogism is available, then the Smith argument can be modeled within $\mathcal{CNR}$ and Goble's criterion is satisfied. A dilemma arises, however, when we consider the fact that (**ds**) is actually logically equivalent to ($\oplus$1). This equivalence is straightforward from (cl).

$$\dfrac{\dfrac{}{\varphi \oplus \psi \longrightarrow \neg\psi \multimap \varphi}(\oplus 1)}{(\varphi \oplus \psi) \otimes \neg\psi \longrightarrow \varphi}(\mathrm{cl}) \qquad \dfrac{\dfrac{}{(\varphi \oplus \psi) \otimes \neg\psi \longrightarrow \varphi}(\mathbf{ds})}{\varphi \oplus \psi \longrightarrow \neg\psi \multimap \varphi}(\mathrm{cl})$$

Thus the dilemma: The disjunctive syllogism, which is an intuitively appealing principle, comes with the paradox of weak augmentation. Hence, there is a choice that needs to be made between the satisfaction of disjunctive syllogism and the paradox of weak augmentation. That is, either we model conditional normative reasoning within $\mathcal{CNR}$(b) or $\mathcal{CNR}$(ab), in which case disjunctive syllogism is satisfied but weak augmentation is derivable, or we model it within $\mathcal{CNR}$ or $\mathcal{CNR}$(a), in which case neither weak augmentation nor disjunctive syllogism is derivable. As a result, there is a trade-off to be made between deductive power and too many paradoxes. In what follows, we propose to sacrifice disjunctive syllogism in order to avoid the paradoxes that are derivable in $\mathcal{CNR}$(b) and $\mathcal{CNR}$(ab). We will provide a justification to discard disjunctive syllogism in section 8.

6 More deductive power without paradoxes

In light of the aforementioned considerations, it would appear that, if one wants to avoid the conditionalization paradox, the paradox of weak augmentation and the unconditionalization paradox, the only choice that is left to model conditional normative reasoning is $\mathcal{CNR}$, as it was initially defined in [20,22]. While weak augmentation is related to ($\oplus$1), which is logically equivalent to (**dm1**) and (**dm3**), the unconditionalization paradox is a consequence of (Ga_f), which is known to imply ($\oplus$2) within $\mathcal{CNR}$, which in turn is logically equivalent to (**dm2**) and (**dm4**) and implies (**lem**). Though rejecting ($\oplus$1) implies the rejection of disjunctive syllogism and some of De Morgan's dualities, one might be tempted to pay that price to avoid the paradox of weak augmentation. With that being said, rejecting all of De Morgan's dualities together with the law of

[9] The notation (**ds**, cut), for example, is used as a notational convention to shorten the notation. It means that we have another branch in the proof-tree with the proper instance of (**ds**) which is used with (cut).

excluded middle in order to avoid the unconditionalization paradox seems to be a price that is a bit excessive to pay.

Fortunately, the framework of monoidal logics provides us with a solution that enables us to keep (**dm2**), (**dm4**) and (**lem**) while avoiding (Ga_f), (Gb_f) and ($\oplus$1).

Definition 6.1 $\mathcal{CNR}_2$ is defined from $\mathcal{CNR}$, ($\oplus$2) and (**lem**).

We adopt the following definition of validity.

Definition 6.2 Let $\mathcal{D}$ be a deductive system. Given a model $\mathcal{M} = \langle \mathbf{S}, v \rangle$ constructed from a pre-ordered set $\mathbf{S} = \langle S, \leq \rangle$ together with a valuation mapping $v : \mathcal{D} \longrightarrow \mathbf{S}$, a proof $\varphi \longrightarrow \psi$ is *valid* with respect to a class $\mathbb{M}$ of models if and only if $v(\varphi) \leq v(\psi)$ for all $\mathcal{M}$.[10]

Theorem 6.3 ([24]) *Neither (**wd**), ($\oplus$1) nor (Ga_f) is derivable within $\mathcal{CNR}_2$.*

Proof. *Let $\mathbf{S} = \langle \{0,1,2,3\}, \leq, \cdot, /, + \rangle$ be such that $\leq$ is a pre-order relation with $0 \leq 1 \leq 2 \leq 3$. The operations on $\mathbf{S}$ are defined by the following tables.*

$\cdot$	0	1	2	3
0	0	0	0	0
1	0	1	2	3
2	0	2	3	3
3	0	3	3	3

$+$	0	1	2	3
0	0	1	2	3
1	1	2	3	3
2	2	3	3	3
3	3	3	3	3

$/$	0	1	2	3
0	3	0	0	0
1	3	1	0	0
2	3	2	1	0
3	3	3	3	3

Define a model $\mathcal{M} = \langle \mathbf{S}, v \rangle$ through a mapping $v : \mathcal{CNR}_2 \longrightarrow \mathbf{S}$ such that:

$$
\begin{aligned}
v(\bot) &= 2 \\
v(1) &= 1 \\
v(0) &= 0 \\
v(\varphi \otimes \psi) &= v(\varphi) \cdot v(\psi) \\
v(\varphi \multimap \psi) &= \frac{v(\psi)}{v(\varphi)} \\
v(\varphi \oplus \psi) &= v(\varphi) + v(\psi)
\end{aligned}
$$

*It is easy to verify using truth tables that the rules and axiom schemas of $\mathcal{CNR}_2$ preserve validity. Hence, if a proof is derivable within $\mathcal{CNR}_2$, then it is valid. However, neither (**wd**), ($\oplus$1) nor (Ga_f) is valid.*

(i) *For (**wd**), take $v(\varphi) = 2$, $v(\psi) = 0$ and $v(\rho) = 1$. Then, $v(\varphi \otimes (\psi \oplus \rho)) = 2 \cdot (0 + 1) = 2$ and $v((\varphi \otimes \psi) \oplus \rho) = (2 \cdot 0) + 1 = 1$, but $2 \not\leq 1$.*

(ii) *For ($\oplus$1), take $v(\varphi) = 1$ and $v(\psi) = 1$. Then, $v(\varphi \oplus \psi) = 1 + 1 = 2$, $v(\neg\psi \multimap \varphi) = \frac{v(\varphi)}{v(\neg\psi)} = \frac{1}{2} = 0$ and $2 \not\leq 0$.*

(iii) *For (Ga_f), take $v(\varphi) = 2$, $v(\psi) = 1$, $v(\rho) = 2$ and $v(\tau) = 3$. Then, $v(\tau) \leq v(\rho \multimap (\varphi \oplus \psi))$ since $v(\rho \multimap (\varphi \oplus \psi)) = \frac{v(\varphi \oplus \psi)}{v(\rho)} = \frac{2+1}{2} = \frac{3}{2} = 3$.*

[10] Note that v takes arrows (not formulas) from $\mathcal{D}$ as arguments.

However, $v(\tau) \nleq v(\varphi \oplus (\rho \multimap \psi))$ *since* $v(\varphi \oplus (\rho \multimap \psi)) = 2 + \frac{1}{2} = 2 + 0 = 2$ *and* $3 \nleq 2$.

□

7 Is this relevant to a logic of norms?

As a result of our analysis, we obtain that $\mathcal{CNR}_2$ is stronger than $\mathcal{CNR}$ but weaker than $\mathcal{CNR}$(a), $\mathcal{CNR}$(b) and $\mathcal{CNR}$(ab). In comparison with Lambek's analysis of substructural logics, we obtain an intermediate system stronger than BL1 but weaker than BL1(a), BL1(b) and BL2. With that being said, $\mathcal{CNR}_2$ is meant to be a logic that can model norms and conditional reasoning. However, one might still object that $\mathcal{CNR}_2$ lacks some deductive power. For instance, one might expect that some principle allowing to copy information, such as (Δ), might be a desirable property for a logic of norms. Indeed, once one knows that some norm or some information φ holds, it does not matter whether or not one iterates that piece of information. Consequently, one might wonder whether or not a principle such as (Δ) can be added to $\mathcal{CNR}_2$ without allowing the derivation of the paradoxes it is meant to avoid.

$$\varphi \longrightarrow \varphi \otimes \varphi \tag{Δ}$$

Definition 7.1 $\mathcal{CNR}_{2\Delta}$ is defined by adding (Δ) to $\mathcal{CNR}_2$.

Theorem 7.2 *Neither (*wd*), (⊕1) nor (*Ga_f*) is derivable within* $\mathcal{CNR}_{2\Delta}$.

Proof. *It suffices to note that* $\mathcal{M}$, *as defined in theorem 6.3, is a model of* $\mathcal{CNR}_{2\Delta}$.

□

One objection against the satisfaction of (Δ) might be that some obligations, when fulfilled, are discharged (e.g. [3]). For instance, if Peter gets a ticket for speeding on the highway, then he ought to pay for the ticket, but this obligation is discharged once Peter has paid for the ticket. In the case of dischargeable obligations, it would thus seem that (Δ) fails. In a sense, it is not because a dischargeable obligation holds once that it holds twice. If Peter owes, say, 150$ for the ticket, then it is not the case that he owes 150$ *and* 150$.

In the context of $\mathcal{CNR}_{2\Delta}$, however, this objection can easily be dealt with. First, to 'pay for the ticket' is an action α, and $\mathcal{AL}$ does not satisfy an axiom schema comparable to (Δ). Accordingly, one cannot duplicate actions within $\mathcal{CNR}_{2\Delta}$. More importantly, a dischargeable obligation is actually a conditional obligation. Looking at the aforementioned example, the obligation to pay for the ticket ($O\alpha$) holds under a context where Peter did not already pay for it ($\neg\varphi$). As such, (Δ) does not allow to duplicate $O\alpha$, which would be arguable, but rather allows to duplicate $\neg\varphi \multimap O\alpha$. This last duplication is perfectly acceptable. If the norm 'Peter ought to pay for the ticket if he did not already pay for it' holds, then we can duplicate this information without obtaining that Peter ought to pay for the ticket *twice*. Consequently, though (Δ) is not acceptable for a logic of *actions*, it is perfectly desirable for a logic of *norms*.

Besides Lambek's analysis of bilinear logics, there are two other well-known substructural logics, namely BCK and relevant logic (cf. [26]). While BCK is characterized by the satisfaction of weakening (K), relevant logic satisfies contraction (W).

$$\frac{\Gamma \vdash \Delta}{\Gamma; \Sigma \vdash \Delta}\ (\mathrm{K}) \qquad \frac{(\Gamma; \Sigma); \Sigma \vdash \Delta}{\Gamma; \Sigma \vdash \Delta}\ (\mathrm{W})$$

These rules can be translated in the language of $\mathcal{CNR}$ as follows.

$$\frac{\varphi \longrightarrow \tau}{\varphi \otimes \psi \longrightarrow \tau}\ (\mathrm{K}_f) \qquad \frac{(\varphi \otimes \psi) \otimes \psi \longrightarrow \tau}{\varphi \otimes \psi \longrightarrow \tau}\ (\mathrm{W}_f)$$

In [22], Peterson argued that the rule ($\otimes$-out) was the real culprit for the problems of augmentation, (factual) detachment of deontic conditionals and deontic explosion.

$$\frac{\varphi \longrightarrow \psi \otimes \rho}{\varphi \longrightarrow \psi}\ (\otimes\text{-out}) \qquad \frac{\varphi \longrightarrow \psi \otimes \rho}{\varphi \longrightarrow \rho}\ (\otimes\text{-out})$$

As it turns out, (K_f) can be proven to be logically equivalent to ($\otimes$-out) in $\mathcal{CNR}$, whereas (W_f) is logically equivalent to (Δ). Hence, while it can be argued that conditional normative reasoning should not be modeled in logical systems comparable to BCK, we obtain that $\mathcal{CNR}_{2\Delta}$ is comparable to a fragment of (multiplicative and classical) relevant logic.

8 Closing remarks

For the sake of the comparison with the literature, let us adopt the following definitions.

Definition 8.1 $\mathcal{CNR}_{Rel}$ is defined by adding (W_f) to $\mathcal{CNR}$(ab).

Definition 8.2 $\mathcal{CNR}_{BCK}$ is defined by adding (K_f) to $\mathcal{CNR}$(ab) .

Definition 8.3 $\mathcal{CNR}_{Mon}$ is defined by adding (W_f) and (K_f) to $\mathcal{CNR}$(ab).

Table 1 summarizes the logical principles satisfied by the different deductive systems presented within this paper as well as the problems and paradoxes that are derivable. The notation $\Leftrightarrow$ indicates logically equivalent principles and a $\boxed{\checkmark}$ indicates the logical principle responsible for the derivation of the problem. A $\surd$ simply indicates that the logical principle is satisfied. As we can see, $\mathcal{CNR}_2$ and $\mathcal{CNR}_{2\Delta}$ are better suited to model conditional normative reasoning when conditional obligations are represented via $\multimap$. In addition to the problems and paradoxes that were mentioned within this paper, it should be emphasized that $\mathcal{CNR}_2$ and $\mathcal{CNR}_{2\Delta}$ also avoid Ross's [27] and Prior's [25] paradoxes (cf. [21]). Accordingly, $\mathcal{CNR}_2$ and $\mathcal{CNR}_{2\Delta}$ strike the right balance between enough deductive power and too many paradoxes, although this comes at the cost of disjunctive syllogism.

From the perspective of substructural logics, the interest of $\mathcal{CNR}_2$ and $\mathcal{CNR}_{2\Delta}$ is that they are intermediate substructural logics. In addition to the fact that $\mathcal{CNR}_2$ falls outside of the scope of Lambek's [17] analysis by being stronger than BL1 but weaker than BL1(a), BL1(b) and BL2, $\mathcal{CNR}_{2\Delta}$ is also

	$\mathcal{CNR}$	$\mathcal{CNR}_2$	$\mathcal{CNR}_{2\Delta}$	$\mathcal{CNR}(a)$	$\mathcal{CNR}(b)$	$\mathcal{CNR}(ab)$	$\mathcal{CNR}_{Rel}$	$\mathcal{CNR}_{BCK}$	$\mathcal{CNR}_{Mon}$
$(\mathsf{CUT}_f)\Leftrightarrow(\mathbf{wd})\Leftrightarrow(\mathsf{Gb}_f)$					$\boxed{\surd}$	$\surd$	$\surd$	$\surd$	$\surd$
(Ga_f)				$\boxed{\surd}$		$\surd$	$\surd$	$\surd$	$\surd$
$(\oplus 1)\Leftrightarrow(\mathbf{ds})\Leftrightarrow(\mathbf{dm1})\Leftrightarrow(\mathbf{dm3})$					$\boxed{\surd}$	$\surd$	$\surd$	$\surd$	$\surd$
$(\oplus 2)\Leftrightarrow(\mathbf{dm2})\Leftrightarrow(\mathbf{dm4})$		$\surd$	$\surd$	$\surd$		$\surd$	$\surd$	$\surd$	$\surd$
(**lem**)		$\surd$	$\surd$	$\surd$		$\surd$	$\surd$	$\surd$	$\surd$
$(\Delta)\Leftrightarrow(\mathrm{W}_f)$			$\surd$				$\surd$		$\surd$
$(\otimes\text{-out})\Leftrightarrow(\mathrm{K}_f)$								$\boxed{\surd}$	$\surd$
Problems:									
Prior								$\surd$	$\surd$
Ross								$\surd$	$\surd$
Factual detachment								$\surd$	$\surd$
Augmentation								$\surd$	$\surd$
Deontic explosion								$\surd$	$\surd$
Weak augmentation					$\surd$	$\surd$	$\surd$	$\surd$	$\surd$
Conditionalization					$\surd$	$\surd$	$\surd$	$\surd$	$\surd$
Unconditionalization				$\surd$		$\surd$	$\surd$	$\surd$	$\surd$
Analogue:	BL1			BL1(a)	BL1(b)	BL2	Relevant logic	BCK	Monotonic *D*-system

Table 1
Systems for conditional normative reasoning

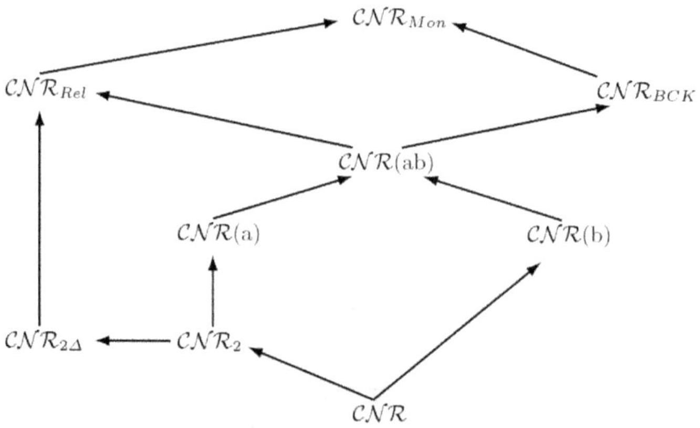

Fig. 1. Relationships between $\mathcal{CNR}$-systems

weaker than (multiplicative and classical) relevant logic (which is an extension of BL2). Figure 1 summarizes the relationships between the logical systems presented within this paper. To conclude, let us mention that our analysis provides us with a reason to discard disjunctive syllogism insofar as adding (**ds**) to $\mathcal{CNR}_2$ or $\mathcal{CNR}_{2\Delta}$ would yield $\mathcal{CNR}$(ab) or $\mathcal{CNR}_{Rel}$, which would allow the recovery of all the paradoxes these systems are meant to avoid.

References

[1] Alchourrón, C., *Detachment and defeasibility in deontic logic*, Studia Logica **57** (1996), pp. 5–18.

[2] Anglberger, A. J. J., N. Gratzl and O. Roy, *Obligation, free choice and the logic of weakest permissions*, The Review of Symbolic Logic **8** (2015), pp. 807–827.

[3] Brown, M. A., *Doing as we ought: Towards a logic of simply dischargeable obligations*, in: M. Brown and J. Carmo, editors, *Deontic Logic, Agency and Normative Systems*, Springer, 1996 pp. 47–65.

[4] Chellas, B. F., "Modal logic: An introduction," Cambridge University Press, 1980.

[5] Chisholm, R., *Contrary-to-duty imperatives and deontic logic*, Analysis **24** (1963), pp. 33–36.

[6] Dunn, J. M., *Star and perp: Two treatments of negation*, Philosophical Perspectives **7** (1993), pp. 331–357.

[7] Girard, J.-Y., *Linear Logic*, Theoretical Computer Science **50** (1987), pp. 1–102.

[8] Goble, L., *A proposal for dealing with deontic dilemmas*, in: A. Lomuscio and D. Nute, editors, *DEON 2004*, Lecture Notes in Computer Science **3065**, Springer, 2004 pp. 74–113.

[9] Goble, L., *Normative conflicts and the logic of 'ought'*, Nos **43** (2009), pp. 450–489.

[10] Gor, R., *Substructural logics on display*, Logic Journal of the IGPL **6** (1998), pp. 451–504.

[11] Grishin, V. N., *On a generalization of the Ajdukiewicz-Lambek system*, in: *Studies in Non-classical Logics and Formal Systems*, Nauka, Moscow, 1983 pp. 315–343, translated by Anna Chernilovskaya in *Symmetric Categorial Grammar*, R. Bernardi and M. Moortgat (eds), 2007.

[12] Grossi, D. and A. J. I. Jones, *Constitutive norms and counts-as conditionals*, in: D. Gabbay, J. Horty, X. Parent, R. van der Meyden and L. van der Torre, editors, *Handbook of Deontic Logic and Normative Systems*, College Publications, 2013 pp. 407–441.

[13] Horty, J., *Non-monotonic foundations for deontic logic*, in: D. Nute, editor, *Defeasible Deontic Logic*, Kluwer Academic Publishers, 1997 pp. 17–44.

[14] Hyland, M. and V. De Paiva, *Full intuitionistic linear logic (extended abstract)*, Annals of Pure and Applied Logic **64** (1993), pp. 273–291.

[15] Jones, A. J. I., *On the logic of deontic conditionals*, Ratio Juris **4** (1991), pp. 355–366.

[16] Lambek, J., *The mathematics of sentence structure*, The American Mathematical Monthly **65** (1958), pp. 154–170.

[17] Lambek, J., *From categorial grammar to bilinear logic*, in: P. Schroeder-Heister and K. Došen, editors, *Substructural Logics*, Studies in Logic and Computation **2**, Oxford Science Publications, 1993 pp. 207–237.

[18] Mac Lane, S., "Categories for the working mathematician," Springer, 1971, 2nd edition.

[19] Ono, H., *Semantics for substructural logics*, in: P. Schroeder-Heister and K. Došen, editors, *Substructural Logics*, Studies in Logic and Computation **2**, Oxford Science Publications, 1993 pp. 259–291.

[20] Peterson, C., *The categorical imperative: Category theory as a foundation for deontic logic*, Journal of Applied Logic **12** (2014), pp. 417–461.

[21] Peterson, C., *Monoidal logics: How to avoid paradoxes*, , **1315**, CEUR Workshop Proceedings, 2014 pp. 122–133.

[22] Peterson, C., *Contrary-to-duty reasoning: A categorical approach*, Logica Universalis **9** (2015), pp. 47–92.

[23] Peterson, C., *A logic for human actions*, in: G. Payette and R. Urbaniak, editors, *Applications of Formal Philosophy*, Springer, 2015 To appear.

[24] Peterson, C., *A comparison between monoidal and substructural logics*, Journal of Applied Non-Classical Logics (2016), in press (doi: 10.1080/11663081.2016.1179528).

[25] Prior, A., *The paradoxes of derived obligation*, Mind **63** (1954), pp. 64–65.

[26] Restall, G., "An introduction to substructural logics," Routledge, 2000.

[27] Ross, A., *Imperatives and logic*, Theoria **7** (1941), pp. 53–71.

[28] Trypuz, R. and P. Kulicki, *On deontic action logics based on Boolean algebra*, Journal of Logic and Computation **25** (2015), pp. 1241–1260.

[29] van der Torre, L. and Y.-H. Tan, *The many faces of defeasibility in defeasible deontic logic*, in: D. Nute, editor, *Defeasible Deontic Logic*, Kluwer Academic Publishers, 1997 pp. 79–121.

[30] van Eck, J., *A system of temporally relative modal and deontic predicate logic and it's philosophical applications*, Logique et Analyse **25** (1982), pp. 249–290.
[31] von Wright, G. H., *Ought to be–ought to do*, in: G. Meggle, editor, *Actions, Norms, Values*, Walter de Gruyter, 1999 pp. 3–10.

Update Semantics for Weak Necessity Modals

Alex Silk

University of Birmingham
a.silk@bham.ac.uk

Abstract

This paper develops an update semantics for weak necessity modals ('ought', 'should'). I start with the basic approach to the weak/strong necessity modal distinction developed in Silk 2012b: Strong necessity modals are given their familiar semantics of necessity, predicating the necessity of the prejacent of the evaluation world. The apparent "weakness" of weak necessity modals results from their bracketing whether the necessity of the prejacent is verified in the actual world. I formalize these ideas within an Update with Centering framework. The meaning of 'Should ϕ' is explained, fundamentally, in terms of how its use updates attention toward possibilities in which ϕ is necessary. The semantics is also extended to deontic conditionals. The proposed analyses capture various contrasting discourse properties of 'should' and 'must'—e.g. in context-sensitivity, entailingness, and force—and provide an improved treatment of largely neglected data concerning information-sensitivity.

Keywords: Modals, weak/strong necessity modals, update semantics.

This paper develops an update semantics for weak necessity modals ('ought', 'should'). I start with a basic approach to the weak/strong necessity modal distinction developed in Silk 2012b (§1). The central idea, on this view, is that the apparent "weakness" of weak necessity modals results from their bracketing whether the necessity of the prejacent is verified in the actual world. Weak necessity modals afford a means of entertaining and planning for hypothetical extensions or minimal revisions of the current context in which relevant considerations (norms, expectations) apply, without needing to settle that those considerations actually do apply. I formalize these ideas within an Update with Centering framework (§2). The conventional meaning of 'Should ϕ' is explained, fundamentally, not in terms of truth-conditions, but in terms of how its use updates attention toward possibilities in which ϕ is necessary. The semantics is also extended to deontic conditionals. The proposed analyses capture various contrasting discourse properties of 'should' and 'must'—e.g. in context-sensitivity, entailingness, and force—and provides an improved treatment of largely neglected data concerning information-sensitivity. I close by considering several alternative static and dynamic implementations (§3).

A word on terminology: Following common practice, I label modals such as 'should'/'ought' "weak necessity" modals, and modals such as 'must'/'have to' "strong necessity" modals. The expressions in each family pattern with one another in linguistically distinctive ways. In invoking these labels I am not assuming that uses of the former modals invariably convey a weaker conversational force, nor am I assuming a particular type of theoretical analysis. For instance, I am not assuming that the modals express different "kinds" of necessity, that they have a structurally analogous semantics, that they comprise a scale of quantificational strength, or even that they stand in an entailment relation. Indeed we will see reasons for questioning each of these claims.

1 Weak and strong necessity modals in context

This section presents the basic approach to the weak/strong necessity modal distinction which I will assume in this paper, as developed in Silk 2012b. I won't be able to fully defend the approach here. The aim is simply to present certain core data to motivate the paper's primary constructive project in §2.

There is robust evidence supporting a distinction in strength between modals such as 'should' and 'ought', on the one hand, and 'must', 'have to', and '(have) got to', on the other.[1] For instance, even holding the readings of the modals fixed, 'Should ϕ' can be followed by 'Must ϕ', but not vice versa, as reflected in (1). Similarly, (2a) is consistent in a way that (2b) is not.

(1) a. I should help the poor. In fact, I must.
b. I must help the poor. #In fact, I should.

(2) a. I should help the poor, but I don't have to.
b. #I must help the poor, but it's not as if I should.

There are also important conversational differences. The relative felicity of 'should' and 'must' depends on standing assumptions in the context. It is this feature of weak and strong necessity modals that I focus on here.

Start with an epistemic case. Suppose we are working on an art project, and I ask where the colored pencils are. Normally you put them in the drawer, but sometimes you accidentally leave them on the shelf. In this context it is more appropriate for you to use 'should' in responding to my question, as in (3).

(3) *Me:* Do you know where the colored pencils are?
You: They should be in the drawer with the crayons.

Suppose, alternatively, that we are looking for the colored pencils together, and you indicate that you have seen something that leads you to conclude that

[1] See, e.g., Sloman 1970; Horn 1972; Wertheimer 1972; Lyons 1977; Woisetschlaeger 1977; Williams 1981; Coates 1983; McNamara 1990; Palmer 1990, 2001; Huddleston & Pullum 2002; von Fintel & Iatridou 2008; Rubinstein 2012; Silk 2012b. I use 'should' as my representative weak necessity modal, and 'must' as my representative strong necessity modal. These modals are typically used "subjectively" (Lyons 1977, 1995), in the sense that they typically present the speaker as endorsing the considerations with respect to which the modal is interpreted; non-endorsing uses (more common with e.g. 'have to', 'supposed to') introduce complications that would be distracting to our discussion here. See Silk 2012b for discussion.

they are in the drawer. Perhaps you noticed that they weren't on the shelf, and this is the only other place you think they could be. In this context it is more natural for you to use 'must', as in (4).

(4) *Me:* Do you know where the colored pencils are?
You: They must be in the drawer with the crayons.

Its following from our evidence (knowledge, information) that the colored pencils are in the drawer depends on today not being one of the atypical days when you accidentally put the colored pencils on the shelf. Using the strong necessity modal 'must' is preferred if, and only if, you know that conditions are indeed normal in this way. What is illuminating is that you can use 'should' even if you aren't in a position to judge that they are. Accepting your 'should' claim doesn't require us to presuppose that your evidence is indefeasible.

Similarly, consider a deontic case.[2] Suppose I am considering whether to fight in the Resistance or take care of my ailing mother. I mention the importance of the value of family, and you agree. But the issue is complex, and we haven't settled whether there might be more important competing values. Sensitive to this, you may find it more appropriate to express your advice that I help my mother by using 'should', as in (5).

(5) *Me:* Family is very important.
You: I agree. You should take care of your mother.

But if we settle that family is of primary importance, as in (6), it can become more natural to use 'must' and for us to accept that I have to help my mother.

(6) *Me:* Family is most important — more important than country.
You: I agree. You must take care of your mother.

My having an obligation to help my mother depends on the value of family being more important (or at least not less important[3]) in my situation than any competing value. Using 'must' is preferred if it is settled in the conversation that this condition obtains. Parallel to the epistemic case, what is illuminating is that you can felicitously express your advice that I help my mother using 'should', advice which I may accept, even if it isn't common ground that this precondition for my having a genuine obligation is satisfied. Accepting your 'should' claim needn't require us to presuppose that the value of family is more important than other potentially competing values.

These cases highlight what I regard as the fundamental difference between the class of weak necessity modals and the class of strong necessity modals. It is typical to gloss epistemic notions of necessity as concerning what follows from one's evidence (knowledge, information), and deontic notions of necessity

---

[2] See Woisetschlaeger 1977, McNamara 1990 for related examples and prescient early discussion. See Rubinstein 2012 and Silk 2012a,b for extensive recent discussion.

[3] I will bracket complications concerning incomparabilities and irresolvable dilemmas.

as concerning what is obligatory.[4] In this sense accepting epistemic 'Should ϕ' needn't commit one to accepting that ϕ is epistemically necessary, and accepting deontic 'Should ϕ' needn't commit one to accepting that ϕ is deontically necessary: we can accept your epistemic 'should' claim in (3) without settling that conditions are normal in the relevant respects, and thus without accepting that our evidence actually implies that the colored pencils are in the drawer; and we can accept your deontic 'should' claim in (5) without settling that family is the most important relevant value, and thus without accepting that I have an actual obligation to help my mother. Whether 'should' or 'must' is preferred depends on context, in the sense of depending on whether one accepts (or is willing to accept) that all preconditions for the prejacent to be necessary are satisfied. If these preconditions are accepted, using 'must' is preferred. But even if they aren't, we can still use 'should'.[5] 'Should ϕ' doesn't conventionally communicate that ϕ is necessary (deontically, epistemically, etc.).[6]

In Silk 2012b I develop these points in what I call a *modal-past approach* to the weak/strong necessity modal distinction. The core of this approach is as follows. There is nothing specially "strong" about strong necessity modals. Strong necessity modals can be given their familiar semantics of necessity: 'Must ϕ' is true iff ϕ is necessary (deontically, epistemically, etc.), and uses of 'Must ϕ' predicate the necessity of ϕ of the actual world. The apparent "weakness" of weak necessity modals derives from their bracketing the assumption that the necessity of ϕ is verified in the actual world. 'Should ϕ' can be accepted without committing that ϕ follows from what the relevant considerations (norms, etc.) imply given the facts, or that the necessity of ϕ is verified throughout the set of live possibilities (the context set, or set of worlds compatible with what is accepted for purposes of conversation; Stalnaker 1978).

This feature of 'should' certainly doesn't mark the only dimension along which necessity modals differ. However, I claim that it does distinguish the *class* of weak necessity modals from the *class* of strong necessity modals. There are various ways of implementing the proposed difference in the formal semantics and pragmatics. Yet even at the present level of abstraction, we can see several respects in which the above approach to weak necessity modals differs from others in the literature. (See Silk 2012b for more detailed comparisons.)

First, many existing accounts of weak necessity modals are developed by considering a limited range of modal flavors; extensions to other readings, to the extent that they are discussed at all, are often strained (e.g., Copley 2006, Swanson 2011, Rubinstein 2012, Charlow 2013, Ridge 2014, Portner & Rubinstein 2016, Yalcin 2016). The approach in this section, by contrast, generalizes

[4] See Lyons 1977, Coates 1983, Palmer 1990, 2001, Huddleston & Pullum 2002, a.m.o.

[5] Of course 'must' may also be appropriate in certain contexts. In (5), if you can be presumed a normative authority on the issue and use 'must', I may accommodate by accepting the precondition for me to have a genuine obligation, namely that the value of family takes precedence. Such contexts notwithstanding (contexts in which the speaker lacks or doesn't want to exercise the relevant authority), 'should' will be preferred (see Silk 2012b).

[6] I will often omit this parenthetical in what follows, but it should be understood.

across readings of the modals (epistemic, deontic, etc.). Second, the approach takes seriously the effects of standing contextual assumptions on the relative felicity of 'should' and 'must'. Explanatory mechanisms for capturing this sort of context-dependence are often lacking in existing accounts (see n. 2 for notable exceptions). Third, nearly all alternative approaches (e.g. comparative possibility/probability analyses, domain restriction analyses, normality-based analyses) agree in treating weak necessity modals as predicating a distinctive *kind* of necessity, namely weak necessity, of the prejacent at the actual world (evaluation world). The intuitive differences in strength among the modals are typically analyzed in terms of asymmetric entailment: 'Must ϕ' $\models$ 'Ought ϕ' $\models$ 'May ϕ'. For instance, domain restriction accounts (e.g., Copley 2006, von Fintel & Iatridou 2008, Swanson 2011, Rubinstein 2012) maintain that accepting 'Should ϕ' requires accepting that ϕ is a necessity; what distinguishes 'should' from 'must' is the logical strength of the relevant notion of necessity. 'Should' is treated as quantifying over a subset of the set of worlds quantified over by 'must'. The present approach, by contrast, diagnoses the apparent "weakness" of weak necessity modals in terms of a failure to presuppose that the relevant worlds in which the prejacent is a necessity are candidates for actuality.

2 Update semantics for weak/strong necessity modals

This section develops one way of implementing the general approach to the weak/strong necessity modal distinction from §1 in an update semantics. The arguments in §1 are by no means conclusive (see Silk 2012b), though I hope they may suffice to motivate the positive project pursued here. I develop the semantics in an Update with Centering framework, adapting Bittner 2011. Alternatives are of course possible (cf. §3). (Hereafter I couch the discussion in terms of deontic readings, though the points generalize to other readings.)

2.1 UC$_\omega$ background

Update with Centering is a dynamic system that represents how informational (Veltman 1996) and attentional (Grosz et al. 1995) states develop in discourse. Update with *Modal* Centering, UC$_\omega$, includes typed discourse referents not only for individuals δ, but also for worlds ω and propositions Ω (sets of worlds ωt) (Bittner 2011; cf. Stone 1999). The meanings of sentences are given in terms of how they update contexts, conceived as informational-attentional states. Such states are represented with sets of sequences of discourse referents. The discourse referents in each sequence are divided between those currently in the center of attention, or *topical* ($\top$), and those currently *backgrounded* ($\bot$). The bottom sublist $\bot$ can be utilized in analyzing grammatical centering, negation, questions, and, I will suggest, modal remoteness/weakness. The discourse referents in each sublist, $\top, \bot$, are *ranked* according to their relative salience or attentional prominence. The column $||$ picks out the set of discourse referents from a given list. For instance, $\top\Omega_1$ is the most salient (leftmost) proposition in the top sublist, and $\top\omega_1||$ is the set of worlds in the most salient world column in the top sublist. (I write $\top a, \bot a$ as short for $\top a_1, \bot a_1$.) Each $\top\bot$-list, i.e. pair $\langle \top, \bot \rangle$ of sublists of discourse referents, is a semantic object of type s, though not a discourse referent. A *context* is a set of $\top\bot$-lists (type st). The *context set* is the topical proposition $\top\Omega = \top\omega||$.

In UC_ω *all* sentences are treated as introducing a possibility, or modal topic, being talked about. (I treat possibilities as propositions.) With simple indicative sentences the possibility being commented on is the context set, typically the most salient possibility in the discourse (cf. Stalnaker 1975, Iatridou 2000, Schlenker 2005). I propose the UC_ω representation of (7) in (8).

(7) Alice is generous.

(8) $^{\top}[\mathbf{x} \mid \mathbf{x} = \mathbf{Alice}]; [\mathbf{w} \mid \mathbf{generous_w}\langle \top\delta \rangle]; [\mathbf{p} \mid \mathbf{p} = \bot\omega||];$
$[\mathbf{w} \mid \mathbf{w} = \bot\omega]; [\bot\omega \in \top\omega||]; [\top\omega = \bot\omega]; {}^{\top}[\mathbf{p} \mid \mathbf{p} = \top\omega||]$

Boxes without variables, $[\ldots]$, are information updates, or tests, which eliminate sequences in the context that don't satisfy the constraint '$\ldots$'. Boxes with variables, $^{\top}[\mathbf{d} \mid \ldots\mathbf{d}\ldots]$ or $[\mathbf{d} \mid \ldots\mathbf{d}\ldots]$, are recentering updates which introduce a discourse referent satisfying '$\ldots\mathbf{d}\ldots$' into the most prominent spot in the center of attention or background, respectively. Following Murray 2014 I use the top sequence in representing the context set, and the bottom sequence for keeping track of possibilities we are considering but not yet committed to.

Suppose our model contains three worlds w_0, w_1, w_2; Alice is generous only in w_1 and w_2; and the input context c_0 consists of two $\top\bot$-lists each of which includes a discourse referent p_0 for the initial context set and some world $w_0, w_1 \in {}^{\{\}}p_0$. Output contexts for the sequence of updates in (8) are as follows, as specified in the subsequent simplified derivation.[7]

(9) $c_0 = {}^{\chi}\{\langle\langle w, p_0\rangle, \langle\rangle\rangle \mid w \in {}^{\{\}}p_0\} = {}^{\chi}\{\langle\langle w_0, p_0\rangle, \langle\rangle\rangle, \langle\langle w_1, p_0\rangle, \langle\rangle\rangle\}$

c_1	c_2	c_3	c_4
$\langle\langle a, w_0, p_0\rangle, \langle\rangle\rangle$	$\langle\langle a, w_0, p_0\rangle, \langle w_1\rangle\rangle$	$\langle\langle a, w_0, p_0\rangle, \langle q, w_1\rangle\rangle$	$\langle\langle a, w_0, p_0\rangle, \langle w_1, q, w_1\rangle\rangle$
	$\langle\langle a, w_0, p_0\rangle, \langle w_2\rangle\rangle$	$\langle\langle a, w_0, p_0\rangle, \langle q, w_2\rangle\rangle$	$\langle\langle a, w_0, p_0\rangle, \langle w_2, q, w_2\rangle\rangle$
$\langle\langle a, w_1, p_0\rangle, \langle\rangle\rangle$	$\langle\langle a, w_1, p_0\rangle, \langle w_1\rangle\rangle$	$\langle\langle a, w_1, p_0\rangle, \langle q, w_1\rangle\rangle$	$\langle\langle a, w_1, p_0\rangle, \langle w_1, q, w_1\rangle\rangle$
	$\langle\langle a, w_1, p_0\rangle, \langle w_2\rangle\rangle$	$\langle\langle a, w_1, p_0\rangle, \langle q, w_2\rangle\rangle$	$\langle\langle a, w_1, p_0\rangle, \langle w_2, q, w_2\rangle\rangle$

c_5	c_6	c_7
$\langle\langle a, w_0, p_0\rangle, \langle w_1, q, w_1\rangle\rangle$		
$\langle\langle a, w_1, p_0\rangle, \langle w_1, q, w_1\rangle\rangle$	$\langle\langle a, w_1, p_0\rangle, \langle w_1, q, w_1\rangle\rangle$	$\langle\langle p_1, a, w_1, p_0\rangle, \langle w_1, q, w_1\rangle\rangle$

$$c_0[\![{}^{\top}[\mathbf{x} \mid \mathbf{x} = \mathbf{A}]]\!]^g := [\![\lambda\mathbf{I}\lambda\mathbf{j}.\, \exists\mathbf{x}\exists\mathbf{i}(\mathbf{j} = (\mathbf{x} \,{}^{\top}\!\!\oplus\, \mathbf{i}) \wedge \mathbf{Ii} \wedge \mathbf{x} = \mathbf{A})]\!]^g(c_0)$$
$$= {}^{\chi}\{\langle\langle a, w, p_0\rangle, \langle\rangle\rangle \mid w \in {}^{\{\}}p_0 \;\&\; a = [\![\mathbf{A}]\!]\} = c_1$$

$$c_1[\![[\mathbf{w} \mid \mathbf{gen_w}\langle \top\delta\rangle]]\!]^g := [\![\lambda\mathbf{I}\lambda\mathbf{j}.\, \exists\mathbf{w}\exists\mathbf{i}(\mathbf{j} = (\mathbf{w} \,{}^{\bot}\!\!\oplus\, \mathbf{i}) \wedge \mathbf{Ii} \wedge \mathbf{gen}(\mathbf{w}, \top\delta_1\mathbf{i})]\!]^g(c_1)$$
$$= {}^{\chi}\{\langle\langle a, w, p_0\rangle, \langle w'\rangle\rangle \mid w \in {}^{\{\}}p_0 \;\&\; a = [\![\mathbf{A}]\!] \;\&\; a \in {}^{\{\}}[\![\mathbf{gen}]\!](w')\} = c_2$$

$$c_2[\![[\mathbf{p} \mid \mathbf{p} = \bot\omega||]]\!]^g := [\![\lambda\mathbf{I}\lambda\mathbf{j}.\, \exists\mathbf{p}\exists\mathbf{i}(\mathbf{j} = (\mathbf{p} \,{}^{\bot}\!\!\oplus\, \mathbf{i}) \wedge \mathbf{Ii} \wedge \mathbf{p} = \bot\omega_1\{\mathbf{I}\})]\!]^g(c_2) = c_3$$

[7] The superscript $^{\chi}$ indicates the characteristic function, and $^{\{\}}$ the characteristic set; variables $\mathbf{i}$, $\mathbf{j}$ are for $\top\bot$-lists (type s), $\mathbf{I}$ for a set of lists (type st), i.e. a context. For space purposes I refer the reader to Murray 2010, Bittner 2011, Silk 2012b: Appendix for general definitions and DRT-style abbreviations (cf. Muskens 1996, Stone 1999).

$$c_3[\![[\mathbf{w} \mid \mathbf{w} = \bot\omega]]\!]^g := [\![\lambda\mathbf{I}\lambda\mathbf{j}.\, \exists\mathbf{w}\exists\mathbf{i}(\mathbf{j} = (\mathbf{w} \,{}^{\bot}\!\oplus\, \mathbf{i}) \wedge \mathbf{Ii} \wedge \mathbf{w} = \bot\omega_{\mathbf{1}}\mathbf{i})]\!]^g(c_3) = c_4$$
$$c_4[\![[\bot\omega \in \top\omega|||]]\!]^g := [\![\lambda\mathbf{I}\lambda\mathbf{j}.\, \mathbf{Ij} \wedge \bot\omega_{\mathbf{1}}\mathbf{j} \in \top\omega_{\mathbf{1}}\{\mathbf{I}\}]\!]^g(c_4) = c_5$$
$$c_5[\![[\top\omega = \bot\omega]]\!]^g := [\![\lambda\mathbf{I}\lambda\mathbf{j}.\, \mathbf{Ij} \wedge \top\omega_{\mathbf{1}}\mathbf{j} = \bot\omega_{\mathbf{1}}\mathbf{j}]\!]^g(c_5) = c_6$$
$$c_6[\![{}^{\top}[\mathbf{p} \mid \mathbf{p} = \top\omega|||]]\!]^g := [\![\lambda\mathbf{I}\lambda\mathbf{j}.\, \exists\mathbf{p}\exists\mathbf{i}(\mathbf{j} = (\mathbf{p} \,{}^{\top}\!\oplus\, \mathbf{i}) \wedge \mathbf{Ii} \wedge \mathbf{p} = \top\omega_{\mathbf{1}}\{\mathbf{I}\})]\!]^g(c_6) = c_7$$

The first update introduces into each top sequence $\top$ an individual discourse referent a for Alice, yielding c_1. The second update introduces the worlds where Alice is generous, w_1 and w_2, into the bottom sequence, yielding c_2. The worlds added to the bottom sequence at this step needn't be in the current context set. The third update introduces a propositional discourse referent q for this set of most prominent worlds in the bottom sequence $\bot\omega|| = {}^{\{\}}q = \{w_1, w_2\}$, yielding c_3. However, the context set isn't yet restricted; the update is a pure attention update. The fourth update represents a commitment to this possibility, by reintroducing the worlds in which it is true into the bottom sequence, yielding c_4. The fifth update represents the proposal to update with the proposition that Alice is generous, by restricting the set of worlds introduced in the fourth update to the worlds in the context set, yielding c_5. The sixth update represents acceptance of the assertion, by checking for each world $\top\omega$ in the context set that it is identical to the most prominent world $\bot\omega$ in its row. The first sequence is ruled out and the context set is restricted to $\{w_1\}$, yielding c_6. The final update centers attention on the new context set by introducing into the top sequence a propositional discourse referent p_1 for it, yielding c_7.

The main features of this sequence of updates are these: First, updates 1–3 introduce the at-issue proposition that Alice is generous into the bottom sequence. This registers this possibility as being under consideration, or on the conversational table, though not yet accepted. Second, I treat all uses of declarative sentences as involving a commitment update such as update 4. Though the commitment update may seem trivial in (8), its importance will become apparent below in distinguishing 'should' and 'must'. This update is distinctive of the version of UC_ω developed here. Third, updates 5–6 occur with all assertions (cf. Murray 2014). The proposal update reflects how in assertions the worlds being talked about are typically the worlds treated as live in the discourse. The success of the assertion registers an attitude of acceptance toward the proposed possibility. Asserting (7) thus both updates information, reflected in the reduction of the context set, and updates attention, reflected in the introduction of a new modal referent as the primary topic.

2.2 Modals in root clauses

Turning to modal sentences, I follow standard ordering semantics in treating modals as contributing a preorder frame $\lesssim^?$, or function from worlds to preorders, where the resolution of ? is tied to the reading of the modal (Lewis 1973, 1981, Kratzer 1981, 1991). The "ideal" of a preordered set, written $\text{MIN}(Q, \lesssim_w)$, is the set of $\lesssim_w$-minimal elements of a modal base Q, the set of Q-worlds that aren't $\lesssim_w$-bettered by any other Q-world. For instance, $\text{MIN}(Q, \lesssim_w^d)$ is the set of worlds in Q that best satisfy the relevant norms in w.[8]

Start with (10) with the strong necessity modal 'must'.

(10) Alice must be generous.

Like with (7), the meaning of (10) is given in terms of how it updates the default modal topic, or context set. The distinctive dynamic contribution of the modal is that it *itself* introduces a topical possibility — here, the possibility that Alice is generous — and then comments on it (cf. Stone 1999, Kaufmann 2000).[9] I propose the UC_ω representation of (10) in (11). As the reader can verify, the input and output updates for (11) are as in (12). (As above, assume an input context c_0 with context set ${}^{\{\}}p_0 = \{w_0, w_1\}$. And assume a model with three worlds w_0, w_1, w_2, such that Alice is generous only in w_1 and w_2, and Alice's being generous is deontically necessary only at w_0 and w_2.)

(11) $$^{\top}[\mathbf{x} \mid \mathbf{x} = \mathbf{Alice}]; [\mathbf{w} \mid \mathbf{generous_w}\langle\top\delta\rangle]; [\mathbf{w} \mid \text{MIN}\{\top\omega||, \precsim^d_{\mathbf{w}}\} \subseteq \bot\omega||]; [\mathbf{p} \mid \mathbf{p} = \bot\omega||]; [\mathbf{w} \mid \mathbf{w} = \bot\omega]; [\bot\omega \in \top\omega||]; [\top\omega = \bot\omega]; {}^{\top}[\mathbf{p} \mid \mathbf{p} = \top\omega||]$$

(12) $$c_0 = {}^{\chi}\{\langle\langle w_0, p_0\rangle, \langle\rangle\rangle, \langle\langle w_1, p_0\rangle, \langle\rangle\rangle\} \qquad c_8 = {}^{\chi}\{\langle\langle p_1, a, w_0, p_0\rangle, \langle w_0, q, w_0, w_1\rangle\rangle, \langle\langle p_1, a, w_0, p_0\rangle, \langle w_0, q, w_0, w_2\rangle\rangle\}$$

As with (8), the first update introduces into each top sequence an individual discourse referent a for Alice, and the second update introduces into each bottom sequence the worlds where the topical individual $\top\delta$ (=Alice) is generous, i.e. w_1 and w_2. The third update reflects the modal's evaluation of this possibility $\bot\omega|| = \{w_1, w_2\}$. The update introduces into the bottom sequence the worlds $\mathbf{w}$ such that every $\precsim^d_{\mathbf{w}}$-minimal world in the topical modality (=the context set $\top\omega||$) is a world where Alice is generous, i.e. w_0 and w_2. The fourth update introduces a propositional discourse referent q for this set of worlds, $\bot\omega|| = {}^{\{\}}q = \{w_0, w_2\}$. This attentional update represents the necessity claim being put on the conversational table. The fifth update represents the speaker's commitment to this possibility, and the sixth update represents the proposal to update the context set with it. The seventh update represents the acceptance of the necessity claim. This update eliminates sequences in which w_1 is the (local) topical world, restricting the context set $\top\omega||$ to $\{w_0\}$. The final update recenters attention on the new context set by introducing a propositional discourse referent $p_1 = {}^{\chi}\{w_0\}$, yielding c_8.

Two remarks: First, as in (8), the update $[\mathbf{w} \mid \mathbf{w} = \bot\omega]$ represents a commitment to the proposition which has been placed on the conversational table, here the deontic necessity claim. Accepting (10) requires accepting that the deontic necessity of Alice's being generous is verified throughout the context set.

[8] The preorder could be determined from a premise set (Kratzerian ordering source) in the usual way: $u \precsim_{P(w)} v := \forall p \in P(w): v \in p \Rightarrow u \in p$. For simplicity I make the limit assumption (Lewis 1973: 19–20) and assume that MIN is well-defined.

[9] Interestingly, it has been argued that a principle use in the development of modals diachronically involves encouraging the hearer to "focus mentally" on the embedded proposition (Van Linden 2012: chs. 6, 8).

Successfully asserting (10) again both updates information, reflected in the reduction of the context set, and updates attention, reflected in the introduction of the modal referent p_1 as the primary modal topic. Second, with 'must' the anaphoric modal base for the relevant norms is resolved to the topical modality $\top\omega||$. This reflects the indicative presupposition that the worlds being talked about are in the context set. We will return to this.

In §1 I argued, following Silk 2012b, that weak necessity modals bracket whether the necessity claim is verified in the actual world. One way of implementing this idea in the present framework is to treat 'should' as having an ordinary semantics of necessity, but as canceling the assumption associated with indicative assertions that the speaker is committed to the at-issue proposition. I propose (14) as a first-pass UC_ω representation of (13).

(13) Alice should be generous.

(14) $^{\top}[\mathbf{x} \mid \mathbf{x} = \mathbf{Alice}]; [\mathbf{w} \mid \mathbf{generous_w}\langle\top\delta\rangle]; [\mathbf{w} \mid \textsc{min}\{?\omega||, \lesssim^d_{\mathbf{w}}\} \subseteq \bot\omega||]; [\mathbf{p} \mid \mathbf{p} = \bot\omega||]; [\mathbf{w} \mid \mathbf{w} = \top\omega]; [\bot\omega \in \top\omega||]; [\top\omega = \bot\omega]; {}^{\top}[\mathbf{p} \mid \mathbf{p} = \top\omega||]$

(15) $c_8 = {}^{\chi}\{\langle\langle p_2, a, w_0, p_0\rangle, \langle w_0, q, w_0, w_1\rangle\rangle, \langle\langle p_2, a, w_0, p_0\rangle, \langle w_0, q, w_2, w_1\rangle\rangle, \langle\langle p_2, a, w_0, p_0\rangle, \langle w_0, q, w_0, w_2\rangle\rangle, \langle\langle p_2, a, w_0, p_0\rangle, \langle w_0, q, w_2, w_2\rangle\rangle, \langle\langle p_2, a, w_1, p_0\rangle, \langle w_1, q, w_0, w_1\rangle\rangle, \langle\langle p_2, a, w_1, p_0\rangle, \langle w_1, q, w_2, w_1\rangle\rangle, \langle\langle p_2, a, w_1, p_0\rangle, \langle w_1, q, w_0, w_2\rangle\rangle, \langle\langle p_2, a, w_1, p_0\rangle, \langle w_1, q, w_2, w_2\rangle\rangle\}$

The first four updates are (nearly) the same as in (11): the deontic necessity claim is placed on the conversational table, as represented by the introduction of a propositional discourse referent q for this possibility into the bottom sequence. (I return to the semantically unspecified modal base $?\omega||$ in update 3 below.) The crucial contrast is in the fifth update: It is the worlds in the context set, rather than the worlds in the at-issue proposition q, that are introduced into the bottom sequence. This update recommits to the topical modality $\top\omega||$, rather than to the necessity claim q. The subsequent updates, which are associated with any assertion, have no effect: There is no restriction of the context set to worlds where it is deontically necessary that Alice is generous, and the output context set p_2 is ${}^{\chi}\{w_0, w_1\} = p_0$, as reflected in (15).

Although the updates in (14) don't directly restrict the context set, they don't have no conversational import. Both (11) and (14) introduce a modal topic for consideration — the possibility q that Alice's being generous is deontically necessary. Updating with (11), with 'must', requires committing to this possibility; it requires committing that norms of generosity take precedence in Alice's situation over other potentially competing considerations. We might not be prepared to restrict the future course of the conversation in this way. If not, using 'must' will be dispreferred. Using 'should' and updating with (14) centers attention on the set of worlds at which Alice's being generous is deontically necessary, but doesn't explicitly require committing that the actual world is among them.[10] The conventional role of weak necessity modals, on

[10] This contrasts with non-assertive discourse moves like questions, which introduce into the

this semantics, isn't to update information. It is to place a necessity claim on the conversational table and center attention on it. Using 'should' allows us to consider the necessity of ϕ as holding, not necessarily in the current context, but in a preferred (normal, desirable) continuation or minimal revision of the context, whatever that might turn out to be. Weak necessity modals afford a means of coordinating on the implications of our values, etc. without having to settle precisely how they weigh against one another in particular circumstances, and while remaining open to new evidence about how they apply. In accepting (13) with 'should' we can leave open the possibility that norms of generosity might ultimately be outweighed or defeated in Alice's situation. We can capture a crucial role for 'should'-claims in discourse and deliberation without treating them as conventionally constraining the context set.

These analyses give precise expression to the informal intuition that 'should' is "weaker" or more tentative than 'must'. In uttering 'Should ϕ' the speaker introduces a claim about the necessity of ϕ but fails to mark her utterance as being about worlds that are candidates for actuality. Yet, as Stalnaker notes, "normally a speaker is concerned only with possible worlds within the context set, since this set is defined as the set of possible worlds among which the speaker wishes to distinguish" (1975: 69). So, using 'should' implicates that one isn't in a position to commit to the prejacent's being necessary throughout the context set. 'Should ϕ' and 'Must ϕ' are ordered not in terms of (e.g.) subset/superset relations in their domains of quantification, as on domain restriction accounts (§1), but in terms of epistemic attitude regarding the proposition that ϕ is a necessity. The basis of the scale between 'should' and 'must' isn't fundamentally logical but epistemic strength (cf. Verstraete 2006, Van Linden & Verstraete 2008). Since 'should' is weaker than 'must' in this way, Grice's first quantity maxim — "Make your contribution as informative as is required" (Grice 1989: 26) — can generate a familiar upper-bounding implicature (Horn 1972). Using 'should' implicates that for all one knows, or is willing to presuppose in the conversation, 'Must ϕ' is false. This implicature has the usual properties of implicatures. For instance, it is cancelable and reinforceable, as in (1a)–(2a). In (1a) the speaker first places the deontic necessity claim on the conversational table (with 'should'), and then commits to it (with 'must').

I noted above that the modal base for 'must' is resolved to the topical modality $\top\omega||$, reflecting the indicative presupposition that the worlds being talked about are in the context set. 'Should' lacks this restriction, as reflected in the semantically unspecified modal base $?\omega||$ in (14). This difference in modal bases helps capture another attested contrast between 'should' and 'must', in entailingness. Uttering 'Should ϕ' is compatible with denying 'ϕ' ((16)); when used with the perfect it even implicates $\neg\phi$ ((17)). Uses of '(Must ϕ) $\wedge$ $\neg\phi$', by contrast, are generally anomalous. There is robust evidence that this holds not only with epistemic readings but also with deontic readings ((18)).[11]

bottom sequence discourse referents for each answer, inducing a partition on the context set.

[11] See esp. Werner 2003, Ninan 2005, Portner 2009; also, a.o., Wertheimer 1972, Harman

(16) a. Alice should be here by now, but she isn't.
b. You should help your mother, but you won't.

(17) We should have given to Oxfam. (*Implicates:* we didn't)

(18) a. #Alice must be here by now, but she isn't.
b. ??You must help your mother, but you won't.

Of course obligations can go unfulfilled. What is interesting is that speakers appear to assume otherwise, at least for purposes of conversation, when expressing obligations with 'must'.

One way of adapting common definitions of truth in dynamic semantics for UC_ω is as follows. Definition 2.1 says that a sentence K is true at w iff given perfect information about w, i.e. an initial context set $\{w\}$, updating with K doesn't lead to the absurd state (cf. van Benthem et al. 1997: 594).[12]

Definition 2.1 (truth, v1). For an $(st)st$ term K and world w:

- Let C_w be the set of contexts c such that $\{(\top j)_1 \mid j \in {}^{\{\}}c\} \neq \{{}^{\chi}\{w\}\}$ and $\{((\top j)_{\omega t})_1 \mid j \in {}^{\{\}}c\} = \{{}^{\chi}\{w\}\}$

i. K is true at w iff for any $c \in C_w$, $\{(\top j)_1 \mid \forall g\colon j \in {}^{\{\}}[\![K]\!]^g(c)\} = \{{}^{\chi}\{w\}\}$

ii. K is false at w iff for any $c \in C_w$, $\{(\top j)_1 \mid \forall g\colon j \in {}^{\{\}}[\![K]\!]^g(c)\} = \emptyset$

This predicts that '(Must ϕ) $\wedge$ $\neg\phi$' is necessarily false: there is no $\neg\phi$-world at which 'Must ϕ' is true, and hence no world at which '(Must ϕ) $\wedge$ $\neg\phi$' is true. Although Definition 2.1 doesn't assign a truth value to (14) (since (14) doesn't recenter the primary modal topic), this definition could be revised to assign truth values to terms that update the primary background item $(\bot j)_1$ to a specific proposition. Replacing $\top$ with $\bot$ throughout Definition 2.1 would predict the possible truth and consistency of '(Should ϕ) $\wedge$ $\neg\phi$'. Note that the semantics does allow for consistent updates with 'Must ϕ. $\ldots\neg\phi$' (at least in that order); observe that w_0 in (12) is a $\neg\phi$-world. However, such sequences are still predicted to be *incoherent* in the sense that no non-empty information state is a fixed-point of an update with it:[13]

Definition 2.2 (coherence). K is coherent iff for some c, $\exists p \in D_{\omega t}\colon {}^{\{\}}p \neq \emptyset$ and $\{((\top j)_{\omega t})_1 \mid j \in {}^{\{\}}c\} = \{((\top i)_{\omega t})_1 \mid \exists g\colon i \in {}^{\{\}}[\![K]\!]^g(c)\} = \{p\}$

2.3 Conditionals and information-sensitivity

This section describes one way of extending the above semantics for 'should' and 'must' in root clauses to deontic conditionals. Simple indicative sentences comment on the topical modality $\top\Omega$, the input context set. Indicative conditionals introduce a *sub*domain of this modality — the set of context-set worlds

1973, Lyons 1977, Coates 1983, Palmer 1990, Myhill 1996, Huddleston & Pullum 2002, Close & Aarts 2010. The point about 'must' holds only for "subjective" uses (Lyons 1977), uses that express the speaker's endorsement of the relevant norms; see Silk 2012b for discussion.

[12] K has a truth value iff it updates the primary topic to a proposition. For a list j, $(\top j)_1$ is the first element in the top sublist, and $((\top j)_{\omega t})_1$ the first type ωt element in the top sublist.

[13] I leave open how this definition might be revised to generalize to non-declarative sentences.

where the antecedent is realized — which is commented on by the consequence clause. I offer (20) as a first-pass UC_ω representation of a simple indicative conditional such as (19) ($K^{\top}$; K' is a topic-comment sequence).

(19) If Alice has a job, she will be generous.

(20) $({}^{\top}[\mathbf{x} \mid \mathbf{x} = \mathbf{Alice}]; [\mathbf{w} \mid \mathbf{job_w}\langle\top\delta\rangle]; [\bot\omega \in \top\omega||]; [\mathbf{p} \mid \mathbf{p} = \bot\omega||])$
${}^{\top}; ([\mathbf{generous}_{\bot\omega}\langle\top\delta\rangle]; [\mathbf{w} \mid \text{MIN}\langle\bot\Omega, \lesssim^{e}_{\mathbf{w}}\rangle \subseteq \bot\omega||]);$
$[\mathbf{w} \mid \mathbf{w} = \bot\omega]; [\bot\omega \in \top\omega||]; [\top\omega = \bot\omega]; {}^{\top}[\mathbf{p} \mid \mathbf{p} = \top\omega||]$

The 'if'-clause introduces a propositional discourse referent into the bottom sequence for the set of worlds in the context set $\top\omega||$ in which Alice has a job, as reflected in the first line of (20). This topical subdomain $\bot\Omega$ forms the modal base of an expectational modal comment in the consequent clause, as reflected in the second line: the first update restricts the topical subdomain to worlds in which Alice is generous (modal anaphora via $\bot\omega$), and the next update introduces worlds $\mathbf{w}$ in which the most $\mathbf{w}$-expected ($\lesssim^{e}_{\mathbf{w}}$-minimal) worlds in the modal base $\bot\Omega$ are worlds in which this possibility ($\bot\omega||$) is realized. The now-familiar updates in the third line represent commitment to this possibility $\bot\omega||$, the proposal to update with it, the acceptance of this proposal, and the recentering of attention on the new topical modality $\top\Omega$.

Our analyses of 'should' and 'must' in root clauses can be integrated into this general treatment of indicative conditionals:

(21) If Alice has a job, she must be generous.

(22) $({}^{\top}[\mathbf{x} \mid \mathbf{x} = \mathbf{Alice}]; [\mathbf{w} \mid \mathbf{job_w}\langle\top\delta\rangle]; [\bot\omega \in \top\omega||]; [\mathbf{p} \mid \mathbf{p} = \bot\omega||])$
${}^{\top}; ([\mathbf{w} \mid \mathbf{generous_w}\langle\top\delta\rangle]; [\text{MIN}\{\bot\omega_2||, \lesssim^{d}_{\bot\omega_2}\} \subseteq \bot\omega||];$
$[\mathbf{w} \mid \text{MIN}\langle\bot\Omega, \lesssim^{e}_{\mathbf{w}}\rangle \subseteq \bot\omega_2||]);$
$[\mathbf{w} \mid \mathbf{w} = \bot\omega]; [\bot\omega \in \top\omega||]; [\top\omega = \bot\omega]; {}^{\top}[\mathbf{p} \mid \mathbf{p} = \top\omega||]$

(23) If Alice has a job, she should be generous.

(24) $({}^{\top}[\mathbf{x} \mid \mathbf{x} = \mathbf{Alice}]; [\mathbf{w} \mid \mathbf{job_w}\langle\top\delta\rangle]; [\bot\omega \in \top\omega||]; [\mathbf{p} \mid \mathbf{p} = \bot\omega||])$
${}^{\top}; ([\mathbf{w} \mid \mathbf{generous_w}\langle\top\delta\rangle]; [\text{MIN}\{?\omega||, \lesssim^{d}_{\bot\omega_2}\} \subseteq \bot\omega||];$
$[\mathbf{w} \mid \text{MIN}\langle\bot\Omega, \lesssim^{e}_{\mathbf{w}}\rangle \subseteq \bot\omega_2||]);$
$[\mathbf{w} \mid \mathbf{w} = \top\omega]; [\bot\omega \in \top\omega||]; [\top\omega = \bot\omega]; {}^{\top}[\mathbf{p} \mid \mathbf{p} = \top\omega||]$

As in (20), the 'if'-clauses in (22)/(24) introduce the set of worlds in the context set $\top\omega||$ in which Alice has a job. In both cases this subdomain is further restricted to worlds in which Alice's being generous is deontically necessary (modal anaphora via $\bot\omega_2$). The comment is that these worlds are the most expected worlds in the subdomain $\bot\Omega$. The proposition "that the most expected context-set worlds where Alice has a job are worlds in which Alice's being generous is deontically necessary" is introduced into the bottom sequence as being under consideration. As with root assertions, the crucial contrast between the 'should' and 'must' conditionals concerns what attitude is taken toward this possibility, as reflected in the first update of the third line. Updating with (24) simply places on the table the possibility that Alice's being generous is deontically necessary, conditional on her having a job. Commitment to the conditional necessity claim isn't required by the conventional meaning of (23).

This account helps capture an apparent contrast between 'should' and 'must' conditionals in information-sensitivity. Consider the Miners Puzzle:

> Ten miners are trapped either in shaft A or in shaft B, but we do not know which. Flood waters threaten to flood the shafts. We have enough sandbags to block one shaft, but not both. If we block one shaft, all the water will go into the other shaft, killing any miners inside it. If we block neither shaft, both shafts will fill halfway with water, and just one miner, the lowest in the shaft, will be killed. (Kolodny & MacFarlane 2010: 115–116)

As has been extensively discussed, there are readings of (25)–(27) on which they appear jointly consistent. (25) seems true, since we don't know which shaft the miners are in, and the consequences will be disastrous if we choose the wrong shaft. (26)/(27) are also natural to accept, since, given that the miners are in shaft A/B, blocking shaft A/B will save all the miners.

(25) We should block neither shaft.

(26) If the miners are in shaft A, we should block shaft A.

(27) If the miners are in shaft B, we should block shaft B.

A wrinkle in the discussions of information-sensitivity is that nearly all examples use weak necessity modals, and little attention is paid to how context affects speakers' judgments. Several authors have observed, first, that using 'must' in the conditionals is generally dispreferred (Charlow 2013; Silk 2013a).

(28) ?If the miners are in shaft A, we must block shaft A.

(29) ?If the miners are in shaft B, we must block shaft B.

Intuitively, the 'should' conditionals say what is best on a condition: given that the miners are in shaft A/B, our blocking shaft A/B is the expectably best action. (26)–(27) don't impose obligations on us conditional on how the world happens to be, unbeknownst to us. By contrast, (28)–(29) do seem to impose such obligations. This is likely part of why many speakers find using 'must' in the conditionals to be dispreferred to using 'should'.

Elsewhere I have argued that deontic conditionals with 'must' (and also with 'may') don't give rise to the same sorts of apparent modus ponens violations as deontic 'should' conditionals, and that the puzzles raised by cases like the Miners Case turn on features peculiar to weak necessity modals (Silk 2013a). Here I only wish to observe how the account in this section sheds light on the apparent contrasts in information-sensitivity between 'should' and 'must' conditionals. That said, the data about the broader spectrum of examples is admittedly less robust than would be desired. More careful assessment of the predicted contrasts among deontic conditionals must be left for future research.

As with (21), updating with (28), using 'must', requires committing (roughly) that our blocking shaft A is deontically necessary at every world in the context set in which the miners are in shaft A. Yet in some of these worlds we don't know that the miners are in shaft A. So, updating with (28) requires accepting that we have an obligation to block shaft A conditional on the miners being in shaft A, independent of whether we learn that they are. Though perhaps one could imagine accepting this in a particularly urgent context, at least for purposes of conversation, doing so would typically be inapt. Hence the general anomalousness of (28). (Likewise for (29).)

Accepting (26)–(27) with 'should', by contrast, requires no such commitments. Suppose we accept information-dependent norms which obligate us to block shaft A/B iff we learn that the miners are in shaft A/B. On the one hand, updating with (25) allows us to entertain the possibility that we won't learn where the miners are and hence will have an obligation to block neither shaft. In accepting (25) with 'should' we can coordinate on a plan for this likely scenario, to block neither shaft, but without needing to settle decisively that we won't end up getting new evidence about where the miners are, as we would need to do if we accepted 'We *must* block neither shaft'. On the other hand, updating with (26)/(27) places on the conversational table the possibility that we will be obligated to block shaft A/B conditional on the miners' being in shaft A/B, and hence — given the information-dependent norms we accept — the possibility that we learn that the miners are in shaft A/B. In accepting (26)–(27) with 'should' we can remaining open to the possibility, however slight, that we might learn which shaft the miners are in, and plan for this contingency. We can capture these points without treating the 'if'-clauses as explicitly reinterpreted as 'if ϕ and we learn it' (as in von Fintel 2012), and without introducing general revisions to the semantics of modals or conditionals (information-dependent preorder frames, selection functions, etc., as in Kolodny & MacFarlane 2010, Cariani et al. 2013, Charlow 2013, Silk 2014; cf. Willer 2010). It is weak necessity modals like 'should', unlike 'must', that play this complex role in conversation, deliberation, and planning.

3 Alternatives: Static and dynamic

One might worry that the account of the joint acceptability of (25)–(27) in §2.3 is a symptom of a general defect in the semantics. If updating with a 'should' sentence only centers attention on a necessity claim, why can't *any* set of 'should' sentences be coherently accepted? I won't attempt to resolve this question here. In this section I simply wish to raise several strategies of reply, so as to introduce certain of the critical empirical and theoretical issues.

'Should ϕ', on the proposed update semantics, places the possibility that ϕ is necessary on the conversational table and centers attention on this possibility. One option is to maintain this as an account of the conventional meaning, and capture ideas about the logic of 'should' sentences in an *extra*-semantic account of rationality constraints on the relevant discourse moves. We might view work in deontic logic on prima facie obligations, weights and priorities, dilemmas, etc. as addressing precisely this issue. Settling on controversial issues about the logic and metaethics arguably isn't required for semantic competence with modals. This line provides a way of situating respective work in logic and linguistic semantics in an overall theory of modality and modal language.[14]

An alternative response is to treat updates with 'Should ϕ' as restricting the context set, but revise what proposition is conventionally placed on the conversational table. For example, one might treat 'Should ϕ' as predicating the necessity of ϕ of a set of worlds that satisfy some (possibly counterfactual) condition, or of a set of (possibly counterfactual) worlds that are minimal/preferred in some contextually relevant sense (most desirable, normal, etc.; cf. Silk 2012a,b). Updating with (30) restricts the context set to worlds **w** such that the relevant **w**-accessible worlds, which may not themselves be in the context set, are worlds at which Alice's being generous is deontically necessary.

(30) $$\begin{aligned}&{}^{\top}[\mathbf{x} \mid \mathbf{x} = \textbf{Alice}]; [\mathbf{w} \mid \textbf{generous}_{\mathbf{w}}\langle \top\delta\rangle]; [\mathbf{w} \mid \textsc{min}\{?\omega||, \lesssim^{d}_{\mathbf{w}}\} \subseteq \bot\omega||]; \\ &[\mathbf{w} \mid \textsc{pref}^{?}_{\mathbf{w}} \subseteq \bot\omega||]; [\mathbf{p} \mid \mathbf{p} = \bot\omega||]; \\ &[\mathbf{w} \mid \mathbf{w} = \bot\omega]; [\bot\omega \in \top\omega||]; [\top\omega = \bot\omega]; {}^{\top}[\mathbf{p} \mid \mathbf{p} = \top\omega||]\end{aligned}$$

(30) explicitly represents an attitudinal comment about the (deontic) necessity claim. The logic of 'should' sentences could then be captured via the logic of the relevant further notion of minimality/preference (normality, desirability).

Whereas the update semantics developed in §2 is essentially dynamic, (30) treats 'should' assertions as updating context like any other assertion. This kind of analysis could be implemented in a static semantics which provides straightforward truth conditions for 'Should ϕ' (cf. Silk 2012b: Def. 2). Either way, insofar as $\textsc{pref}^{?}_{w}$ needn't be included in the context set, we capture the idea from §1 that 'Should ϕ' brackets whether ϕ is actually necessary (epistemically, deontically, etc.). Yet even if informal ideas about the contrasting discourse properties of 'should' and 'must' *could* be implemented in a static or dynamic semantics, this leaves open whether the ideas are best explained in terms of truth. Thorough investigation of grammatical and discourse differences among necessity modals, as well as general philosophical reflection on the explanatoriness of static vs. dynamic frameworks, is needed.[15]

4 Conclusion

Let's recap. Following Silk 2012b, I argued that the common semantic core of weak necessity modals is that they bracket whether the prejacent is necessary (deontically, epistemically, etc.) in the actual world. To implement this idea I developed an update semantics for weak and strong necessity modals. An account of deontic conditionals was also integrated into a more general update semantics for conditionals. These analyses carve out an important role for expressions of weak necessity in discourse, deliberation, and planning.

The data considered here certainly aren't the only data that must be explained by an overall theory of necessity modals. Elsewhere (Silk 2012a,b) I argue that the proposed approach to the weak/strong necessity modal distinction

[14] For discussion of related methodological issues, see Forrester 1989, Silk 2013b, 2015.

[15] See e.g. Starr 2010, Lassiter 2011, 2012, Iatridou & Zeijlstra 2013, Rubinstein 2014.

also captures the morphosyntactic properties of expressions of weak necessity cross-linguistically (von Fintel & Iatridou 2008), and the contrasting logical properties and illocutionary properties of weak and strong necessity modals. There are also contrasts in interactions with comparatives, quantifiers, modifiers, neg-raising, and polarity effects, among others (n. 15). Though I focused here on what distinguishes the *classes* of weak vs. strong necessity modals, I bracketed differences among weak necessity modals and among strong necessity modals (see Silk 2012b, 2015 and references therein). Moreover our discussion highlighted how phenomena with weak and strong necessity modals interact with general issues concerning context-sensitivity, assertion, the roles of truth and discourse function in linguistic theorizing, and relations among logic, semantics, and pragmatics in an overall theory of modals. These interactions afford rich possibilities for future research.

References

van Benthem, Johan, Reinhard Muskens & Albert Visser. 1997. Dynamics. In Johan van Benthem & Alice ter Meulen (eds.), *Handbook of logic and language*, 587–648. Amsterdam: Elsevier.

Bittner, Maria. 2011. Time and modality without tenses or modals. In Monika Rathert & Renate Musan (eds.), *Tense across languages*, 147–188. Tübingen: Niemeyer.

Cariani, Fabrizio, Stefan Kaufmann & Magdalena Kaufmann. 2013. Deliberative modality under epistemic uncertainty. *Linguistics and Philosophy* 36(3). 225–259.

Charlow, Nate. 2013. What we know and what to do. *Synthese* 190. 2291–2323.

Charlow, Nate & Matthew Chrisman (eds.). 2016. *Deontic modality*. New York: Oxford University Press.

Close, Joanne & Bas Aarts. 2010. Current change in the modal system of English: A case study of *must, have to* and *have got to*. In Ursula Lenker, Judith Huber & Robert Mailhammer (eds.), *The history of English verbal and nominal constructions*, 165–181. Amsterdam: John Benjamins.

Coates, Jennifer. 1983. *The semantics of the modal auxiliaries*. London: Croom Helm.

Copley, Bridget. 2006. What should *should* mean? MS, CNRS/Université Paris 8.

von Fintel, Kai. 2012. The best we can (expect to) get? Challenges to the classic semantics for deontic modals. Paper presented at the 2012 Central APA, Symposium on Deontic Modals.

von Fintel, Kai & Sabine Iatridou. 2008. How to say *ought* in foreign: The composition of weak necessity modals. In Jacqueline Guéron & Jacqueline Lecarme (eds.), *Time and modality*, 115–141. Springer.

Forrester, James William. 1989. *Why you should: The pragmatics of deontic speech*. Hanover: Brown University Press.

Grice, Paul. 1989. *Studies in the ways of words*. Cambridge: Harvard University Press.

Grosz, Barbara J., Arivind K. Joshi & Scott Weinstein. 1995. Centering: A framework for modeling the local coherence of discourse. *Computational Linguistics* 21(2). 203–225.

Harman, Gilbert. 1973. Review of Roger Wertheimer, *The significance of sense: Meaning, modality, and morality*. *The Philosophical Review* 82. 235–239.

Horn, Laurence R. 1972. *On the semantic properties of logical operators in English*: University of California, Los Angeles Ph.d. thesis.

Huddleston, Rodney & Geoffrey K. Pullum (eds.). 2002. *The Cambridge grammar of the English language*. Cambr: Cambridge University Press.

Iatridou, Sabine. 2000. The grammatical ingredients of counterfactuality. *Linguistic Inquiry* 31(2). 231–270.

Iatridou, Sabine & Hedde Zeijlstra. 2013. Negation, polarity, and deontic modals. *Linguistic Inquiry* 44(4). 529–568.

Kaufmann, Stefan. 2000. Dynamic context management. In Martina Faller, Stefan Kaufmann & Marc Pauly (eds.), *Formalizing the dynamics of information*, 171–188. Stanford: CSLI Publications.

Kolodny, Niko & John MacFarlane. 2010. Ifs and oughts. *Journal of Philosophy* 115–143.

Kratzer, Angelika. 1981. The notional category of modality. In Hans-Jürgen Eikmeyer & Hannes Rieser (eds.), *Words, worlds, and contexts: New approaches in word semantics*, 38–74. Berlin: de Gruyter.

Kratzer, Angelika. 1991. Modality/Conditionals. In Arnim von Stechow & Dieter Wunderlich (eds.), *Semantics: An international handbook of contemporary research*, 639–656. New York: de Gruyter.

Lassiter, Daniel. 2011. *Measurement and modality: The scalar basis of modal semantics*: New York University dissertation.

Lassiter, Daniel. 2012. Quantificational and modal interveners in degree constructions. In Anca Chereches (ed.), *Proceedings of SALT 22*, 565–583. Ithaca: CLC Publications.

Lewis, David. 1973. *Counterfactuals*. Cambridge: Harvard University Press.

Lewis, David. 1981. Ordering semantics and premise semantics for counterfactuals. *Journal of Philosophical Logic* 10(2). 217–234.

Lyons, John. 1977. *Semantics*, vol. 2. Cambridge: Cambridge University Press.

Lyons, John. 1995. *Linguistic semantics: An introduction*. Cambridge: Cambridge University Press.

McNamara, Paul. 1990. *The deontic quadecagon*: University of Massachusetts Ph.d. thesis.

Murray, Sarah E. 2010. *Evidentiality and the structure of speech acts*: Rutgers University Ph.d. thesis.

Murray, Sarah E. 2014. Varieties of update. *Semantics and Pragmatics* 7. 1–53.

Muskens, Reinhard. 1996. Combining Montague semantics and discourse representation. *Linguistics and Philosophy* 19. 143–186.

Myhill, John. 1996. The development of the strong obligation system in American English. *American Speech* 71(4). 339–388.

Ninan, Dilip. 2005. Two puzzles about deontic necessity. In Jon Gajewski, Valentine Hacquard, Bernard Nickel & Seth Yalcin (eds.), *MIT working papers in linguistics: New work on modality*, vol. 51, 149–178.

Palmer, F.R. 1990. *Modality and the english modals.* New York: Longman 2nd edn.

Palmer, F.R. 2001. *Mood and modality.* Cambridge: Cambridge University Press 2nd edn.

Portner, Paul. 2009. *Modality.* Oxford: Oxford University Press.

Portner, Paul & Aynat Rubinstein. 2016. Extreme and non-extreme deontic modals. In Charlow & Chrisman (2016).

Ridge, Michael. 2014. *Impassioned belief.* Oxford University Press.

Rubinstein, Aynat. 2012. *Roots of modality*: University of Massachusetts Amherst dissertation.

Rubinstein, Aynat. 2014. On necessity and comparison. *Pacific Philosophical Quarterly* 95. 512–554.

Schlenker, Philippe. 2005. The lazy Frenchman's approach to the subjunctive (Speculations on reference to worlds and semantic defaults in the analysis of mood). In Twan Geerts, Ivo van Ginneken & Haike Jacobs (eds.), *Romance languages and linguistic theory 2003: Selected papers from Going Romance 17*, 269–309. Nijmegen: John Benjamins.

Silk, Alex. 2012a. Modality, weights, and inconsistent premise sets. In Anca Chereches (ed.), *Proceedings of SALT 22*, 43–64. Ithaca: CLC Publications.

Silk, Alex. 2012b. Weak and strong necessity. MS, University of Michigan. `http://goo.gl/xY9roN`.

Silk, Alex. 2013a. Deontic conditionals: Weak and strong. In *Proceedings of the 19th international congress of linguists*, Département de Linguistique de l'Université de Genève. `http://goo.gl/xSGbcJ`.

Silk, Alex. 2013b. Modality, weights, and inconsistent premise sets. MS, University of Birmingham.

Silk, Alex. 2014. Evidence sensitivity in weak necessity deontic modals. *Journal of Philosophical Logic* 43(4). 691–723.

Silk, Alex. 2015. What normative terms mean and why it matters for ethical theory. In Mark Timmons (ed.), *Oxford studies in normative ethics*, vol. 5,

296–325. Oxford: Oxford University Press.

Sloman, Aaron. 1970. 'Ought' and 'better'. *Mind* 79. 385–394.

Stalnaker, Robert. 1975. Indicative conditionals. In Stalnaker (1999) 63–77.

Stalnaker, Robert. 1978. Assertion. In Stalnaker (1999) 78–95.

Stalnaker, Robert. 1999. *Context and content: Essays on intentionality in speech and thought.* Oxford: Oxford University Press.

Starr, William. 2010. *Conditionals, meaning, and mood*: Rutgers University Ph.d. thesis.

Stone, Matthew. 1999. Reference to possible worlds. RuCCS Report 49, Rutgers University.

Swanson, Eric. 2011. On the treatment of incomparability in ordering semantics and premise semantics. *Journal of Philosophical Logic* 40. 693–713.

Van Linden, An. 2012. *Modal adjectives: English deontic and evaluative constructions in synchrony and diachrony.* Berlin: Mouton de Gruyter.

Van Linden, An & Jean-Christophe Verstraete. 2008. The nature and origins of counterfactuality in simple clauses: Cross-linguistic evidence. *Journal of Pragmatics* 40. 1865–1895.

Veltman, Frank. 1996. Defaults in update semantics. *Journal of Philosophical Logic* 25. 221–261.

Verstraete, Jean-Christophe. 2006. The nature of irreality in the past domain: Evidence from past intentional constructions in Australian languages. *Australian Journal of Linguistics* 26(1). 59–79.

Werner, Tom. 2003. *Deducing the future and distinguishing the past: Temporal interpretation in modal sentences in English*: Rutgers University Ph.d. thesis.

Wertheimer, Roger. 1972. *The significance of sense: Meaning, modality, and morality.* Ithaca: Cornell University Press.

Willer, Malte. 2010. *Modality in flux*: University of Texas, Austin dissertation.

Williams, Bernard. 1981. Practical necessity. In *Moral luck*, 124–131. Cambridge: Cambridge University Press.

Woisetschlaeger, Erich Friedrich. 1977. *A semantic theory of the English auxiliary system*: MIT dissertation.

Yalcin, Seth. 2016. Modalities of normality. In Charlow & Chrisman (2016).

Coarse Deontic Logic (short version)

Frederik Van De Putte [1]

Ghent University
Blandijnberg 2, B-9000 Ghent, Belgium
frederik.vandeputte@ugent.be

Abstract

In recent work, Cariani has proposed a semantics for *ought* that combines two features: (i) it invalidates Inheritance in a principled manner; (ii) it allows for coarseness, which means that $ought(\varphi)$ can be true even if there are specific ways of making φ true that are (intuitively speaking) impermissible. We present a group of multi-modal logics based on Cariani's proposal. We study their formal properties and compare them to existing approaches in the deontic logic literature — most notably Anglberger et al.'s logic of obligation as weakest permission, and deontic stit logic.

Keywords: Deontic logic; contrastivism; modal inheritance; Ross paradox; deontic STIT logic; coarseness

1 Introduction

Contrastivism about "ought" says that claims using this modality can only be understood relative to a (usually implicit) contrast class. [2] So according to this view, "you ought to take the bus" is shorthand for "given the set of alternatives $\mathcal{A}$ under consideration, you ought to take the bus". Here $\mathcal{A}$ may consist of various ways of getting somewhere (say, the university).

In recent work, Cariani has proposed a formal semantics which starts from a contrastivist reading of ought [5]. This proposal is interesting for at least two reasons. First, it gives a principled account of why Inheritance [3] fails in cases like the Ross paradox, which makes it more insightful than most existing semantics for non-normal modalities. [4] Second, it allows for what Cariani calls *coarse* ought-claims, which means that $ought(\varphi)$ can be true even if there are specific ways of making φ true that are (intuitively speaking) impermissi-

---

[1] We are indebted to Mathieu Beirlaen and three anonymous referees for incisive comments on previous versions.

[2] See [20, footnote 1] for some key references to contrastivism in deontic logic.

[3] By *Inheritance* we mean here: from $ought(\varphi)$ and $\varphi \vdash \psi$, to infer $ought(\psi)$. This property is also often called *monotony*.

[4] As Cariani [5, p. 537] remarks, such semantics are "often purely algebraic", in the sense that they just translate rules for *ought* into conditions on neighbourhood functions.

ble.[5] This unusual combination – coarseness *without* Inheritance – is possible precisely because of the way the alternatives are modeled: rather than single worlds, they are (mutually exclusive) *sets* of worlds.

Before one can argue for or against Cariani's proposal, one has to study the logics obtained from it. We do this here. In Section 2, we present Cariani's proposal, both informally and in terms of a possible-worlds semantics. We discuss the most salient properties of the resulting logic. Next, we consider variants of this semantics that are defined over the same modal language (Section 3). Section 4 provides a map of the various logics obtained and presents their axiomatization.[6] Finally, we show how they relate to existing work in the deontic logic field, and where one can draw on this link in order to solve existing problems and puzzles (Section 5).

Preliminaries We use $p, q, \ldots$ for arbitrary propositional variables. The boolean connectives are denoted by $\neg, \vee, \wedge, \supset, \equiv$ (only the first two are primitive) and occasionally we will use the falsum and verum constants ($\bot$, resp. $\top$). $\varphi, \psi, \ldots$ are metavariables for formulas and $\Gamma, \Delta, \ldots$ for sets of formulas. *ought* refers to operators proposed as formal counterparts of the natural language "ought". Given an expression of the type $ought(\varphi)$, φ is the *prejacent* of this formula.

2 Cariani's Semantics

In this section, we introduce and illustrate Cariani's semantics for *ought*. We first present the semantics informally in our own terms, after which we indicate the relation with Cariani's original presentation (Section 2.1). Next, we define a formal semantics which implements Cariani's ideas (Section 2.2) and discuss the most salient properties of the resulting logic (Section 2.3).

2.1 Cariani's proposal, informally

Our Version Cariani's *ought* is defined in terms of various more basic concepts. To spell these out, we need three parameters:

(a) a set of (mutually exclusive) *alternatives* or *options* $\mathcal{A}$
(b) a set $\mathcal{B} \subseteq \mathcal{A}$ of "optimal" or "best" options
(c) a set $\mathcal{I} \subseteq \mathcal{A}$ of "impermissible" options

For instance, in a context where we are deliberating about how Lisa ought to get to the university, her options may be represented by the following set:

$$\mathcal{A}_{\mathsf{ex}} = \{\text{walk}, \text{bike}, \text{bus}, \text{car}\}$$

indicating that she may walk to the university, drive her bike, take the bus, or drive by car. Some of these options may be optimal – e.g. biking or taking the

[5] We explain and illustrate Cariani's notion of coarseness in Section 2.1.

[6] In the full version of this paper [23] (available on request), we show that our axiomatizations are sound and (strongly) complete, and we establish the finite model property for each of the logics.

bus. Driving may well be impermissible (since she may not yet have obtained her driver's licence) and walking may be suboptimal (since given the distance, she risks getting late) but nevertheless permissible. So we have:

$$\mathcal{B}_{\mathsf{ex}} = \{\text{bike}, \text{bus}\}$$

$$\mathcal{I}_{\mathsf{ex}} = \{\text{car}\}$$

Each of the options in $\mathcal{A}_{\mathsf{ex}}$ can be carried out in many different ways; e.g. Lisa may drive her bike in a blue dress or in a green dress; she may drive her bike in a hazardous way or very cautiously. In Cariani's terms, this means the alternatives are *coarse-grained.* In other words, they correspond to generic action-types or general properties (sets of worlds in a Kripke-model), in contrast to action-tokens or maximally specific descriptions of a state of affairs (worlds in a Kripke-model).[7]

This explains at once how it is possible, in Cariani's framework, that there are (intuitively) impermissible instances of an optimal (or permissible) alternative. Even if Lisa ought to drive her bike or take the bus, this does not imply that every way of doing so is normatively ok. Indeed, relative to a more fine-grained set of alternatives, it may turn out that some ways of driving her bike are impermissible. Mind that the framework does not explicitly represent the impermissibility of such specific actions – hence, they are only impermissible "intuitively speaking". The point is exactly that, by choosing one specific level of granularity in a certain context, we decide to leave those more specific (impermissible) actions out of the picture. Once we make them explicit, the level of granularity changes, and with it the truth of any given *ought*-claim.[8]

Since options are coarse-grained, they do not fix every property of the world. Still, some propositions are fixed by taking one option rather than the other. If Lisa takes her bike, she is definitely not taking the bus or driving her car. In general, we say that an option $X \in \mathcal{A}$ *guarantees* a proposition φ iff following that option ensures that φ is the case.

We are now ready to spell out an informal version of Cariani's proposal. That is, where φ is a proposition, *ought*(φ) is true (relative to $\mathcal{A}, \mathcal{B}, \mathcal{I}$) iff each of the following hold:

(i) φ is *visible*, i.e. for all $X \in \mathcal{A}$: X guarantees φ or X guarantees $\neg\varphi$
(ii) φ is *optimal*, i.e. for all $X \in \mathcal{B}$: X guarantees φ
(iii) φ is *strongly permitted*, i.e. for all $X \in \mathcal{A}$ that guarantee φ, $X \notin \mathcal{I}$.

For instance, in our example, it is true that Lisa ought to ride her bike or take the bus. It is false that she ought to ride her bike, take the bus or take the car, since taking the car is impermissible. It is equally false that she ought

[7] See [5, pp. 544-545] for a more detailed discussion of the link between action types/tokens and Cariani's semantics.

[8] This of course raises the question how oughts concerning such more fine-grained $\mathcal{A}'$ relate to the coarse-grained $\mathcal{A}$ – we return to this point in Section 5.

to ride her bike or take the bus in a green dress, since that proposition is not visible.[9]

This shows us at once that Inheritance is invalid on Cariani's semantics. It is in fact blocked in two different ways – see (i) and (iii) above. As a result, also the Ross paradox is blocked: "you ought to mail the letter" may be true while "you ought to mail the letter or burn it" is false. This will either be the case because burning the letter is invisible, or if we do take it to be a visible option, because it is impermissible.

Ranking and threshold In Cariani's original proposal, instead of $\mathcal{B}$ and $\mathcal{I}$, a "ranking" of $\mathcal{A}$ is used together with a "threshold" t on that ranking. The idea is that the "best" options are those that are maximal (according to the ranking), and the impermissible ones are those that are below the threshold. Although Cariani is not very explicit about the formal properties of his ranking and threshold, it seems that his ranking is a modular pre-order, in the sense that it distinguishes different layers of "ever better" options.[10] In other words, it can be defined as a function $r : \mathcal{A} \to \mathbb{R}$, where intuitively, X is better than X' (for $X, X' \in \mathcal{A}$) iff $r(X) > r(X')$. The threshold is then simply a $t \in \mathbb{R}$, such that whenever $r(X) < t$, X is impermissible.

It is easy enough to check that, once such an r and t are given, we can obtain $\mathcal{B}$ and $\mathcal{I}$ from them as follows: (i) $\mathcal{B} = \{X \in \mathcal{A} \mid r(X) = \mathsf{max}_<(\{r(Y) \mid Y \in \mathcal{A}\})\}$, and (ii) $\mathcal{I}$ is the set of all $X \in \mathcal{A}$ such that $r(X) < t$. Hence our simplified version of Cariani's semantics is at least as general as his original version.

Given fairly weak assumptions, we can also show the converse. That is, consider an arbitrary $\langle \mathcal{A}, \mathcal{B}, \mathcal{I} \rangle$ and suppose that each of the following hold:

(D) $\mathcal{B} \neq \emptyset$
(C$\cap$ $\mathcal{B} \cap \mathcal{I} = \emptyset$

In other words, there are best options, and every best option is permissible. Define the function $r : \mathcal{A} \to \{1, 2, 3\}$ as follows:

(1) if $X \in \mathcal{B}$, then $r(X) = 3$
(2) if $X \in \mathcal{I}$, then $r(X) = 1$
(3) if $X \in \mathcal{A} \setminus (\mathcal{B} \cup \mathcal{I})$ then $r(X) = 2$

Let $t = 2$. It can easily be checked that (i) and (ii) hold. So if we assume (D) and (C$\cap$), the two formats are equivalent (deontically speaking).

In the current section, we will leave restrictions (D) and (C$\cap$) aside. In Section 3.1 we consider variants of our base logic in which these restrictions are added to the semantics.

[9] As the reader may note, "Lisa ought to take her bike, take the bus, or walk to the university" is also true in our example, which might strike one as odd. We return to this point in Section 3.2.

[10] At least it is in all the examples he gives. Also, this seems to be presupposed by the way he uses the notion of a threshold, viz. as a single member X of $\mathcal{A}$ such that any option below X is impermissible.

2.2 The formal semantics of CDL^c

Our language $\mathcal{L}$ is obtained by closing the set of propositional variables $\mathcal{S} = \{p, q, \ldots\}$ under the Boolean connectives and the modal operators U (necessary/holds in every possible world), A (is guaranteed by the chosen alternative), B (is best/is guaranteed by all optimal alternatives), and P (is strongly permitted).

Two comments are in place here. First, Cariani does not explicitly mention the operators U and A. However, both are fairly natural modalities in this context. U is just a global modality – see [11] for a systematic study. A expresses the concept of being guaranteed by a given option, which Cariani uses in the semantic clause of his *ought*-operator. Moreover, adding both modalities to the language allows us to obtain a sound and complete axiomatization of the logic – see Section 4. [11]

Second, rather than taking it as primitive as Cariani does, we treat "is visible", V, as a defined operator:

$$\mathsf{V}\varphi =_{\mathsf{df}} \mathsf{U}(\mathsf{A}\varphi \vee \mathsf{A}\neg\varphi)$$

Likewise, O (Cariani's *ought*) is a defined operator:

$$\mathsf{O}\varphi =_{\mathsf{df}} \mathsf{V}\varphi \wedge \mathsf{B}\varphi \wedge \mathsf{P}\varphi$$

The following two definitions make the informal semantics from Section 2.1 exact: [12]

Definition 2.1 A **CDL^c**-*frame* is a tuple $F = \langle W, \mathcal{A}, \mathcal{B}, \mathcal{I}\rangle$, where W is a non-empty set, $\mathcal{A} \in \wp(\wp(W))$ is a partition of W, $\mathcal{B} \subseteq \mathcal{A}$ is the set of *best options* in $\mathcal{A}$, and $\mathcal{I} \subseteq \mathcal{A}$ is the set of *impermissible options* in $\mathcal{A}$.

A **CDL^c**-*model* M is a **CDL^c**-frame $\langle W, \mathcal{A}, \mathcal{B}, \mathcal{I}\rangle$ augmented with a valuation function $v : \mathcal{S} \to \wp(W)$.

Since $\mathcal{A}$ is a partition of W, all worlds are by definition a member of some alternative in the contrast class. In other words, we exclude the possibility that some members of W are simply irrelevant for the deontic claims that are at stake. We leave the investigation of such a possibility for another occasion.

In line with the preceding, the members of $\mathcal{A}$ are interpreted as action types or options a given agent faces, whereas the members of W represent action tokens (specific ways of carrying out a given action or option). Formulas are evaluated relative to a given $w \in W$, in accordance with Definition 2.2. This means that in general, whether or not a formula is true may depend on the alternative that is chosen and on the specific way it is carried out or materializes. However, for purely normative claims, this is not the case (see our discussion of the property of Uniformity in Sections 2.3 and 3.3).

Definition 2.2 Let $M = \langle W, \mathcal{A}, \mathcal{B}, \mathcal{I}, v\rangle$ be a **CDL^c**-model and $w \in W$. Where $w \in W$, let X^w denote the $X \in \mathcal{A}$ such that $w \in X$.

---

[11] It remains an open question whether one can obtain such an axiomatization without these modalities, and with V primitive.

[12] **CDL** is shorthand for "Coarse Deontic Logic". The superscript c refers to Cariani.

(SC1) $M, w \models \varphi$ iff $w \in v(\varphi)$ for all $\varphi \in \mathcal{S}$
(SC2) $M, w \models \neg\varphi$ iff $M, w \not\models \varphi$
(SC3) $M, w \models \varphi \vee \psi$ iff $M, w \models \varphi$ or $M, w \models \psi$
(SC4) $M, w \models \mathsf{U}\varphi$ iff $M, w' \models \varphi$ for all $w' \in W$
(SC5) $M, w \models \mathsf{A}\varphi$ iff $M, w' \models \varphi$ for all $w' \in X^w$
(SC6) $M, w \models \mathsf{B}\varphi$ iff for all $X \in \mathcal{B}$, for all $v \in X$, $M, v \models \varphi$
(SC7) $M, w \models \mathsf{P}\varphi$ iff for all $X \in \mathcal{A}$ s.t. (for all $v \in X$, $M, v \models \varphi$), $X \notin \mathcal{I}$

Note that $\mathsf{V}\varphi$ means (by our definition) that at every world w in the current model, either φ is guaranteed or $\neg\varphi$ is guaranteed. Since $\mathcal{A}$ is a partition of W, this is equivalent to saying that every option either guarantees φ or guarantees $\neg\varphi$, which corresponds to Cariani's original semantics for "is visible".

As usual, $\Gamma \Vdash_{\mathbf{CDL^c}} \varphi$ iff for all $\mathbf{CDL^c}$-models M and every world w in the domain of M, if $M, w \models \psi$ for all $\psi \in \Gamma$, then $M, w \models \varphi$.

2.3 Properties of CDL^{c}

It can be easily verified that each of U, A and B are normal modal operators in $\mathbf{CDL^c}$. In fact, both U and A are $\mathbf{S5}$-modalities. Second, P is a non-normal but classical modality (in the sense of Chellas [6]), which means it satisfies at least replacement of equivalents. As a result, also the defined operators V and O are classical.

Now for some more distinctive properties. Each of the following hold for $\Vdash = \Vdash_{\mathbf{CDL^c}}$:

$$\mathsf{O}(\varphi \wedge \psi) \nVdash \mathsf{O}\varphi \tag{1}$$

$$\mathsf{O}\varphi, \mathsf{O}\psi \Vdash \mathsf{O}(\varphi \wedge \psi) \tag{2}$$

$$\mathsf{O}\varphi, \mathsf{O}\psi \Vdash \mathsf{O}(\varphi \vee \psi) \tag{3}$$

$$\mathsf{O}\varphi, \mathsf{P}\psi \nVdash \mathsf{O}(\varphi \vee \psi) \tag{4}$$

$$\mathsf{O}\varphi, \mathsf{P}(\varphi \vee \psi) \nVdash \mathsf{O}(\varphi \vee \psi) \tag{5}$$

$$\mathsf{O}\varphi, \mathsf{P}\psi, \mathsf{V}\psi \Vdash \mathsf{O}(\varphi \vee \psi) \tag{6}$$

$$\mathsf{V}\varphi, \mathsf{V}\psi \Vdash \mathsf{V}(\varphi \vee \psi) \tag{7}$$

$$\mathsf{V}\varphi, \mathsf{V}\psi \Vdash \mathsf{V}(\varphi \wedge \psi) \tag{8}$$

$$\Vdash \mathsf{P}(\varphi \vee \psi) \supset (\mathsf{P}\varphi \wedge \mathsf{P}\psi) \tag{9}$$

$$\not\Vdash (\mathsf{P}\varphi \wedge \mathsf{P}\psi) \supset \mathsf{P}(\varphi \vee \psi) \tag{10}$$

$$\Vdash (\mathsf{P}\neg\mathsf{A}\neg\varphi \wedge \mathsf{P}\neg\mathsf{A}\neg\psi) \supset \mathsf{P}(\neg\mathsf{A}\neg\varphi \vee \neg\mathsf{A}\neg\psi) \tag{11}$$

Let us comment on these properties one by one. That O does not satisfy *Inheritance* – see (1) – was already explained above. Quite surprisingly, *Aggregation* (2) holds for O. In a context where the possibility of deontic conflicts is omitted, this is often considered a nice feature. It follows from the fact that the three operators B, V, and P are each aggregative – witness (8) and (9). For similar reasons, *Weakening* (3) also holds in $\mathbf{CDL^c}$ – this follows by the normality of B, (7), and (10).

Both Aggregation and Weakening deserve our attention here. As shown in [4], these properties fail on what is perhaps the most well-known contrastive semantics for *ought*, viz. the actualist semantics from [17], which has been worked out and axiomatized by Goble [8,9].

(4) and (5) tell us that, contrary to what one might expect, neither $\mathsf{P}\psi$ nor $\mathsf{P}(\psi \vee \varphi)$ suffice in order to derive $\mathsf{O}(\varphi \vee \psi)$ from $\mathsf{O}\varphi$.[13] The reason is that neither of those propositions warrant that $\varphi \vee \psi$ is visible, which is required for $\mathsf{O}(\varphi \vee \psi)$ to hold. Only if we add $\mathsf{V}\psi$ do we obtain a restricted form of Inheritance that is $\mathbf{CDL^c}$-valid – see (6).

Together with replacement of equivalents, (9) entails that P is "downward closed": whatever is stronger than something that is permitted, is itself also permitted. To see why this is so, note that $\mathsf{P}(\varphi \vee \psi)$ expresses that guaranteeing $\varphi \vee \psi$ implies that one is choosing a permissible option. Hence *a fortiori* guaranteeing φ (resp. ψ) is sufficient for permissibility. By the definition of O, this also means that $\mathsf{O}(\varphi \vee \psi) \Vdash_{\mathbf{CDL^c}} \mathsf{P}\varphi, \mathsf{P}\psi$: that $\varphi \vee \psi$ ought to be implies that $\varphi \vee \psi$ is strongly permitted, which in turn implies that both φ and ψ are strongly permitted. We return to this property in Section 3.2.

In view of (10), P is not an operator of "free choice permission" in the strict sense of [25]. To see why (10) holds, recall our example. Here, "Lisa takes the car in a green dress" (*car* $\wedge$ *green*) is permissible in a vacuous way, since there is simply no option which guarantees that proposition. Likewise, "Lisa takes the car, but not in a green dress" (*car* $\wedge \neg$*green*) is permissible. However, *car* (which is equivalent to the disjunction of both propositions) is not permissible.[14]

(11) shows that for the more specific case where φ and ψ are of the form $\neg\mathsf{A}\neg\tau$, we do get the converse of (9). If it is permissible that (a) one leaves open

---

[13] Snedegar [20, pp. 217-218] refers to Goble [10, Note 49] who rejects such a rule. However, in Goble's case, the P-operator is one of weak permission, i.e. $\mathsf{P} =_{\mathsf{df}} \neg\mathsf{O}\neg$. Besides that, Goble's main concern is to accommodate deontic conflicts, a target which Cariani explicitly rules out – as Snedegar acknowledges.

[14] In view of this example, P seems to express only part of the meaning of "is permitted". A more plausible operator of (strong) permission can be defined as $\mathsf{P}^v = \mathsf{P}\varphi \wedge \mathsf{V}\varphi$. Note that $(\mathsf{P}^v\varphi \wedge \mathsf{P}^v\psi) \Vdash_{\mathbf{CDL^c}} \mathsf{P}^v(\varphi \vee \psi)$, but $\mathsf{P}^v(\varphi \vee \psi) \not\Vdash_{\mathbf{CDL^c}} \mathsf{P}^v\varphi \wedge \mathsf{P}^v\psi$. We leave the investigation of such definable operators for future work.

the possibility that φ, and it is also permissible that (b) one leaves open the possibility that ψ, then it is permissible that (c) one leaves open the possibility that φ or one leave open the possibility that ψ. Indeed, whenever (c) holds, either (a) or (b) hold and hence one is definitely taking one of the permissible options.

Other interesting validities concern the interaction between the alethic modalities U, A and the deontic modalities B, P, and O. These are of two types:

$$\text{where } \nabla \in \{\mathsf{B}, \mathsf{P}, \mathsf{O}\} : \Vdash \nabla\varphi \equiv \nabla\mathsf{A}\varphi \tag{12}$$

$$\text{where } \nabla \in \{\mathsf{B}, \mathsf{P}, \mathsf{O}\} : \Vdash \nabla\varphi \equiv \mathsf{U}\nabla\varphi \tag{13}$$

Contrast-sensitivity, (12), expresses that the deontic modalities really apply to alternatives $X \in \mathcal{A}$, rather than worlds $w \in W$. For instance, $\mathsf{B}\varphi$ can only be true if φ is true in all worlds that belong to an optimal alternative; but that is the same as saying that all optimal alternatives guarantee φ. This property is therefore essential for Cariani's constrastive approach.

Uniformity, (13), expresses that deontic claims are either always settled true or settled false (to use terminology from [2]). It follows from the fact that $\mathcal{B}$ and $\mathcal{I}$ are independent of the world w one happens to be at in a model. We return to this property in Section 3.3.

3 Some Variants

We now consider variants of the **CDL**$^{\mathbf{c}}$-semantics and motivate each of them independently. This will be useful in Section 5, where we compare Cariani's construction to existing work in deontic logic.

3.1 Conditions (D) and (C$\cap$)

We first return to the conditions mentioned at the end of Section 2.1. (D) corresponds to the requirement in Standard Deontic Logic that the accessibility relation is *serial*, and hence, that there is at least one "ideal" or "optimal" world. It can be moreover easily checked that (D) is expressed by the familiar axiom schema $\mathsf{B}\varphi \supset \neg\mathsf{B}\neg\varphi$ within **CDL**$^{\mathbf{c}}$. This axiom schema (along with the failure of the T-schema, $\mathsf{B}\varphi \supset \varphi$) is traditionally seen as the distinctive feature of deontic logics.

Although it is a much debated property in the context of deontic logic in general, (D) does seem to have some intuitive power in the present context. After all, the idea is that we start from a fixed set of alternatives, one particular ranking r, and one threshold t.[15] Finiteness of $\mathcal{A}$ already entails (D). But even if we allow for a possibly infinite number of options, it seems sensible to say that we only consider finitely many of those as viable options, such that a ranking on them will always yield a non-empty set of best alternatives.

[15] As Cariani notes, one may generalize the entire setting to cases with multiple rankings and threshold functions; that seems to be his preferred way of allowing for deontic conflicts.

(C∩) is more difficult to interpret in the present context. It states that every best option is permissible. Interestingly, this condition is not definable in the language of **CDL**$^{\mathbf{c}}$. In fact, imposing it onto the semantics has no impact on the resulting logic.[16] This means in turn that, once we assume (D), and as far as the consequence relation is concerned, there really is no difference between Cariani's original semantics and our reformulation of it.

3.2 Putting the threshold at optimality

Bronfman & Dowell note that Cariani's use of a set of alternatives (as a set of sets of worlds) and a ranking on them does not conflict *per se* with the standard approach in modal logic [3, p. 6]. The distinctive feature of Cariani's semantics, according to them, is the use of the permissibility threshold in order to block Inheritance. It is this feature that they attack.

To understand their argument, we should briefly rehearse the pragmatic defense of Inheritance for *ought*. This defense says, roughly speaking, that although affirming $ought(\varphi \vee \psi)$ is rather pointless in cases where we also know $ought(\varphi)$, the former expression is nevertheless true whenever the latter is. It is much like affirming "John is either Dutch or Italian" when we actually know that John is Dutch: not maximally helpful, but also not plainly false or mistaken. What *is* false is the Gricean implicature that follows when we *only* state $ought(\varphi \vee \psi)$, viz. that $\varphi \vee \psi$ is the most specific necessary condition for optimality.

Cariani rejects this defense of Inheritance, since it cannot account for the way *ought* behaves in embeddings [5, pp. 549]. Such behavior, he argues, can only be explained by the following principle:

(Implicated) $ought(\varphi \vee \psi)$ communicates that one has two ways of doing as one ought, viz. by making φ true or by making ψ true.

In contrast, Cariani's account covers (Implicated) well: as we saw in Section 2.3, $\mathsf{O}(\varphi \vee \psi) \Vdash_{\mathbf{CDL^c}} \mathsf{P}\varphi, \mathsf{P}\psi$.

However, Bronfman & Dowell rightly remark that (Implicated) gives counterintuitive results when applied to Cariani's own semantics. That is, by taking an option that is suboptimal but permissible, the agent is also doing as (s)he ought – at least if (Implicated) holds. Let us illustrate this with our running example. The options *bus* and *bike* are the only two optimal ones. However, since *walk* is permissible, $ought(bus \vee bike \vee walk)$ comes out true. But, given (Implicated), this means that by walking to the university, Lisa is doing as she ought.

Bronfman & Dowell suggest that, if one really wants to satisfy Cariani's requirement, one should put the threshold at optimality.[17] There are two

---

[16] See [23] where these claims are proven.

[17] There remains a problem though. Suppose that "Lisa ought to go to the supermarket" is true. Since the semantics satisfies replacement of equivalents, it follows that "Lisa ought to either go to the supermarket and pay for whatever she buys or go to the supermarket and steal something." Given (Implicated), it follows that by going to the supermarket and stealing something, Lisa is doing as she ought. So whatever refinement one proposes of

ways to implement this suggestion. The first is to change the semantic clause for P, such that $M, w \models \mathsf{P}\varphi$ iff, whenever $X \in \mathcal{A}$ is such that $M, w' \models \varphi$ for all $w' \in X$, then $X \in \mathcal{B}$. This means that $\mathcal{I}$ becomes superfluous in the semantics of the logic.

Secondly, one may leave the semantic clause for P unaltered, but treat $\mathcal{I}$ simply as the set of all suboptimal alternatives. This means that we impose the following frame condition on $\mathbf{CDL^c}$-models:

(C+) $\mathcal{I} = \mathcal{A} \setminus \mathcal{B}$

The advantage of this second approach – which we will follow in the remainder – is that it allows for a smooth comparison with Cariani's original proposal. Note that (C+) is equivalent to the conjunction of condition (C$\cap$) (see Section 2.1) and the following:

(C$\cup$) $\mathcal{I} \cup \mathcal{B} = \mathcal{A}$

Henceforth, let M be a $\mathbf{CDL^{bd}}$-model iff it is a $\mathbf{CDL^c}$-model that satisfies (C+); we denote the associated consequence relation by $\Vdash_{\mathbf{CDL^{bd}}}$.

Obviously, $\mathbf{CDL^{bd}}$ is an extension of $\mathbf{CDL^c}$. But exactly what additional validities (in our language $\mathcal{L}$) do we get from imposing this condition? Each of (1)-(10) from Section 2.3 hold also for $\Vdash \; = \; \Vdash_{\mathbf{CDL^{bd}}}$, and hence not much seems to change to the deontic part of the language.

However, once we consider the interaction with U, we do get an important additional feature: if two ought-claims are both true, their prejacents have the same extension in the model. Following [7], we call this *Uniqueness*:

$$\mathsf{O}\varphi, \mathsf{O}\psi \Vdash_{\mathbf{CDL^{bd}}} \mathsf{U}(\varphi \equiv \psi) \tag{14}$$

This property fails for $\mathbf{CDL^c}$ – witness our example: both $ought(bike \vee bus)$ and $ought(bike \vee bus \vee walk)$ are true, but one is obviously more specific than the other. Note that by $\mathbf{S5}$-properties of U, Uniqueness entails both Aggregation and Weakening for O.

Even if condition (C+) is well-motivated, *Uniqueness* may be hard to swallow from the viewpoint of natural language. One morning John may have to ensure that he gets to the office in time (p), but also that his children get to school in time (q). So $ought(p)$ and $ought(q)$ both seem true in this scenario. But John would be rather lucky if making q true would at once ensure that p also holds (or vice versa).

Still, this kind of critique misses the point behind Cariani's semantics. That is: once we fix a set of alternatives and a way to compare them, then (usually) there is no doubt that one or several of those alternatives are optimal. So in the above scenario, we are really looking at different sets of alternatives, or in more technical terms, different partitions of one and the same set of possible worlds.

Cariani's (or Kratzer's) semantics, pragmatic factors will anyway have to be called for at some point. (This example is a variant of Hansson's "vegetarian's free lunch" [12, p. 218].)

Another question is whether, if we do allow for several such partitions of the given W, there should be some interaction between the related *ought*-claims. We will not be able to tackle this important issue in the present paper and postpone it for future work.

3.3 Rejecting Uniformity

As we just saw, there are reasons for strengthening $\mathbf{CDL^c}$ in various ways. There are however also reasons for weakening $\mathbf{CDL^c}$, in the sense that it is no longer assumed that optimality and permissibility are uniform throughout a model. That is, rather than taking $\mathcal{B}$ and $\mathcal{I}$ as sets of alternatives, one may think of them as functions, taking as their argument worlds $w \in W$ (or alternatives $X \in \mathcal{A}$), and mapping these to sets of alternatives. This means in turn that the validities mentioned in (13) — see page 263 — are denied.

To motivate such a weakening, we can point to various arguments that have been put forth in the literature. First, from the viewpoint of game theory, it has been argued that which action of a given agent α is best, may depend on the actions other agents perform; hence, it will also depend on the specific world one happens to be at. See e.g. [1, Section 4.2] where this point is discussed and linked to some properties of the deontic operators.

Second, in [26], Wansing attacks specific constructions of deontic logic based on a branching-time framework, in which the truth of "obligation reports" (say, claims about what ought to be, what is best, what one ought to do, etc) depend only on the moment m of evaluation. This means that such claims are either true at all moment-history pairs m/h, or false at all m/h. In the present, more abstract framework, moments correspond to the entire set W, whereas moment/history-pairs correspond to single worlds $w \in W$.

Wansing's arguments for this claim are of two kinds: on the one hand, he says that certain obligations are simply of such a type that they depend on future contingents. For instance, if "you ought to give the prize to the winner of this race" is true, then depending on who actually wins (say a or b), it may be true that "you ought to give the prize to a" – but this will of course not be settled true. The other argument is more intricate, as it concerns the so-called Restricted Complement thesis from [2]. As Wansing shows, this thesis together with Uniformity trivializes nested ought-claims of the type "John ought to see to it that it is forbidden for Mary to eat the cake."

Third and last, Uniformity is typically rejected by actualist theories of *ought*. In contrast to possibilists, actualists argue that what ought to be depends on what is actually the case (now or in the future), rather than on what can be (or may become) the case.[18] Of course, the temporal dimension is not explicit in the simple $\mathbf{CDL^c}$-models we considered so far. Nevertheless, the fact that we abstract from the temporal dimension in our models seems a sufficient reason to remain neutral about those properties that would become

[18] See e.g. [15, Section 7.4.3] where the two views are briefly discussed and linked to two different notions of *ought* in stit logic. A more unified theory that encompasses both these notions is presented in [16].

problematic, once we add time back in.

4 Coarse Deontic Logics

Let us take stock. We first generalize Definition 2.1 from Section 2.2:

Definition 4.1 A **CDL**-*frame* is a tuple $F = \langle W, \mathcal{A}, \mathcal{B}, \mathcal{I} \rangle$, where W is a non-empty set, $\mathcal{A} \in \wp(\wp(W))$ is a partition of W, $\mathcal{B} : W \to \wp(\mathcal{A})$ maps every $w \in W$ to the set of w-*best options* in $\mathcal{A}$, and $\mathcal{I} : W \to \wp(\mathcal{A})$ maps every $w \in W$ to the set of w-*impermissible options* in $\mathcal{A}$.

The definition of a model and the semantic clauses remain the same, with the exception of the following:

(SC6') $M, w \models \mathsf{B}\varphi$ iff for all $X \in \mathcal{B}(w)$, for all $v \in X$, $M, v \models \varphi$
(SC7') $M, w \models \mathsf{P}\varphi$ iff for all $X \in \mathcal{A}$ s.t. (for all $v \in X$, $M, v \models \varphi$), $X \in \mathcal{A} \setminus \mathcal{I}(w)$

Table 1 gives a sound and (strongly) complete axiomatization of **CDL**. The first six axioms and rules in this table are standard. The axioms (G_A), (G_B) and (G_P) follow from the fact that U is a global modality. (C_B), (C_P), (P1) and (P2) were already discussed in Section 2.3. Finally, (EQ_P) is a strengthened version of replacement of equivalents for P.

(CL)	any complete axiomatization of classical propositional logic
(MP)	from $\varphi, \varphi \supset \psi$ to infer ψ
(NEC_U)	from $\vdash \varphi$, to infer $\vdash \mathsf{U}\varphi$
(K_B)	$\mathsf{B}(\varphi \supset \psi) \supset (\mathsf{B}\varphi \supset \mathsf{B}\psi)$
($\mathrm{S5}_\mathsf{A}$)	**S5** for U
($\mathrm{S5}_\mathsf{A}$)	**S5** for A
(G_A)	$\mathsf{U}\varphi \supset \mathsf{A}\varphi$
(G_B)	$\mathsf{U}\varphi \supset \mathsf{B}\varphi$
(G_P)	$\mathsf{U}\varphi \supset \mathsf{P}\neg\varphi$
(C_B)	$\mathsf{B}\varphi \equiv \mathsf{B}\mathsf{A}\varphi$
(C_P)	$\mathsf{P}\varphi \equiv \mathsf{P}\mathsf{A}\varphi$
(P1)	$\mathsf{P}(\varphi \vee \psi) \supset (\mathsf{P}\varphi \wedge \mathsf{P}\psi)$
(P2)	$(\mathsf{P}\neg\mathsf{A}\neg\varphi \wedge \mathsf{P}\neg\mathsf{A}\neg\psi) \supset \mathsf{P}(\neg\mathsf{A}\neg\varphi \vee \neg\mathsf{A}\neg\psi)$
(EQ_P)	$\mathsf{U}(\varphi \equiv \psi) \supset (\mathsf{P}\varphi \equiv \mathsf{P}\psi)$

Table 1
Axiomatization of **CDL**.

Table 2 provides an overview of the conditions on **CDL**-frames we have considered so far, and the axioms (if any) that correspond to these frame conditions. Where (C1), ..., (Cn) are frame conditions from Table 2, say M is an $\mathbf{CDL}_{\mathbf{C1},\ldots,\mathbf{Cn}}$-model iff M is an **CDL**-model and M obeys these conditions. We use $\Vdash_{\mathbf{CDL}_{\mathbf{C1},\ldots,\mathbf{Cn}}}$ to refer to the associated semantic consequence relation. Note that $\mathbf{CDL}^{\mathbf{c}} = \mathbf{CDL}_{\mathrm{U}_\mathsf{B},\mathrm{U}_\mathsf{P}}$ and $\mathbf{CDL}^{\mathbf{bd}} = \mathbf{CDL}_{\mathrm{U}_\mathsf{B},\mathrm{U}_\mathsf{P},\mathrm{C}+}$.

Not all of these conditions are independent. As noted, (C+) is equivalent to the conjunction of (C$\cup$) and (C$\cap$). In view of axiom (G_A), (U_B) implies (A_B), and (U_P) implies (A_P). Also, (C+) implies that (U_B) and (U_P) are equivalent,

(U_B)	for all $w, w' \in W$, $\mathcal{B}(w) = \mathcal{B}(w')$	$\mathsf{B}\varphi \equiv \mathsf{U}\mathsf{B}\varphi$
(U_P)	for all $w, w' \in W$, $\mathcal{I}(w) = \mathcal{I}(w')$	$\mathsf{P}\neg\mathsf{A}\varphi \equiv \mathsf{U}\mathsf{P}\neg\mathsf{A}\varphi$
(A_B)	where $w, w' \in X$, $\mathcal{B}(w) = \mathcal{B}(w')$	$\mathsf{B}\varphi \equiv \mathsf{A}\mathsf{B}\varphi$
(A_P)	where $w, w' \in X$, $\mathcal{I}(w) = \mathcal{I}(w')$	$\mathsf{P}\neg\mathsf{A}\varphi \equiv \mathsf{A}\mathsf{P}\neg\mathsf{A}\varphi$
(D)	for all $w \in W$, $\mathcal{B}(w) \neq \emptyset$	$\mathsf{B}\varphi \supset \neg\mathsf{B}\neg\varphi$
(C$\cup$)	for all $w \in W$, $\mathcal{B}(w) \cup \mathcal{I}(w) = W$	$(\mathsf{B}\varphi \wedge \mathsf{P}\neg\mathsf{A}\varphi) \supset \mathsf{U}\varphi$
(C$\cap$)	for all $w \in W$, $\mathcal{B}(w) \cap \mathcal{I}(w) = \emptyset$	-
(C+)	for all $w \in W$, $\mathcal{B}(w) = \mathcal{A} \setminus \mathcal{I}(w)$	-

Table 2
Frame conditions and axioms for **CDL**.

and that (A_B) and (A_I) are equivalent. We leave it to the reader to verify that this exhausts the dependencies between each of the frame conditions from Table 2.

Whereas most of the frame conditions are modally expressible, (C+) and (C$\cap$) are not. This means that there is no formula φ which is globally valid on all and only those **CDL**-frames that satisfy (C+). One can nevertheless give a sound and complete, Hilbert-style axiomatization of the logics in question, by adding the associated axioms from Table 2 to **CDL**. One can moreover show that all these logics satisfy the finite model property, and hence are decidable. We refer to [23] where each of these claims are spelled out in exact terms and proven.

5 Related Work

CDL and its extensions bear close resemblances to existing work in deontic logic. In fact, leaving some specific modeling choices aside, one could say that they are just a combination of two well-known constructions in the field. We only explain both of these here in a nutshell; a more detailed and exact comparison of the respective logics is left for future work.

5.1 Deontic necessity and sufficiency

The idea of combining a notion of necessity and sufficiency for modeling *ought* was proposed fairly recently in [1,19] under the name "obligation as weakest permission". The idea is that what one ought to do is that which is implied by every strongly permitted proposition, where a proposition is strongly permitted iff it is sufficient for optimality. The resulting *ought*-operator satisfies the same basic properties as our O in $\mathbf{CDL_{C+}}$ does – Uniqueness, and hence also Aggregation and Weakening. Likewise, it does not satisfy Inheritance, Uniformity, and the rule of Necessitation.

In [24], richer logics are studied in which both deontic necessity and sufficiency are expressible, which can be traced back to an extended abstract by van Benthem [22]. As shown in [7, Section 3], the deontic action logic from [21] is a fragment of van Benthem's system, and hence belongs to the same family of logics.

The main difference between the aforementioned logics and the **CDL**-family is that the former speak about the optimality (permissibility) of single worlds or action-tokens, whereas the latter speak about sets of worlds or action-types. As a result, we can also express the additional condition that φ is visible whenever $ought(\varphi)$ is true. Also, because of this feature, the logic of obligation as weakest permission and its relatives do not allow for coarseness.

5.2 Deontic stit logics

Deontic logics for optimal actions, which are conceived as *sets* of worlds, are at least as old as Horty's [14], which he further developed in the 2001 book [15].[19] Roughly speaking, $ought_\alpha(\varphi)$ is true at a world w in a model of Horty's most basic semantics if and only if α sees to it that φ whenever it takes one of its best options at w. Horty further distinguishes between two ways to determine what the best options are; one is called *dominance act utilitarianism* and satisfies Uniformity; the other is called *orthodox act utilitarianism* and invalidates Uniformity.[20] Both satisfy the (D)-axiom (see Table 2). Horty's $ought_\alpha$-operators are hence much like the B-operator of $\mathbf{CDL_{D,U_B}}$ (resp. $\mathbf{CDL_D}$), with the obvious difference that they refer explicitly to an agent or group of agents. Horty's systems lack an operator for strong permission (our P).

Apart from the usual benefits – the transfer of insights and results from one system to the other –, there is one particular sense in which this link can be highly useful. In [20], Snedegar considers the problem of *coarsening inferences*, i.e. inferences that involve sets of alternatives that differ in their degree of coarseness. Snedegar's question then is: how do ought-claims relative to $\mathcal{A}$ relate to ought-claims relative to a finer partition $\mathcal{A}'$?

In view of the preceding, this question is analogous to asking how the obligations of a group of agents relate to the obligations of subgroups of that group, within the framework of deontic stit-logic.[21] Indeed, the alternatives that are available to the group correspond exactly to a partition that refines the partition corresponding to the alternatives available to a subgroup.

6 Summary and Outlook

The main contribution of this paper consists in the formal study of different variants of Cariani's semantics for *ought*. Spelling out these variants in turn allowed us to point at links with existing work in deontic logic, most particularly the logic of obligation as weakest permission and deontic stit logic.

Many issues remain unsettled, such as a more exact comparison of these systems. As explained, the link with deontic stit logic suggests possible solu-

[19] In Horty's stit-based semantics, the points of evaluation are moment-history pairs rather than worlds, and the sets of worlds are rather sets of histories. There is however a one-to-one correspondence between such models and more regular Kripke-models – see e.g. [13, Section 2.1].

[20] In [16], Horty proposes a way to unify both accounts and hence overcome semantic ambiguity w.r.t. "the right action(s)".

[21] Horty discusses this relation in 6.2 of his book, showing that dominant act utilitarianism differs from orthodox act utilitarianism in this respect. See e.g. [18] for formal results on this matter.

tions to the problem of coarsening inferences; in future work we want to study this relation in more detail. Also, it is an open question whether deontic stit logic can be enriched with an operator for strong permission, and in particular, how such an operator will behave for group obligations.

References

[1] Albert J.J. Anglberger, Norbert Gratzl, and Olivier Roy. Obligation, free choice, and the logic of weakest permissions. *Review of Symbolic Logic*, 8(4):807–827, 2015.

[2] N. Belnap, Perloff M., Xu M., and Bartha P. *Facing the Future: Agents and Choice in Our Indeterminist World.* Oxford University Press, 2001.

[3] Aaron Bronfman and Janice Dowell. The language of reasons and oughts. In: Star, D. (ed.) Oxford Handbook of Reasons. Oxford University Press, Oxford (forthcoming).

[4] Fabrizio Cariani. Logical consequence for actualists (and other contrastivists). Unpublished manuscript, April 15, 2015.

[5] Fabrizio Cariani. "Ought" and resolution semantics. *Noûs*, 47(3):534–558, 2013.

[6] Brian Chellas. *Modal Logic: an Introduction.* Cambridge : Cambridge university press, 1980.

[7] Huimin Dong and Olivier Roy. Three deontic logics for rational agency in games. In *Studies in Logic*, volume 8, pages 7–31. College Publications, 2015.

[8] Lou Goble. A logic of "good", "should", and "would": Part I. *Journal of Philosophical Logic*, 19(2):169–199, 1990.

[9] Lou Goble. A logic of "good", "should", and "would": Part II. *Journal of Philosophical Logic*, 19(3):253–76, 1990.

[10] Lou Goble. Normative conflicts and the logic of "ought". *Noûs*, 43(3):450–489, 2009.

[11] Valentin Goranko and Solomon Passy. Using the universal modality: Gains and questions. *Journal of Logic and Computation*, 2(1):5 – 30, 1992.

[12] Sven Ove Hansson. *The Varieties of Permission*, chapter 3, pages 195–240. College Publications, 2013.

[13] Andreas Herzig and François Schwarzentruber. Properties of logics of individual and group agency. In *Advances in Modal Logic*, 2008.

[14] John F. Horty. *Deontic Logic, Agency and Normative Systems: ΔEON '96: Third International Workshop on Deontic Logic in Computer Science, Sesimbra, Portugal, 11 - 13 January 1996*, chapter Combining Agency and Obligation (Preliminary Version), pages 98–122. Springer London, London, 1996.

[15] John F. Horty. *Agency and Deontic Logic.* Oxford University Press, 2001.

[16] John F. Horty. Perspectival act utilitarianism. In Patrick Girard, Olivier Roy, and Mathieu Marion, editors, *Dynamic Formal Epistemology*, volume 351 of *Synthese Library*, pages 197–221. Springer Netherlands, 2011.

[17] Frank Jackson. On the semantics and logic of obligation. *Mind*, 94:177–195, 1985.

[18] Barteld Kooi and Allard Tamminga. Moral conflicts between groups of agents. *Journal of Philosophical Logic*, 37(1):1–21, February 2008.

[19] Olivier Roy, Albert J.J. Anglberger, and Norbert Gratzl. The logic of obligation as weakest permission. In Thomas Agotnes, Jan Broersen, and Dag Elgesem, editors, *Deontic Logic in Computer Science*, volume 7393 of *Lecture Notes in Computer Science*, pages 139–150. Springer Berlin Heidelberg, 2012.

[20] Justin Snedegar. Deontic reasoning across contexts. In Fabrizio Cariani, Davide Grossi, Joke Meheus, and Xavier Parent, editors, *Deontic Logic and Normative Systems*, volume 8554 of *Lecture Notes in Computer Science*, pages 208–223. Springer International Publishing, 2014.

[21] Robert Trypuz and Piotr Kulicki. A systematics of deontic action logics based on boolean algebra. *Logic and Logical Philosophy*, 18(3-4):253–270, 2010.

[22] Johan van Benthem. Minimal deontic logics (abstract). *Bulletin of the Section of Logic*, 8(1):36–41, 1979.

[23] Frederik Van De Putte. Coarse Deontic Logic (Extended Version). In preparation.

[24] Frederik Van De Putte. That will do: Logics of deontic necessity and sufficiency. Under review, 2016.

[25] Georg Henrik von Wright. *An Essay in Deontic Logic and the General Theory of Action.* North-Holland Publishing Company, 1968.

[26] Heinrich Wansing. Nested deontic modalities: Another view of parking on highways. *Erkenntnis*, 49(2):185–199, 1998.

Deontic Logic as a Study of Conditions of Rationality in Norm-related Activities

Berislav Žarnić [1]

Faculty of Humanities and Social Sciences
University of Split
Split, Croatia

Abstract

The program put forward in von Wright's last works defines deontic logic as "a study of conditions which must be satisfied in rational norm-giving activity" and thus introduces the perspective of logical pragmatics. In this paper a formal explication for von Wright's program is proposed within the framework of set-theoretic approach and extended to a two-sets model which allows for the separate treatment of obligation-norms and permission norms. The three translation functions connecting the language of deontic logic with the language of the extended set-theoretical approach are introduced, and used in proving the correspondence between the deontic theorems, on one side, and the perfection properties of the norm-set and the "counter-set", on the other side. In this way the possibility of reinterpretation of standard deontic logic as the theory of perfection properties that ought to be achieved in norm-giving activity has been formally proved. The extended set-theoretic approach is applied to the problem of rationality of principles of completion of normative systems. The paper concludes with a plaidoyer for logical pragmatics turn envisaged in the late phase of Von Wright's work in deontic logic.

Keywords: Deontic logic, logical pragmatics, reinterpretation of standard deontic logic, G.H. von Wright.

1 Von Wright's reinterpretation of deontic logic

The foundational role and crucial influence of Georg Henrik von Wright (1916–2003) in the development of deontic logic is beyond dispute. [2] Recently, Bulygin [3] has has divided von Wright's work in deontic logic in four phases: 1) dogmatic phase of 1950s marked by ignoring the fact that norms do not have truth-value; 2) eclectic phase of "Norm and Action" introducing the distinction between logic of norms and logic of norm propositions; 3) sceptic phase marked

---

[1] This work has been supported in part by Croatian Science Foundation's funding of the project IP-2014-09-9378 and by the grant of Faculty of Humanities and Social Sciences, University of Split.

[2] Although usually called "the founding father of deotic logic", Von Wright preferred and used the term 'midwife' to denote his role in the development of the discipline.

by the thesis that logic of norms is impossible; 4) logic without truth-phase with the reinterpretation of deontic logic as the study of rationality conditions of the norm-giving activity. Von Wright's reinterpretation of deontic logic given in his later works (from 1980s onwards) has remained a non-formalized manifesto which so far has not received a fuller elaboration. In this paper the reinterpretation will be understood as the turn towards logical pragmatics. An exemplar programmatic statement is given in the following quote.

> Deontic logic, one could also say, is neither a logic of norms nor a logic of norm-propositions but a study of conditions which must be satisfied in rational norm-giving activity. [13, p.111]

Von Wright's reinterpretation of deontic logic developed gradually and has introduced important conceptual distinctions and theses, among which the following stand out: the distinction between prescriptive and descriptive use of deontic sentences [10]; the thesis that relation between permission and absence of prohibition is not conceptual but normative in character [12]; this relation is one among "perfection properties" of normative system that the norm-giver ought to achieve in norm-giving activity [14,15]. These theses are mutually supporting. A normative systems can come into existence thanks to the prescriptive use of language . The logical properties of real normative systems can be described using the language of the "logic of norm-propositions". Some logical properties are "perfection-properties" of a normative-system. The absence of a certain perfection-property does not deprive a normative system of its normative force. In the prescriptive use of language the norm-giver ought to achieve some perfection properties of the normative system. Deontic logic is a study of logical perfection properties; properties which act as the normative source of requirements to which the norm-giver is subordinated.

Von Wright's reinterpretation of deontic logic can be formally explicated within the set-theoretic approach. The set-theoretic approach has been introduced into the logic of normative systems by Alchourrón and Bulygin [1]. Within this framework deontic sentences are treated as claims on membership in the set of consequences $Cn(\mathcal{N})$ of "explicitly commanded propositions" $\mathcal{N}$. Thus, $\mathrm{O}\varphi$ in their approach means $\varphi \in Cn(\mathcal{N})$, while $\mathrm{P}\varphi$ is explicated as $\neg\varphi \notin Cn(\mathcal{N})$. More recently a refinement and generalization of the set-theoretic approach has been developed by Broome [2] where the set of requirements is equated with the value of a code function, which takes as its arguments a normative source, an actor and a situation. The major point of divergence within the set-theoretic approaches lies in the properties that are assigned to sets of norms or requirements [17]. It is in accord with the approach proposed by Von Wright to treat real norm-sets, the one corresponding to obligation-norms and the other to permission-norms, as simple sets consisting just of affirmed and negated propositional contents of explicitly promulgated norms, not presupposing any a priori given properties. Rather, it is the question of compliance with second-order normativity whether a real normative system posses desirable logical properties and approximates an ideal system. In the approach of this paper it is neither assumed that a norm-set is deductively closed nor that it is closed under equivalence.

If permission and obligation are not interdefinable, then two types of consistency should be distinguished. External consistency deals with the relation between obligation-norms and permission norms: $\neg(\mathrm{O}\varphi \wedge \mathrm{P}\neg\varphi)$. Internal consistency deals with obligation-norms alone: $\neg(\mathrm{O}\varphi \wedge \mathrm{O}\neg\varphi)$. According to Von Wright, the set of obligation-norms ought to have perfection properties.

> ... classic deontic logic, on the descriptive interpretation of its formulas, pictures a gapless and contradiction-free system of norms. A factual normative order *may* have these properties, and it may be thought desirable that it *should* have them. But can it be a *truth of logic* that a normative order has ("must have") these "perfection"-properties? [14, p.20]

Perfection properties produce "normative demands on normative systems" and define "rationality conditions of norm-giving activity". If the relation between permission and absence of prohibition is not a conceptual relation, then an addition to Von Wright's outline is required. It is not sufficient to determine perfection properties of the set of obligation-norms. Perfection properties of the set corresponding to permission-norms must be taken into account, too, as well as perfection relations between obligation-norms and permission-norms, like external consistency. The needed extension of the set-theoretic approach can be obtained by the addition of the set related to permission norms. The formal explication of the relation between standard deontic logic and the theory of normative system perfection properties requires a provision of the translation function from the language of standard deontic logic without iterated operators to the language of extended set-theoretic approach. The translation function should reveal the fact that axioms of standard deontic logic are descriptions of an ideal normative system, a system endowed with "perfection properties". These are the properties that a normative system ought have, as von Wright noted, and, as will be argued here, these are the properties to which the norm-giver and the norm-recipient relate in their corrective activities when faced with an imperfect normative system.

2 Perfection properties of a normative system

According to the extended Von Wright's reinterpretation of deontic logic the norm-giver and the norm-recipient relate to the ideal concepts of obligation and permission.

Definition 2.1 Let $\mathcal{L}_{pl}$ be the language of propositional logic. A set $\mathcal{N} \subseteq \mathcal{L}_{pl}$ is called norm-set and contains contents of obligation-norms. A set $\overline{\mathcal{N}} \subseteq \mathcal{L}_{pl}$ is called counter-set and contains negated contents of permission-norms. A normative system is the pair $\langle \mathcal{N}, \overline{\mathcal{N}} \rangle$.

The ideal concepts of obligation and permission can be explicated by pointing out the "perfection properties" of their corresponding sets, namely, of the norm-set and the counter-set. Since the filter structure and the weak-ideal structure of the respective sets will be later recognized as responsible for their perfection properties these terms must be introduced. The first one is a well-know concept while the second will be introduced here.

2.1 Filter and weak ideal

It is well-known fact that the set of truth-sets of sentences belonging to a consistent and deductively closed set of sentences exemplifies a "filter" structure [4]. A filter F is a set of subsets of a given set W satisfying the following conditions [6, p.73]: (i) $\emptyset \notin F$, (ii) $W \in F$, (iii) if $X \in F$ and $Y \in F$, then $X \cap Y \in F$, (iv) for all $X, Y \subseteq W$, if $X \in F$ and $X \subseteq Y$, then $Y \in F$. In classical propositional logic the set of sets of valuations $\{[\![\varphi]\!] \mid \varphi \in Cn(T)\}$ is a filter if $Cn(T)$ is consistent, where $T \subseteq \mathcal{L}_{pl}$ is a theory, $Cn(T) = \{\varphi \mid T \vdash_{pl} \varphi\}$ is its deductive closure, and $[\![\varphi]\!] = \{v \mid v(\varphi) = \mathrm{t}\}$ is the truth-set of φ. The properties of a filter can be reformulated in terms of logical syntax. In particular, reformulated condition (iii) expresses the closure under conjunction; reformulated condition (iv) expresses closure under entailment, i.e., if $\varphi \in Cn(T)$ and φ entails ψ, then $\psi \in Cn(T)$.

On the other hand, the set-theoretic structure corresponding to the "counter-theory", $\mathcal{L} - Cn(T)$ is closed in the opposite direction: if $[\![\varphi]\!]$ corresponds to some $\varphi \in (\mathcal{L} - Cn(T))$, then so does any subset of it. In the syntactic reformulation: if $\varphi \in (\mathcal{L} - Cn(T))$ and ψ entails φ, then $\psi \in (\mathcal{L} - Cn(T))$. The structure of an 'ideal' is a particular kind of structure, which can be found in some but not all sets of truth-sets of sentences in counter-theories. An ideal I is defined as a set of subsets of a given set W satisfying the following conditions [6, p.73]: (i) $\emptyset \in I$, (ii) $W \notin I$, (iii) if $X \in I$ and $Y \in I$, then $X \cup Y \in I$, (iv) for all $X, Y \subseteq W$, if $X \in I$ and $Y \subseteq X$, then $Y \in I$. Conditions can be reformulated in syntactic terms. In particular, condition (iv) can be reformulated as the 'closure under implicants'.

The complement of a filter needs not be an ideal, but if a theory is complete and consistent, its corresponding filter will be maximal, and its complement will be an ideal. The complement of any filter shares an essential property of the structure of an ideal, namely the property of "closure under implicants" as the first item in Proposition 2.2 shows.

Proposition 2.2 *If $S = W - F$ and F is a filter, then*

(i) *if $[\![\varphi]\!] \subseteq [\![\psi]\!]$ and $[\![\psi]\!] \in S$, then $[\![\varphi]\!] \in S$,*

(ii) *if $[\![\varphi]\!] \cap [\![\psi]\!] \in S$, then $[\![\varphi]\!] \in S$ or $[\![\psi]\!] \in S$,*

In this paper the question of logical structure of "counter-theory" will play an important role in the determination of perfection properties of the counter-set. Therefore, the new notion of weak ideal will be introduced.

Definition 2.3 A structure S is a weak ideal iff (i) $[\![\varphi]\!] \subseteq [\![\psi]\!]$ and $[\![\psi]\!] \in S$, then $[\![\varphi]\!] \in S$, and (ii) if $[\![\varphi]\!] \cap [\![\psi]\!] \in S$, then $[\![\varphi]\!] \in S$ or $[\![\psi]\!] \in S$.

Syntactic conditions corresponding to a weak ideal structure are: (i) inclusion of at least one conjunct for each conjunction contained, and (ii) closure under implicants, respectively.

Proposition 2.4 *Let $T \subseteq \mathcal{L}_{pl}$ and $Cn(T) \neq \mathcal{L}_{pl}$. The set $\{[\![\varphi]\!] \mid \varphi \in Cn(T)\}$ is a filter. The set $\{[\![\varphi]\!] \mid \varphi \in \mathcal{L}_{pl} - Cn(T)\}$ is a weak ideal.*

3 Translations, theorems of standard deontic logic and ideal normative systems

A formal explication of Von Wright's reinterpretation of deontic logic asks for the establishment of a connection between the theorems of standard deontic logic and properties of ideal normative systems. For this purpose the translation function has been introduced in [17] connecting theorems standard deontic logic with the properties of the norm-set. Now additional translations function will be introduced, connecting deontic theorems also with the perfection properties of the counter-set and perfection relations between the norm-set and the counter-set.

Definition 3.1 Language $\mathcal{L}_{\mathrm{sdl}}$ is a deontic language without iterated modalities: $\varphi ::= p \mid \mathrm{O}\varphi \mid \mathrm{P}\varphi \mid \neg\varphi \mid (\varphi \wedge \varphi) \mid$, where p is a sentence of language $\mathcal{L}_{pl}$ of propositional logic. The definitions of deontic modality F and of truth-functional connectives are standard.

Definition 3.2 Language $\mathcal{L}_{ns}$ is the language of the norm-set and counter-set membership within the extended set-theoretic approach: $\varphi ::= p \mid \ulcorner p \urcorner \in \mathcal{N} \mid \ulcorner p \urcorner \in \overline{\mathcal{N}} \mid \neg\varphi \mid (\varphi \wedge \varphi) \mid$ where $p \in \mathcal{L}_{pl}$.[3]

Definition 3.3 Functions $\tau^{+}, \tau^{-}, \tau^{*} : \mathcal{L}_{sdl} \mapsto \mathcal{L}_{ns}$ translate formulas of the deontic language $\mathcal{L}_{sdl}$ without iterated modalities to the language $\mathcal{L}_{ns}$ of the extended set-theoretic approach.

$$\begin{aligned}
\tau^{+}(\mathrm{O}\varphi) &= \ulcorner\varphi\urcorner \in \mathcal{N}\\
\tau^{+}(\mathrm{P}\varphi) &= \ulcorner\neg\varphi\urcorner \notin \mathcal{N}\\
\tau^{+}(\mathrm{P}\neg\varphi) &= \ulcorner\varphi\urcorner \notin \mathcal{N}\\
\tau^{-}(\mathrm{O}\varphi) &= \ulcorner\varphi\urcorner \notin \overline{\mathcal{N}}\\
\tau^{-}(\mathrm{P}\varphi) &= \ulcorner\neg\varphi\urcorner \in \overline{\mathcal{N}}\\
\tau^{-}(\mathrm{P}\neg\varphi) &= \ulcorner\varphi\urcorner \in \overline{\mathcal{N}}\\
\tau^{*}(\mathrm{O}\varphi) &= \tau^{+}(\mathrm{O}\varphi)\\
\tau^{*}(\mathrm{P}\varphi) &= \tau^{-}(\mathrm{P}\varphi)
\end{aligned}$$

For $\star = +, -, *$

$$\begin{aligned}
\tau^{\star}(\varphi) &= \varphi \text{ if } \varphi \in \mathcal{L}_{pl}\\
\tau^{\star}(\neg\varphi) &= \neg\tau^{\star}(\varphi)\\
\tau^{\star}(\varphi \wedge \psi) &= \tau^{\star}(\varphi) \wedge \tau^{\star}(\psi)
\end{aligned}$$

[3] "Quine quotes", $\ulcorner \ldots \urcorner$, will be omitted at most places in the subsequent text for the ease of reading and writing.

POSTULATES OF STANDARD DEONTIC LOGIC	NORM-SET PROPERTIES	COUNTER-SET PROPERTIES
(D) $O\varphi \rightarrow P\varphi$	consistency $\tau^{+}(D) = \varphi \in \mathcal{N} \rightarrow \neg\varphi \notin \mathcal{N}$	completeness $\tau^{-}(D) = \varphi \notin \overline{\mathcal{N}} \rightarrow \neg\varphi \in \overline{\mathcal{N}}$
(2*) $(O\varphi \wedge O\psi) \rightarrow O(\varphi \wedge \psi)$	closure under conjunction $\tau^{+}(2^*) = (\varphi \in \mathcal{N} \wedge \psi \in \mathcal{N}) \rightarrow (\varphi \wedge \psi) \in \mathcal{N}$	having at least one conjunct for each conjunction contained $\tau^{-}(2^*) = (\varphi \wedge \psi) \in \overline{\mathcal{N}} \rightarrow (\varphi \in \overline{\mathcal{N}} \vee \psi \in \overline{\mathcal{N}})$
(Rc) $\dfrac{\vdash_{pl} \varphi \rightarrow \psi}{O\varphi \rightarrow O\psi}$	deductive closure $\tau^{+}(O\varphi \rightarrow O\psi) = \varphi \in \mathcal{N} \rightarrow \psi \in \mathcal{N}$ if $\vdash_{pl} \varphi \rightarrow \psi$	"closure under implicants" $\tau^{-}(O\varphi \rightarrow O\psi) = \psi \in \overline{\mathcal{N}} \rightarrow \varphi \in \overline{\mathcal{N}}$ if $\vdash_{pl} \varphi \rightarrow \psi$
	RELATIONAL PROPERTIES	
(D*) $O\varphi \rightarrow \neg P \neg\varphi$	external consistency $\tau^{*}(D^*) = \varphi \in \mathcal{N} \rightarrow \varphi \notin \overline{\mathcal{N}}$	
(Com) $O\varphi \vee P\neg\varphi$	"gaplessness" $\tau^{*}(Com) = \varphi \in \mathcal{N} \vee \varphi \in \overline{\mathcal{N}}$	

Table 1
Perfection properties come in non-equal pairs where each member is charcterized by the same axiom or rule.

On counter-sets Although counter-intuitive at the first glance, the adequate metaphor for permitting is that of putting the negation of the content into the counter-set. This corresponds to the standard definition "φ is permitted iff it is not obligatory that $\neg\varphi$" in the following way: since $\neg\varphi$ cannot go into the norm-set it must be placed into the counter-set. The perfection properties are different for different sets since "ideal concepts" of obligation and permission have different logical structure. For example, having a contradictory pair is an imperfection property of the norm-set, but for the counter-set this is neither a perfection nor an imperfection property. Similarly, completeness is a perfection property for permissions but not for obligations: it is indifferent whether $\varphi \in \mathcal{N} \vee \neg\varphi \in \mathcal{N}$ holds, whereas $\varphi \in \overline{\mathcal{N}} \vee \neg\varphi \in \overline{\mathcal{N}}$ ought to hold. This model, as will be shown, can account for the fact that perfection properties come in pairs, one for obligations, another for permissions, both of which are characterized by the same theorems of standard deontic logic, as shown in Table 1. The difference in logical structure of the two sets is also visible from the following facts: A perfect counter-set can have a contradictory pair of (negations) of permission-norm contents, which means that a certain state of affairs is optional. This fact does not cause an "explosion" since the principle *ex contradictione quodlibet* does not hold for the ideal counter-set.

The proposed two-sets model bears resemblance to the relation between a theory T and its counter-part $\mathcal{L} - Cn(T)$. The counter-theory has logical properties such as "closure under the implicant" (if $\psi \in \mathcal{L} - Cn(T)$ and φ entails

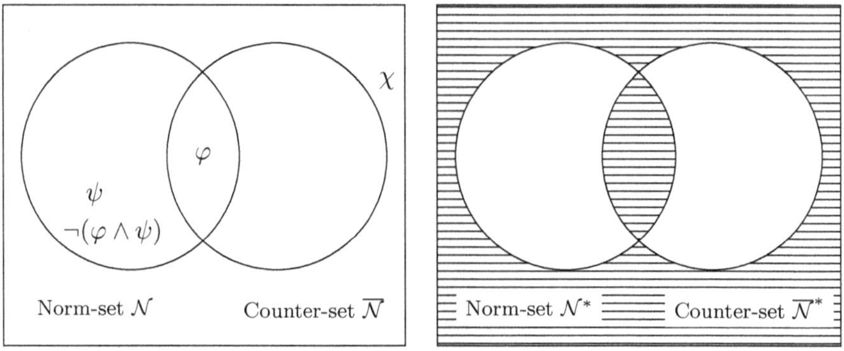

Fig. 1. A comparison between imperfect system $\langle\mathcal{N},\overline{\mathcal{N}}\rangle$ and system $\langle\mathcal{N}^*,\overline{\mathcal{N}}^*\rangle$, which is endowed with some perfection properties. System $\langle\mathcal{N},\overline{\mathcal{N}}\rangle$ is internally inconsistent under deductive closure; it is externally inconsistent as the presence of φ at the intersection shows and makes both $\mathrm{O}\varphi$ and $\mathrm{P}\neg\varphi$ true; it is also incomplete since $\chi \notin \mathcal{N} \cup \overline{\mathcal{N}}$. System $\langle\mathcal{N}^*,\overline{\mathcal{N}}^*\rangle$ is externally consistent and complete.

ψ, then $\varphi \in \mathcal{L} - Cn(T)$). The perfection properties of the descriptive theory have been well investigated within the logic of natural sciences. For example, the completeness of a theory, if attainable, counts as a perfection property, but the completeness of the (obligation) norm-set is not its perfection property. The mismatch holds also on the side of "counter-sets": the completeness of the descriptive counter-part $\mathcal{L} - Cn(T)$ is an indifferent property, while in the realm of normativity it is a perfection property of the "counter-set" representing permission-norms. The construction is different, too: there is no "exclusion" part in building a theory since rejecting a sentence equals accepting its negation. This need not be the case with normative systems, whose obligation and permission parts are separately built. These facts shows that deontic logic as the study of "rationality conditions of norm-giving activity" or "perfection properties of normative systems" is a sui generis logic. If one accepts, together with von Wright, the central position of the phenomenon of normativity in humanities and social sciences, then deontic logic plays the prominent role in the philosophy of the science of man by revealing the logical basis of its methodological autonomy.

4 Deontic logic as the theory of ideal normative systems

As Aristotle famously wrote in *Nicomachean Ethics*, "it is possible to fail in many ways ... while to succeed is possible only in one way". The same goes for constructing a normative system by prescriptive use of language: there are many imperfect normative systems in reality, but only one ideal system, the one, as will be proved here, described by standard deontic logic; compare Table 1.

An ideal normative system (*ins*) is internally (IntC) and externally consistent (ExtC), its obligation norm-set is closed under conjunction (2*) and entailment (Rc), and it is complete (Comp). An ideal normative system is characterized by the following axioms and rules:

(1) $\vdash_{ins} \top \in \mathcal{N}$

(2*) $\vdash_{ins} (\varphi \in \mathcal{N} \wedge \psi \in \mathcal{N}) \rightarrow \varphi \wedge \psi \in \mathcal{N}$

(IntC) $\vdash_{ins} \varphi \in \mathcal{N} \rightarrow \neg\varphi \notin \mathcal{N}$

(ExtC) $\vdash_{ins} \varphi \in \mathcal{N} \rightarrow \varphi \notin \overline{\mathcal{N}}$

(Comp) $\vdash_{ins} \varphi \in \mathcal{N} \vee \varphi \in \overline{\mathcal{N}}$

(Rc) $\dfrac{\vdash_{pl} \varphi \rightarrow \psi}{\vdash_{ins} \mathrm{O}\varphi \rightarrow \mathrm{O}\psi}$

Other properties of an ideal normative system are consequences of these axioms and rules. In particular, perfection properties of the counter-set, closure under implicant and inclusion of at least one conjunct for each conjunction contained, can be derived in $\vdash_{ins}$. The three translation functions when applied to theorems of standard deontic logic (*sdl*) yield the following descriptions:

- if $\vdash_{sdl} \varphi$, then $\tau^{+}(\varphi)$ describes a perfection property of the (obligation) norm-set;
- if $\vdash_{sdl} \varphi$, then $\tau^{-}(\varphi)$ describes a perfection property of the (permission) counter-set;
- if $\vdash_{sdl} \varphi$ and both O and P occur in φ, then $\tau^{*}(\varphi)$ describes a perfection relation between the norm-set and the counter-set.

Since translations of axioms and rules of standard deontic logic yield truths about an ideal normative system, they can be understood as the theory of ideal normative system thus confirming Von Wright's conjecture.

Theorem 4.1 *If* $\vdash_{sdl} \varphi$, *then* $\vdash_{ins} \tau^{+}(\varphi)$, $\vdash_{ins} \tau^{-}(\varphi)$, *and* $\vdash_{ins} \tau^{*}(\varphi)$.

Proof. All axioms and rules of standard deontic logic can be derived in $\vdash_{ins}$. Therefore, any step of a proof in $\vdash_{sdl}$ can be reproduced within $\vdash_{ins}$. □

For the purpose of illustration the proofs for $\vdash_{ins} \tau^{+}(6.11)$, $\vdash_{ins} \tau^{-}(\mathrm{KD})$, and $\vdash_{ins} \tau^{*}(\mathrm{DD}')$ are given in the Appendix.[4] It should be noted that having a norm-set with contingent content, i.e., $\mathcal{N} \cap \{\varphi \mid \nvdash_{pl} \varphi \text{ or } \nvdash_{pl} \neg\varphi\} \neq \emptyset$ does not count as a perfection property of a normative system. Therefore, a nihilistic normative system in which any contingent state of affairs is permitted and none prohibited counts as an instance of an ideal normative system.

Von Wright's "pilgrim's progress" [13] from standard deontic logic to the position he held in his later works in 1990s may look as a circle, but the ending point is not the same. The theorems from 1950s still remain as theorems in 1990s deontic logic, but their position and character has been changed. They cease to be theorems of the "logical syntax" of deontic language, and become the theorems of the "logical pragmatics" of deontic language use. What had been previously understood as a conceptual relation, later becomes a normative

[4] The notations for theorems (6.11), (KD), and (DD′) are taken over from [5].

relation; a norm for the norm-giving activity, and not the logic of the norms being given.

4.1 Rationality of sealing principles

Completeness ("gaplessness") of normative system is a perfection property that is hardly achievable for any non-nihilistic system. Von Wright gives a vivid definition of the problem and possible ways of solving it.

> What is the difference "in practice" between a state of affairs not being prohibited and its being permitted? Suppose there is a code of norms in which there is no norm Pp. Now someone makes it so that p. What should be the law-giver's reaction to this, if any? Could he say: "You were not permitted to do this and you must not do that which you are not permitted to do"? He could say this, making it a meta-norm that everything not-permitted is thereby forbidden. "Logically" this would be just as possible, even though perhaps less reasonable, as to have a meta-norm permitting everything which is not forbidden. But one can also think of some "middle way" between these two principles, a meta-norm to the effect that if something is not permitted by the existing norms of a code one must, as we say, 'ask permission' of the law-giver to do it. [12, p.280]

According to von Wright there are three principles by use of which a normative system can be completed: 1. ($\neg$F $\rhd$ P) everything not forbidden is permitted, 2. ($\neg$P $\rhd$ F) everything not permitted is forbidden, 3. normative gaps are filled in communication between the norm-recipient and the norm-giver. The third principle will be left aside because of its complexity. Using the two-sets model of normative system it can be shown why the first principle is to be preferred over the second one, i.e., why the first principle is "more reasonable". In addition to this it can be proved that the mere "logical possibility" of the second mode ($\neg$P $\rhd$ F) of filling normative gaps is not a sufficient condition of its rationality, according to von Wright's own criterion of rationality of norm-giving activity.

Definition 4.2 A norm-system $\langle \mathcal{N}, \overline{\mathcal{N}} \rangle$ is gapless iff $\{\varphi, \neg\varphi\} \subseteq \mathcal{N} \cup \overline{\mathcal{N}}$ for all doable states of affairs φ and $\neg\varphi$, i.e., $\mathcal{L}_{\text{doable}} = \mathcal{N} \cup \overline{\mathcal{N}}$.

The notion of "doable state of affairs" is taken over from von Wright's works. The notion of doability introduces complex problems of logic of action. Here the set of sentences describing doable states of affairs will be simplified and identified with the set of contingent sentences, $\mathcal{L}_{\text{doable}} = \mathcal{L}_{pl} - \{\varphi \mid \vdash_{pl} \varphi \text{ or } \vdash_{pl} \neg\varphi\}$.

The easy way of making a normative system complete is by applying the principle *everything which is not forbidden is permitted.* The way of filling in the gaps is straightforward, consisting in adding the missing sentences to the counter set and thus obtaining its extension $\overline{\mathcal{N}}^*$, as formula (1) shows.

$$\overline{\mathcal{N}}^* = \overline{\mathcal{N}} \cup \{\varphi \mid \varphi \notin Cn(\mathcal{N})\} \tag{1}$$

A completion of the normative system under the principle *everything which is not permitted is forbidden* is not a functional relation. In this mode the process is under-determined and so does not result in a unique system. The completion proceeds in two steps, each of which includes a choice.

The first step In the first step the counter-set must be completed in the view of perfection relations and properties. Also, the perfection-relation between the obligation norm set and its counter-set ought to be preserved if present and so their intersection must remain empty. This means that it will be expanded to achieve perfection-properties of being closed under implicants and under the rule of having at least one conjunct for each member conjunction. Since the last condition has the disjunctive consequent there may be different ways of performing the closure. Therefore, the weak-ideal expansion of a counter-set results in a set of sets.

Definition 4.3 The minimal weak-ideal closure $\mathrm{WI}(\overline{\mathcal{N}})$ of a counter-set is the set of the smallest sets a satisfying the following conditions:

(i) a includes $\overline{\mathcal{N}}$: $\overline{\mathcal{N}} \subseteq a$,

(ii) if $\psi \in \overline{\mathcal{N}}$ and φ entails ψ and $\varphi \in \mathcal{L}_{\text{doable}}$, then $\varphi \in a$,

(iii) a satisfies one of the following conditions:
  (a) if $\varphi \wedge \psi \in \overline{\mathcal{N}}$, $\psi \notin \overline{\mathcal{N}}$, $\varphi \notin Cn(\mathcal{N})$ and $\varphi \in \mathcal{L}_{\text{doable}}$, then $\varphi \in a$,
  (b) if $\varphi \wedge \psi \in \overline{\mathcal{N}}$, $\varphi \notin \overline{\mathcal{N}}$, $\psi \notin Cn(\mathcal{N})$ and $\psi \in \mathcal{L}_{\text{doable}}$, then $\psi \in a$.

Definition 4.4 Function γ picks an arbitrary member of the set $\mathrm{WI}(\overline{\mathcal{N}})$ of weak-ideal sets: $\gamma(\overline{\mathcal{N}}) \in \mathrm{WI}(\overline{\mathcal{N}})$.

Example 4.5 Let $\mathcal{L}_{\text{doable}} = \{p, q\}$. Let $\mathcal{N} = \emptyset$. Let the only norm be the norm-permission $\mathrm{P}(\neg p \vee \neg q)$. It follows that: $p \wedge q \in \overline{\mathcal{N}}$; $\mathrm{WI}(\overline{\mathcal{N}}) = \{\{p \wedge q, p, p \wedge \neg q\}, \{p \wedge q, q, \neg p \wedge q\}\}$

The second step The second step in a completion of normative system is also under-determined and complex in itself. It consists of two phases. In each of the two phases lists of sentence are being used in the construction. Lists will be understood as lists of equivalence classes $[\varphi] = \{\psi \mid \vdash_{pl} \psi \leftrightarrow \varphi\}$ $[\varphi_1], \ldots, [\varphi_n], \ldots$.

(i) In the first phase the obligation norm-set and its counter-set are closed under appropriate relations by taking into account "partially placed" sentences, i.e., those where only one sentence from a pair of contradictory sentences belongs to the closure of the system.

$$\overline{\mathcal{N}}_0 \in \mathrm{WI}(\gamma(\overline{\mathcal{N}}) \cup \{\varphi \mid \vdash_{pl} \varphi \leftrightarrow \neg\psi \text{ and } \psi \in Cn(\mathcal{N})\})$$
$$\mathcal{N}_0 = Cn(\mathcal{N} \cup \{\varphi \mid \vdash_{pl} \varphi \leftrightarrow \neg\psi \text{ and } \psi \in \overline{\mathcal{N}}_0\})$$

(ii) In the second phase "unplaced sentences" are being added in an iterative manner to the system. "Unplaced sentences" are those where no sentence from a pair of contradictory sentences belongs to the system.

$$\langle \mathcal{N}_{n+1}, \overline{\mathcal{N}_{n+1}} \rangle =$$
$$= \begin{cases} \langle Cn(\mathcal{N}_n \cup \{\varphi_n\}), \overline{\mathcal{N}}_n \cup \{\neg\varphi_n\}\rangle, & \text{if } \mathcal{N}_n \cup \{\varphi\} \text{ is consistent,} \\ \langle Cn(\mathcal{N}_n \cup \{\neg\varphi_n\}), \overline{\mathcal{N}}_n \cup \{\varphi_n\}\rangle, & \text{otherwise.} \end{cases}$$
$$\langle \mathcal{N}^*, \overline{\mathcal{N}}^* \rangle = \langle \bigcup_{0 \geq i} \mathcal{N}_i, \bigcup_{0 \geq i} \overline{\mathcal{N}}_i \rangle$$

There is no preferred ordering of unplaced sentences. The outcome of the iterative process depends on the chosen ordering. In most cases the resulting systems are radically different.

The systems completed by the application of the principle *everything not permitted is forbidden* do not necessarily end in one and the same "ideal state of things".

Example 4.6 Let $\mathcal{N} = \{p \vee q\}$ and $\overline{\mathcal{N}} = \emptyset$. Expansion of the counter-set by partially placed sentences and weak-ideal closure of the counter-set together yield the following set of sets: $\{\{\neg p \wedge \neg q, \neg p, \neg p \wedge q\}, \{\neg p \wedge \neg q, \neg q, p \wedge \neg q\}\}$. So, a choice must be made. Consequently expansion of the obligation-set with respect to the counter set depends on the set being chosen, and thus it yields either $\mathcal{N}_0^{'} = \{p \vee q, p, p \vee \neg q\}$ or $\mathcal{N}_0^{''} = \{p \vee q, q, \neg p \vee q\}$. Finally, the expansion by unplaced sentences depends on the list used in the construction. Suppose that List 1 is given by: $[q], \ldots$, and List 2 by:$[\neg p], \ldots$. Then $\mathcal{N}_1^{'} = Cn(\mathcal{N}_0^{'} \cup \{q\})$, while $\mathcal{N}_1^{''} = Cn(\mathcal{N}_0^{''} \cup \{\neg p\})$. Therefore, $(p \wedge q) \in \mathcal{N}_1^{'}$, and $\neg(p \wedge q) \in \mathcal{N}_1^{''}$ The completion results in incompatible ideal states as translations show: $\mathrm{O}(p \wedge q)$ w.r.t. $\mathcal{N}^{*'}$, while $\mathrm{F}(p \wedge q)$ w.r.t. $\mathcal{N}^{*''}$.

A critique of Von Wright: how many ideal states? Von Wright did not consider completion under the principle *everything not permitted is forbidden* as not rational, but only as less reasonable then the completion under the principle *everything not forbidden is permitted* .

> Generally speaking: a legal order and, similarly, any coherent code or system of norms may be said to envisage what I propose to call an ideal state of things when no obligation is ever neglected and everything permitted is sometimes the case. If this ideal state is not logically possible, i.e., could not be factual, the totality of norms and the legislating activity which has generated it do not conform to the standards of rational willing. Deviations from these standards sometimes occur — and when they are discovered steps are usually taken to eliminate them by 'improved' legislation. [11, p.39]

If a normative system is completed under the principle *everything not permitted is forbidden*, then, if consistent, it can "envisage more then one ideal state", each equally acceptable as the other. Thus, there will be no unique ideal state with respect to obligation-norms. If intending a unique ideal state is essential to rational willing on the side of the norm-giver, then the the principle *everything not permitted is forbidden* is not only "less reasonable", as von Wright claimed, but also not (instrumentally) rational.

5 Concluding remarks and further research

The term 'pragmatics' indicates the study of language-use: the norm-giver is engaged in the prescriptive *use of language* while constructing a normative system; the norm-recipient *uses a system constructed by language use* as the basis of her/his normative reasoning. The term 'social' indicates that more than one language-user (or social role) should be taken into account: the (role of) norm-giver, the (role of) norm-recipient, the (role of) norm-evaluator. Social pragmatics of deontic logic studies the norms that apply to norm-related activities of social actor roles. These norms can be properly called 'second-order norms' since they cover the activities that are related to a normative-system. There is a difference between second-order norms which require construction (envisaging) a logical possible description of an ideal state (e.g., consistency norms) and second-order norms which are related to the will of the norm-giver. In the latter case if the aim is to construct a description of exactly one ideal state, then the second-order norm *everything not permitted is forbidden* is not acceptable since it might end in a multitude of equally valid ideal states. The language of modal deontic logic can be (and perhaps should be) understood as the language in which perfection properties of a normative system are being described. The norm-giver and the norm-recipient are related both to the actual normative-system, which may be imperfect, and to its, possibly missing, perfection properties (from which second-order norms spring). Logic has sometimes been understood as the *ethics of thinking*. Von Wright's reinterpretation of deontic logic prompts us to understood logic also as the *ethics of language use*. In understanding deontic logic the perspectives of different social roles of should be taken into account as well as the purpose of norm giving activity. In this way deontic logic ceases to be a "zero-actor logic" and becomes the logic of language use which requires the presence of "users". This fact redefines deontic logic as a research which necessarily includes the stance of *logical pragmatics*.

This paper is a continuation of the previous research [18] in which the extension of the pragmatic reading of deontic axioms has been introduced with respect to the difference of roles of the norm-subject and the norm-applier, but without separate treatment of permission-norms. Further research should extend the logical pragmatics approach and address the interrelated topics of normative reasoning based on an inconsistent normative system, the problem of conditional norms, and, at the most general level, the determination of the source of the second-order norms of norm-giving activity and provision of an adequate logical framework for their formalization. A sketch of possible directions of further research follows.

A normative vacuum does not appear if the norm-recipient is subordinated to an inconsistent normative system, in which there is no way out of the normative conflict on the basis of the metanormative principles on the priority order over norms. On the other hand, the norm-recipient cannot reason using classical logic since it would lead to the logical "explosion" (on the side of the norm-set). The only remaining option is logic revision. In the view of perfection properties, some postulates of logic revision can be outlined. The first condition that a logic change ought to satisfy is to restore coherence (=non-explosiveness) of the set whose logic is being changed. Secondly, the change of logic ought to preserve desirable logical properties. The two conditions of the logic revision, restoration condition and preservation condition, resemble the content contraction, but the difference lies in the fact that instead of consistency it is the coherence that is being restored, and, instead of maximal preservation of the content, it is the desirable logical properties that are being saved. So, the norm-recipient faced with an inconsistent normative system ought to adopt an inconsistency-tolerant logic under which the normative properties will be preserved, namely, closure under entailment and adjunction of the norm-set together with correlated properties of the counter-set (closure under implicants and closure under having at least one conjunct for each conjunction). Is there such a logic? The deontic dialetheic logic of G. Priest [8] seems to be adequate for the purpose.

The set-theoretic approach must be refined in order to capture the problem of conditional norms, which requires a more refined treatment of interaction between "is" and "ought". The application of the generalized treatment of a code of requirements as a three-place functions introduced by Broome [2]. It has been proved in [17] that "for each world-relative code there is a realization equivalent world-absolute code", or, in other words, that the narrow-scope and wide-scope reading have the same effects. The approach should be extended so to include also "necessary condition conditionals" having the form $\mathrm{O}\varphi \to \psi$ and investigate perfection properties in this respect.

The third topic for the further research has a philosophical character because of its high level of generality. The relevant theoretical basis for this line of research can be found in dynamic logic as a logic of effects of language use, developed by J. van Benthem [9] and the vast group of related researchers. The essential formula of the theory has the form $[C]E$ and it describes communicative act C by its effect E. The inclusion of the actor's identity in the C-part has been introduced by Ju and Liu [7], while Yamada [16] has added deontic effects to the E-part of the formula. Within this framework obligations of the norm-giver in the prescriptive use of deontic language can be captured by the formula $[g : \underline{!\Delta_r\varphi}]\mathbf{D}_g g : \underline{!\Delta'_g\psi}$, where g and r are the norm-giver and the norm-recipient, respectively, ! indicates the prescriptive use of deontic sentence, Δ and Δ' stand for deontic operators of the first-order, while $\mathbf{D}$ stands for a deontic-operator of the second order. For example, in this perspective to the perfection property of external consistency there corresponds a second-order norm type forbidding creation of an externally inconsistent system, which can be formalized by the formula $[g : \underline{!\mathrm{O}_r\varphi}]\mathbf{F}_g\ g : \underline{!\mathrm{F}_r\varphi}$. A hypothesis worth considering is the one stating that the use of language in creation of a normative system is subordinated to the requirements of the second-order normativity, the normativity of language use, which ought to be studied within logical pragmatics of deontic logic. [5]

[5] The author wishes to thank anonymous reviewers for helpful criticism and suggestions.

Appendix

Proposition .1 $\vdash_{ins} \tau^{+}(\mathrm{O}\varphi \wedge \mathrm{P}\psi) \to \mathrm{P}(\varphi \wedge \psi)$

Proof. $\tau^{+}(\mathrm{O}\varphi \wedge \mathrm{P}\psi) \to \mathrm{P}(\varphi \wedge \psi) = (\varphi \in \mathcal{N} \wedge \neg\psi \notin \mathcal{N}) \to \neg(\varphi \wedge \psi) \notin \mathcal{N}$

1	$\varphi \in \mathcal{N} \wedge \neg\psi \notin \mathcal{N}$	
2	$\neg(\varphi \wedge \psi) \in \mathcal{N}$	
3	$\varphi \in \mathcal{N}$	1/ Elim$\wedge$
4	$(\varphi \wedge \neg(\varphi \wedge \psi)) \in \mathcal{N}$	3, 4/ 2*
5	$(\varphi \wedge \neg(\varphi \wedge \psi)) \to \neg\psi$	$\vdash_{pl}$
6	$\neg\psi \in \mathcal{N}$	4, 5/ Rc
7	$\neg\psi \notin \mathcal{N}$	1/ Elim$\wedge$
8	$\neg(\varphi \wedge \psi) \notin \mathcal{N}$	2–7/ Intro$\neg$
9	$(\varphi \in \mathcal{N} \wedge \neg\psi \notin \mathcal{N}) \to \neg(\varphi \wedge \psi) \notin \mathcal{N}$	1–8/ Intro$\to$

□

Proposition .2 $\vdash_{ins} \tau^{-}(\mathrm{O}(\varphi \to \psi) \to (\mathrm{O}\varphi \to \mathrm{O}\psi))$

Proof. $\tau^{-}(\mathrm{O}(\varphi \to \psi) \to (\mathrm{O}\varphi \to \mathrm{O}\psi)) = (\varphi \to \psi) \notin \overline{\mathcal{N}} \to (\varphi \notin \overline{\mathcal{N}} \to \psi \notin \overline{\mathcal{N}})$

1	$(\varphi \to \psi) \notin \overline{\mathcal{N}}$	
2	$\varphi \notin \overline{\mathcal{N}}$	
3	$\varphi \in \mathcal{N}$	2/ Comp
4	$(\varphi \to \psi) \in \mathcal{N}$	1/ Comp
5	$(\varphi \wedge (\varphi \to \psi)) \in \mathcal{N}$	3, 4/ 2*
6	$(\varphi \wedge (\varphi \to \psi)) \to \psi$	$\vdash_{pl}$
7	$\psi \in \mathcal{N}$	5, 6/ Rc
8	$\psi \notin \overline{\mathcal{N}}$	7/ ExtC
9	$\varphi \notin \overline{\mathcal{N}} \to \psi \notin \overline{\mathcal{N}}$	2–8/ Intro$\to$
10	$(\varphi \to \psi) \notin \overline{\mathcal{N}} \to (\varphi \notin \overline{\mathcal{N}} \to \psi \notin \overline{\mathcal{N}})$	1–9/ Intro$\to$

□

Proposition .3 $\vdash_{ins} \tau^{*}(\mathrm{O}\varphi \to \mathrm{P}\varphi)$

Proof. $\tau^*(\mathrm{O}\varphi \to \mathrm{P}\varphi) = \varphi \in \mathcal{N} \to \neg\varphi \in \overline{\mathcal{N}}$

1	$\quad \varphi \in \mathcal{N}$	
2	$\quad\quad \neg\varphi \notin \overline{\mathcal{N}}$	
3	$\quad\quad \neg\varphi \in \mathcal{N}$	2/ Comp
4	$\quad\quad \neg\varphi \notin \mathcal{N}$	1/ IntC
5	$\quad \neg\varphi \in \overline{\mathcal{N}}$	2–4/ Elim¬
6	$\varphi \in \mathcal{N} \to \neg\varphi \in \overline{\mathcal{N}}$	1–5/ Intro→

□

References

[1] Alchourrón, C. E. and E. Bulygin, *The expressive conception of norms*, in: R. Hilpinen, editor, *New Studies in Deontic Logic: Norms, Actions, and the Foundations of Ethics*, Springer Netherlands, Dordrecht, 1981 pp. 95–124.

[2] Broome, J., "Rationality Through Reasoning," The Blackwell / Brown Lectures in Philosophy, Wiley-Blackwell, 2013.

[3] Bulygin, E., *G.H. von Wright's and C.E. Alchourron's struggle with deontic logic*, Paper presented at "The Human Condition Conference in Honour of Georg Henrik von Wright's Centennial Anniversary".

[4] Dunn, J. M. and G. Hardegree, "Algebraic Methods in Philosophical Logic," Oxford University Press, Oxford, 2001.

[5] Hilpinen, R. and P. McNamara, *Deontic logic: A historical survey and introduction*, in: D. Gabbay, J. Horty, X. Parent, R. van der Meyden and L. van der Torre, editors, *Handbook of Deontic Logic and Normative Systems*, College Publications, 2013 pp. 3–136.

[6] Jech, T., "Set Theory: The Third Millennium Edition," Springer, 2003.

[7] Ju, F. and F. Liu, *Prioritized imperatives and normative conflicts*, European Journal of Analytic Philosophy **7** (2011), pp. 35–58.

[8] Priest, G., "In Contradiction : A Study of the Transconsistent," Oxford University Press, Oxford, 2006, expanded edition, first edition in 1987 by Martinus Nijhoff Pub.

[9] van Benthem, J., "Logical Dynamics of Information and Interaction," Cambridge University Press, Cambridge, 2011.

[10] von Wright, G. H., "Norm and Action: A Logical Enquiry," Routledge and Kegan Paul, London, 1963.

[11] von Wright, G. H., *Is and ought*, in: M. C. Doeser and J. N. Kraay, editors, *Facts and Values: Philosophical Reflections from Western and Non-Western Perspectives*, Springer, Dordrecht, 1986 pp. 31–48.

[12] von Wright, G. H., *Is there a logic of norms?*, Ratio Juris **4** (1991), pp. 265–283.

[13] von Wright, G. H., *A pilgrim's progress*, in: *The Tree of Knowledge and Other Essays*, Brill, Leiden, 1993 pp. 103–113.

[14] von Wright, G. H., *Deontic logic — as I see it*, in: P. McNamara and H. Prakken, editors, *Norms, Logics and Information Systems: New Studies in Deontic Logic and Computer Science*, IOS Press, Amsterdam, 1999 pp. 15–25.

[15] von Wright, G. H., *Deontic logic: a personal view*, Ratio Juris **12** (1999), pp. 26–38.

[16] Yamada, T., *Acts of requesting in dynamic logic of knowledge and obligation*, European Journal of Analytic Philosophy **7** (2011), pp. 59–82.

[17] Žarnić, B., *A logical typology of normative systems*, Journal of Applied Ethics and Philosophy **2** (2010), pp. 30–40.
[18] Žarnić, B. and G. Bašić, *Metanormative principles and norm governed social interaction*, Revus: Journal for Constitutional Theory and Philosophy of Law (2014), pp. 105–120.

Author Index

www.ingramcontent.com/pod-product-compliance
Lightning Source LLC
Chambersburg PA
CBHW060228050426
42448CB00009B/1346
* 9 7 8 1 8 4 8 9 0 2 1 5 2 *